Micromanufacturing Processes

Micromanufacturing Processes

Edited by
V. K. Jain

CRC Press
Taylor & Francis Group
Boca Raton London New York

CRC Press is an imprint of the
Taylor & Francis Group, an **informa** business

CRC Press
Taylor & Francis Group
6000 Broken Sound Parkway NW, Suite 300
Boca Raton, FL 33487-2742

First issued in paperback 2017

No claim to original U.S. Government works
Version Date: 20120501

ISBN 13: 978-1-138-07642-6 (pbk)
ISBN 13: 978-1-4398-5290-3 (hbk)

Library of Congress Cataloging-in-Publication Data

Micromanufacturing processes / editor, V.K. Jain.
    p. cm.
    Summary: "In addition to covering all processes in a single source, Micromanufacturing Processes also deals with various aspects of micromanufacturing. This book presents the principles, basic micro tools, and developments in micromanufacturing processes and systems. It also includes measurement techniques as well as research trends. Ideal as a reference for researchers or for undergraduate students, the text addresses several different types of micromanufacturing processes, such as micromachining, microforming, microcasting, microjoining, nanofinishing, and micrometrology. Figures and technical data illustrate the processes and concepts that are discussed by the author"-- Provided by publisher.
    Includes bibliographical references and index.
    ISBN 978-1-4398-5290-3 (hardback)
    1. Micromachining. 2. Manufacturing processes. I. Jain, V. K. (Vijay Kumar), 1948-

TJ1191.5.M55255 2012
670--dc23
        2012015444

Visit the Taylor & Francis Web site at
http://www.taylorandfrancis.com

and the CRC Press Web site at
http://www.crcpress.com

Dedicated

to

Kanti Jain (wife)

and

Aanya Jain (granddaughter)

Contents

Section I Introduction

Section II Micromachining

Traditional Micromachining

Advanced Micromachining

Section III Nanofinishing

Section IV   Microjoining

Section V   Microforming

Section VI   Miscellaneous

Foreword

It is with great pleasure that I read the draft copy of the book *Micromanufacturing Processes* by Professor V.K. Jain of the Indian Institute of Technology, Kanpur, India. I am no stranger to his publications and presentations at Indian and international conferences. He is an erudite scholar and a voracious writer. This book contains 18 chapters written mostly by Indian researchers, including himself, and a few international scientists, including one from the National Institute of Standards and Technology, Gaithersburg, USA, on dimensional metrology for meso-, micro-, and nanomanufacturing and another on biomachining from the famous National Tsing Hua University, Hsinchu, Republic of China. Quite fittingly, the book opens with challenges faced not only in micromanufacturing but also in meso- and nanomanufacturing. India has indeed made a modest start in these areas, which is evident from the many chapters in this book. The impact that the Indian Institutes of Technology have had on Indian graduate education is evident from the vast number of IIT PhD scholars who are coauthors of the chapters.

The areas covered in this book are microturning and grinding, magnetorheological finishing (MRF), abrasive flow machining, laser macro- and microwelding, and electron beam macro- and microwelding. This book will be a valuable companion to many advanced books on manufacturing processes, including my book, *Precision Engineering* (published by McGraw Hill in 2008). The book's creative work includes new derivations for unconventional machining techniques, and researchers will find it a good source of reference. The book has many citations to The International Academy for Production Engineering (CIRP) work, and CIRP's focus group on microproduction engineering will find this new book a valuable asset.

Professor V.C. Venkatesh, DSc, PhD

Department of Mechanical Engineering
University of Nevada, Las Vegas, Nevada

Preface

This introductory book on micromanufacturing is the result of the combined efforts of more than 45 eminent professors and researchers. All these authors are actively engaged in teaching, research, or development activities in the specific areas about which they have written. The basic objective of this book is to acquaint readers with the principles, basic machine tools, and latest developments in micromanufacturing processes. This book deals not only with various micromanufacturing processes but also with measurement techniques and other essential topics, making it a good textbook for undergraduate and postgraduate students as well as a reference book for researchers. This book has a large number of cross-references that will help readers with further in-depth study of a particular area of specialization.

Nowadays, meso (1–10 mm) and micro (1–999 μm) manufacturing are emerging as important technologies, especially in areas where miniaturization yields economic and technical benefits, namely, the aerospace, automotive, optical, and biomedical industries. With time, miniaturization of machines and devices is leading to further demand for parts with dimensions of the order of a few micrometers (1 mm = 10^{-6} m) to a few hundred nanometers (1 nm = 10^{-9} m). The demand of industries for micromanufacturing of various types of materials (metallic, ceramics, and plastics) is also increasing by the day. Some examples of the products that require micromanufacturing are microholes in optical fibers, micronozzles for high-temperature jets, and micromolds. Scientists and researchers are engaged in developing nanofeatured products such as nanoelectromechanical systems (NEMS). Hence, it is quite safe to say that there is a need for manufacturing processes that are capable of dealing with atomic and molecular dimensions.

This book deals with processes that come under the category of micromanufacturing processes. The book is divided into six major parts: *Introduction, Micromachining, Nanofinishing, Microjoining, Microforming,* and *Miscellaneous.*

Section I has two chapters, the first one being an introduction, and the second, on meso-, micro-, and nanomanufacturing. Chapter 1 gives an overview of various micromanufacturing processes including applications, working principles, challenges in scaling down a macromanufacturing process to a micromanufacturing process, and some recent developments in each category of process. The need for product realization from new material in new configurations having micrometer sizes with complex features poses new challenges in manufacturing concepts and procedures. Chapter 2 elaborates on the challenges being faced in micro- and nanomanufacturing processes.

Section II deals with micromachining processes, which have been described under two heads: traditional micromachining and advanced micromachining. Traditional micromachining has a chapter on microturning (Chapter 3), which discusses the need for a thorough understanding of the various areas of the microturning process including the basic differences between microturning and macroturning processes, critical requirements of microturning machines, and mechanisms underlying material removal. Chapter 4, on microgrinding, deals with the mechanics of grinding applicable to microgrinding. The results of single-grit experiments have been effectively used to model grinding-specific energy. The second area of micromachining (that is, advanced micromachining) is presented in two chapters (Chapters 5 and 6); the first one is on biomachining and the second focuses on ion beam machining. In biomachining, the role of chemolithotrophic bacteria has been discussed along with micromachining of metals by culture supernatants. The use of *Acidithiobacillus thiooxidans* in biomachining of metals has also been discussed. In the case of focused ion beam machining or deposition, apart from a description of the working principle and setup, micro- and nanodevices fabricated using focused ion beams have been illustrated.

Section III gives an overview of the three well-known nanofinishing processes, namely, magnetorheological finishing, magnetic abrasive finishing, and abrasive flow finishing. The first process is capable of producing, under certain conditions, even sub-nano surface finish on silicon wafers. Magnetic abrasive finishing, the second process, is capable of producing surface roughness (Ra) of less than 10 nm, while the third process (abrasive flow finishing) is capable of producing a surface finish with Ra as good as 20 nm. The principle, experimental setup, mechanics of nanofinishing, and the applications there of, have been elaborated.

Section IV deals with two microjoining processes, namely, laser microwelding and electron beam microwelding. Both these processes can achieve joining in the microdomain. Chapter 11 on electron beam welding also deals with design aspects of electron beam welding machines.

Section V has three chapters on microforming processes. One of these chapters (Chapter 12) deals with micro- and nanostructured surfaces developed by nano plastic forming and roller imprinting. A combination of nano plastic forming, coating, and roller imprinting (NPF-CRI process) is used to rapidly modify the surface of a material. Chapter 13 describes the design and development of a miniature extrusion machine and its performance during extrusion of microgears, micropins, and microcondensers. The final chapter in this section (Chapter 14) discusses the applications of laser beam for microbending. The governing differential equations for laser microbending are developed, and the application of finite element analysis is illustrated briefly. Microbending uses laser for producing small bend angles on small and large-sized sheets. It also deals with the physics underlying the laser microbending process.

Section VI has four chapters. Chapter 15 deals with the dimensional metrology for micro/nanoscale manufacturing. It deals with the philosophy of Lord Kelvin, "If you can not measure it, you can not improve it." Chapter 16 deals with the soft lithography technique, micromolding. It discusses the process in detail and gives some interesting examples of the products obtained with this technique. Chapter 17 gives an overview of the fabrication of microelectronic devices. It deals with the backbone of the present era, that is, precise manufacturing of a circuit element on a single-crystal Si wafer. The final chapter describes the model for integrated wafer surface evolution for chemical mechanical planarization.

I would like to individually thank all the authors of this book for efforts put in toward writing their chapters.

Finally, I would highly appreciate critical suggestions from the readers of this book to improve its quality in the next edition.

V.K. Jain

Editor

Dr. V.K. Jain completed his BE (mechanical) from Vikram University, Ujjain, India, and ME (production) and PhD from University of Roorkee, Roorkee, India (now, Indian Institute of Technology Roorkee). He has served as a visiting professor at the University of California, Berkeley (U.S.) and the University of Nebraska–Lincoln (U.S.). Currently, he is a professor at the Indian Institute of Technology Kanpur (India). He has also served on the faculty at other Indian institutions, namely, Malviya Regional Engineering College, Jaipur, India; Birla Institute of Technology, Pilani, India; and Motilal Nehru Regional Engineering College, Allahabad, India.

Dr. Jain has a number of medals and best paper awards to his credit. He has written five books and 11 chapters for different books by international publishers. He has been appointed as full editor of two international journals and associate editor of three international journals. He has also worked as a guest editor for 12 special issues of different international journals. He has served as a member of the editorial board of 12 international journals.

Dr. Jain has guided 15 PhD theses and 90 MTech/ME theses. He has more than 270 publications to his credit. He has delivered more than 26 keynote lectures in different conferences/workshops/universities. He has eight Indian patents and one U.S. patent to his credit.

Contributors

S. Aravindan
Department of Mechanical Engineering
Indian Institute of Technology Delhi
New Delhi, India

R. Balasubramaniam
Precision Engineering Division
Bhabha Atomic Research Centre
Mumbai, India

A.V. Bapat
Beam Technology Development Group
Bhabha Atomic Research Centre
Mumbai, India

Ashraf F. Bastawros
Department of Aerospace Engineering and
Department of Mechanical Engineering
Iowa State University
Ames, Iowa

Abhijit Chandra
Department of Mechanical Engineering and
Department of Aerospace Engineering
Iowa State University
Ames, Iowa

Jei-Heui Chang
Department of Power Mechanical
    Engineering
National Tsing Hua University
Hsinchu, Taiwan, Republic of China

Manas Das
Department of Mechanical Engineering
Indian Institute of Technology Guwahati
Guwahati, India

R. Das
Department of Mechanical Engineering
Indian Institute of Technology Guwahati
Guwahati, India

Deepak
Department of Materials Science and
    Engineering
Indian Institute of Technology Kanpur
Kanpur, India

V. Dey
Department of Mechanical Engineering
Indian Institute of Technology Kharagpur
Kharagpur, India

U.S. Dixit
Department of Mechanical Engineering
Indian Institute of Technology Guwahati
Guwahati, India

K. Easwaramoorthy
Beam Technology Development Group
Bhabha Atomic Research Centre
Mumbai, India

S. Ghosh
Department of Mechanical Engineering
Indian Institute of Technology Delhi
New Delhi, India

Micayla Haugen
Department of Mechanical Engineering
Iowa State University
Ames, Iowa

Hong Hocheng
Department of Power Mechanical
    Engineering
National Tsing Hua University
Hsinchu, Taiwan, Republic of China

Umesh U. Jadhav
Department of Power Mechanical
    Engineering
National Tsing Hua University
Hsinchu, Taiwan, Republic of China

V.K. Jain
Department of Mechanical Engineering
Indian Institute of Technology Kanpur
Kanpur, India

S.C. Jayswal
Department of Mechanical Engineering
Madan Mohan Malaviya Engineering
    College
Gorakhpur, India

S.N. Joshi
Department of Mechanical Engineering
Indian Institute of Technology Guwahati
Guwahati, India

Pavan Karra
Department of Mechanical Engineering
Iowa State University
Ames, Iowa

Monica Katiyar
Department of Materials Science and
    Engineering
Indian Institute of Technology Kanpur
Kanpur, India

G.U. Kulkarni
Chemistry and Physics of Materials Unit and
Department of Science & Technology Unit
    on Nanoscience
Jawaharlal Nehru Centre for Advanced
    Scientific Research
Bangalore, India

Vishwas N. Kulkarni
Department of Physics
Indian Institute of Technology Kanpur
Kanpur, India

V. Hemanth Kumar
Department of Mechanical Engineering
Indian Institute of Technology Guwahati
Guwahati, India

T. Matsumura
Department of Mechanical Engineering
Tokyo Denki University
Tokyo, Japan

Shawn P. Moylan
Engineering Laboratory
National Institute of Standards and
    Technology
Gaithersburg, Maryland

D.K. Pratihar
Department of Mechanical Engineering
Indian Institute of Technology Kharagpur
Kharagpur, India

B. Radha
Chemistry and Physics of Materials Unit
    and
Department of Science & Technology Unit
    on Nanoscience
Jawaharlal Nehru Centre for Advanced
    Scientific Research
Bangalore, India

V. Radhakrishnan
Indian Institute of Space Science and
    Technology
Trivandrum, India

Nitul S. Rajput
Department of Physics
Indian Institute of Technology Kanpur
Kanpur, India

J. Ramkumar
Department of Mechanical Engineering
Indian Institute of Technology Kanpur
Kanpur, India

P.V. Rao
Department of Mechanical Engineering
Indian Institute of Technology Delhi
New Delhi, India

M. Ravisankar
Department of Mechanical Engineering
Indian Institute of Technology Guwahati
Guwahati, India

Neeraj Shukla
Department of Physics
Indian Institute of Technology Kanpur
Kanpur, India

Ajay Sidpara
Department of Mechanical Engineering
Indian Institute of Technology Kanpur
Kanpur, India

D.K. Singh
Department of Mechanical Engineering
Madan Mohan Malaviya Engineering
    College
Gorakhpur, India

Vinod Kumar Suri
Precision Engineering Division
Bhabha Atomic Research Centre
Mumbai, India

N.J. Vasa
Department of Engineering Design
Indian Institute of Technology Madras
Chennai, India

Vikram Verma
Department of Materials Science and
    Engineering
Indian Institute of Technology Kanpur
Kanpur, India

Xiaoping Wang
Department of Mechanical Engineering
Iowa State University
Ames, Iowa

K. Willy
Department of Mechanical and Control
    Engineering
Tokyo Institute of Technology
Tokyo, Japan

A. Yamanaka
Department of Mechanical and Control
    Engineering
Tokyo Institute of Technology
Tokyo, Japan

M. Yoshino
Department of Mechanical and Control
    Engineering
Tokyo Institute of Technology
Tokyo, Japan

Section I

Introduction

1

Micromanufacturing: An Introduction

V.K. Jain and Ajay Sidpara
Indian Institute of Technology Kanpur

M. Ravisankar and Manas Das
Indian Institute of Technology Guwahati

CONTENTS

1.1 Introduction

Microproducts are widely used owing to their compactness, low material requirement, low power consumption, high sensitivity, and many other advantages over macroproducts. Micromanufacturing is becoming an inevitable part of household life and various industries such as electronics, telecommunication, medicine, defense, and automotives. It is one of the fundamental technologies that lead the market of miniaturized products, and which involves material removal, material deposition, or constant volume process (by changing the shape and size of the material). This chapter provides a brief introduction to the different micromanufacturing processes (microcutting, microforming, micromolding/casting, and microjoining) and their applications, which are widely used for the production of microcomponents.

1.2 Miniaturization and Applications

Micromanufacturing is a set of processes or techniques used to fabricate microcomponents or microsystems, or to create microfeatures on macro/microparts. Today, the demand for microcomponents/microsystems [microelectromechanical system (MEMS), microreactors, fuel cells, micromechanical devices, and micromedical components] is continuously increasing. Microcomponents are extensively used in vehicles, aircraft, telecommunication, IT industries, home appliances, medical devices, and implants. Currently, the field of miniature technologies is a fast-growing market in a broad spectrum of applications.

Presently, market demand exists not only for small parts but also for small and complex features on large components such as cooling holes in a turbine blade, microfins on a large surface for increasing heat transfer, and similar others. Hence, the objective of miniaturization is either to create small parts with better capabilities or to create small features on a large component to enhance its functionality. The mobile phone is a good example of miniaturization. Its size has gone down from being fairly large to being less than palm sized. In addition, today's mobile phones have many functional capabilities that a computer/laptop has—for example, internet surfing, sending e-mail, banking, and ticket booking.

Miniaturization of products or components carries many merits, including reduced energy and material consumption during manufacturing, lightness and portability, increased selectivity and sensitivity, use of intelligent materials with structures at the nanoscale level, minimally invasive techniques, exploitation of new effects through the breakdown of continuum theory in the microdomain, and cost/performance advantages.

The emerging miniaturization technologies are potential technologies of the future. Miniaturization of various devices/parts involves many areas of science and engineering: physics, chemistry, material science, computer science, ultraprecision engineering (Venkatesh and Izman, 2007), fabrication processes, and equipment design. Hence, manufacturing of these parts has received great attention from researchers in recent years, and nowadays, it is attracting both venture capital and federal research support in developed and developing countries. It has also opened up various new interdisciplinary research areas.

Demand for microparts is consistently increasing in medical applications because the human body has a limited capacity to accommodate "foreign" devices. Therefore, the components of various repair and pain-relief devices should be as small as feasible. *Physicians* also look for devices that are less intrusive so that there is low probability of infection, so as to promote faster healing. *Aerospace and automobile* industries are at the forefront in the application of microproducts. These industries are keen on introducing microsensors, complex fuel injection systems, safety devices, microfluidics for flow control, micromotors, and microactuators in their existing and new products. The *electronics industry* is trying to go beyond the limits imposed by lithographic circuit manufacturing techniques. Huge sums are invested in R&D to find alternative methods for cramming more circuitry and memory elements into a computer. Some researchers are attempting to create small machines that may be introduced into the human body to carry out repairs while others are trying to produce molecular-sized computer memories.

Microfeatures/parts can be produced in two ways, by modifying an existing machine (scaling down some units to handle small parts) to handle a micromanufacturing task or by developing a completely new machine tool specifically for micromanufacturing applications. In the case of modifying a large machine tool for micromanufacturing, it is important to emphasize that there is a limit to which a conventional machine tool can be scaled down. Beyond certain dimensions, the factors that can be ignored in conventional machining suddenly play a vital role in the quality of the miniaturized parts to be made. Deformation of the workpiece and tool, inaccuracies in the machine tool, vibration, temperature, tool offset, rigidity, and chip removal are some of the important considerations in micromanufacturing because these factors have a significant influence on the ultimate size of the component being produced.

Part handling and fixturing is a challenge for microcomponents, and hence special attention is necessary to ensure that the part does not get damaged during this process. Vacuum is more commonly used in the transportation and holding of microparts. Furthermore, inspection and measurement are also difficult because of the small dimensions of the products. Conventional probes and gauges are often too large to be used for the inspection of microparts. Measuring systems must use air, light, or other noncontact scanning methods to evaluate a part's features or dimensions. 3D metrology for micro/nanodevices and features is becoming an essential element of micromanufacturing. In view of the challenges associated with the handling and fixturing of micro/nanoparts, online measurement is becoming a prerequisite for micromanufacturing.

1.3 Classification

Many researchers have classified micromanufacturing technologies in various ways. Masuzawa (2000) has classified micromachining technologies according to their working principle. Madou (2001) has listed miniaturization techniques with their respective characteristics, organizing them as traditional or nontraditional methods and lithographic or nonlithographic methods. Other authors have classified them into various categories such as removal, deposition, and molding technologies (Smith, 2000). A classification based on the tools (solid or beam) or masks (anisotropic and isotropic) used by the process has been proposed (Brinksmeier et al., 2002). Figure 1.1 shows a simplified classification of micromanufacturing processes similar to the one reported by Masuzawa (2000). In the present chapter, an introduction to some of the micromanufacturing processes such as microcutting, microforming, micromolding, microcasting, and microjoining is presented.

1.4 Subtractive Processes

Micromachining is the removal of material in the form of chips or debris with sizes (or cross-section or at least one dimension) in the range of microns (greater than 1 μm and smaller than 999 μm). In other words, it can be considered as the process of creating

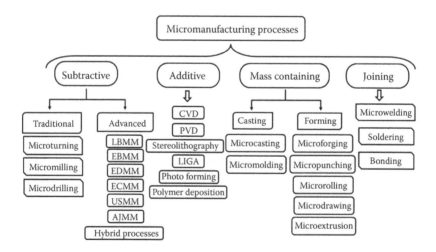

FIGURE 1.1
Classification of micromanufacturing processes. LBMM, laser beam micromachining; EBMM, electron beam micromachining; EDMM, electric discharge micromachining; ECMM, electrochemical micromachining; USMM, ultrasonic micromachining; AJMM, abrasive jet micromachining; CVD, chemical vapor deposition; PVD, physical vapor deposition.

microfeatures on macro- or microcomponents. Subtractive-type micromanufacturing processes are classified into two classes (Figure 1.1), traditional and advanced.

1.4.1 Traditional Micromachining

Traditional machining processes (turning, milling, and drilling) have gone through continuous improvements in the design of machine tools. To some extent, they can also be applied for micromachining by appropriately designing the cutting tool and optimizing the machining conditions. They are characterized by mechanical interaction of a sharp and well-defined geometry tool with the workpiece. This mechanical interaction between a tool and the workpiece results in the removal of material in the form of micro/nanochips from the workpiece, creating the desired microfeature that eventually results in the desired part. Several types of cutting processes are suitable for micromachining. Typical examples of traditional microcutting are drilling of microholes, milling of microgrooves and micro 3D shapes, and turning of micropins. The most attractive advantage of traditional microcutting technologies is the possibility to machine 3D microstructures characterized by a high aspect ratio and comparatively high geometric complexity.

For traditional microturning and other operations, the mini (tabletop) machine should be designed and fabricated (Honegger et al., 2006a,b). Use of the normal machine tools by scaling down the tool and machining conditions would not satisfactorily serve the desired purpose. Efforts are being made in this direction in different universities/institutes, R&D houses, and industries. Traditional microcutting processes involve critical issues: cutting force must be as low as possible, rigidity of the machine tool should be high enough to minimize machining errors, tool edge radius must be smaller than the dimension of the feature to be created, tool size (e.g., in microdrilling) must be smaller than the dimensions of the feature to be created, and tool material hardness should be higher than the workpiece hardness.

1.4.1.1 Microdeburring

The necessary condition for microcutting processes is the availability of ultraprecision machines in order to control tolerances and chip thickness. An important issue in microcutting processes is burr removal (Alting et al., 2003). Traditional methods of burr removal are difficult to apply because the dimensions of the machined features are in microns. Therefore, either a special technique for burr removal or burr-free machining strategies have to be developed. Magnetic abrasive deburring (MADe) is one such method that has been successfully used for deburring of microholes (Madarkar, 2007). Figure 1.2 shows a part of an edge of a microhole with burrs before deburring and no burrs after deburring. Very hard or brittle materials are difficult to machine, and brittle fracture/chipping of the workpiece cannot be suppressed completely. However, brittle materials can be cut in the ductile regime when very small depths of cut are used.

1.4.1.2 Microtool Fabrication

Fabrication of microtools for different types of operations (turning, milling, and drilling) is another important issue that should be addressed before conventional processes can be applied to the micromanufacturing domain. A major hurdle is the relatively high cutting force, which influences machining accuracy and the practical limits of the size that can be machined owing to the deflection of either tool or workpiece, or both. During the

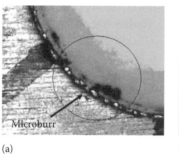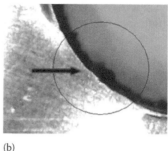

(a)   (b)

FIGURE 1.2
Drilled hole edge (a) before and (b) after deburring by MAF.

interaction between the tool and the workpiece material, no thermally activated diffusion should take place in order to maintain the shape and size of the tool and the workpiece (Alting et al., 2003).

Microdrilling has been widely used in various applications. Drilling of microholes in most metals, plastics, and their composites can be comparatively easily done. One typical example is microdrilling in laminated printed circuit boards (PCBs). Straightness and uniformity of the size of the tool are important issues in microdrilling. It is needless to emphasize the importance of the correct location and orientation of the microtool to avoid defective holes. A microhole drilled in soda lime glass is shown in Figure 1.3 (Park et al., 2002).

1.4.1.3 Diamond Turning

Optical lenses have also found many applications owing to progress in micromanufacturing. Small size and precision are the prime requirements for a lens when it is used in the microdomain. Micro diamond turning, also known as *single-point diamond turning (SPDT)*, is one of the most suitable machining processes for micro-optical device fabrication (Sze-Wei et al., 2007). SPDT is widely used to manufacture high-quality aspheric optical elements from glass, crystals, metals, acrylic, and other materials (Balasubramanium and Suri, 2010).

Lenses of some complex shapes are made by the molding process. Nickel–phosphorus alloy is the commonly used material for mold making. These molds are precisely prepared by SPDT. Some of these molds are used to make precision contact lenses, including aspherical lenses to correct special vision problems. Intraocular lenses are also used as

FIGURE 1.3
Microdrilling on soda lime glass (thickness = 0.13 mm, diameter = 0.1 mm, rpm = 30,000, cutting speed = 9.42 m/min, feed = 0.25 μm/rev).

a replacement for the cornea and inserted within the eye. Optical elements produced by SPDT are used in optical assemblies of telescopes, TV projectors, cameras, cell phones, DVD players, missile guidance systems, scientific research instruments, and numerous other systems and devices. Tool alignment is an essential precondition for achieving the desired quality in SPDT.

1.4.1.4 Micromilling

Micromilling is widely used for fabrication of grooves, cavities, and 3D concave and convex shapes. Some micromanufacturing processes lack the ability to make 3D structures from some materials, such as, silicon and polymer, and they are not suitable to produce 3D geometries for small and medium lot sizes. Micromilling is a promising approach to overcome some of the limitations of common microfabrication techniques. Micromilling of grooves of micron-sized width in brass, Al 6061, steel, and PMMA by means of 2, 4, and 5 cutting edge end mills have been described by Adams et al. (2001). Micromilling of bipolar fuel cell plates using titanium- and diamond-coated cutting tools has been reported by Jackson et al. (2006). Micromilling with tungsten carbide tools can be used for machining various materials such as steel, aluminum, and brass for making miniaturized components. Micromilling is also used for making molds for micromolding purposes (Brinksmeier et al., 2004).

1.4.2 Advanced Micromachining Processes

Owing to the development of new materials that are difficult to machine and stringent requirements of the parts used in hi-tech industries, many advanced micromachining processes (also known as *nontraditional* or *unconventional micromachining processes*) have come into existence, which can machine these materials economically, efficiently and with high accuracy. Consequently, advanced machining processes (AMPs) provide effective solutions to the problems imposed by the increasing demand for high-strength temperature-resistant alloys and the requirements of parts with intricate and complex shapes (Jain, 2010). These materials are so hard as to defy machining by traditional methods. These processes are advanced in the sense that they do not employ a conventional or traditional tool for material removal. Instead, they directly utilize some form of energy for micromachining. These processes are classified into different groups on the basis of their working principle as shown in Figure 1.4.

Mechanical micromachining processes such as ultrasonic micromachining (USMM) (Jadoun, 2010), abrasive jet micromachining (AJMM) (Ko, 2010), abrasive water jet micromachining (AWJMM) (Miller, 2004), and water jet micromachining (WJMM) utilize the kinetic energy (KE) of either abrasive particles or a water jet or both to remove material from a workpiece. The performance of these processes depends on hardness, strength, and other mechanical properties of the workpiece material. In the case of brittle materials, material removal takes place by brittle fracture, whereas in the case of ductile materials, it is because of the deformation process. The mechanism of material removal is also influenced by the angle at which an abrasive particle or an abrasive water jet hits the workpiece surface. Figure 1.5 illustrates the mechanism of material removal followed in these mechanical micromachining processes while machining ductile material or brittle material. The strain hardening property of the workpiece also plays an important role in these mechanical micromachining (M^3) processes.

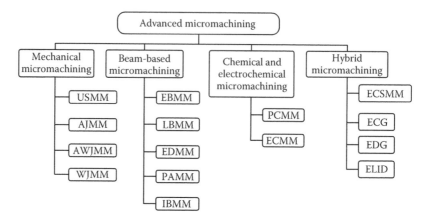

FIGURE 1.4
Classification of advanced micromachining processes: AWJMM, abrasive water jet micromachining; WJMM, water jet micromachining; PAMM, plasma arc micromachining; PCMM, photochemical micromachining; ECSMM, electrochemical spark micromachining; ELID, electrolytic in-process dressing; ECG, electrochemical grinding; EDG, electric discharge grinding.

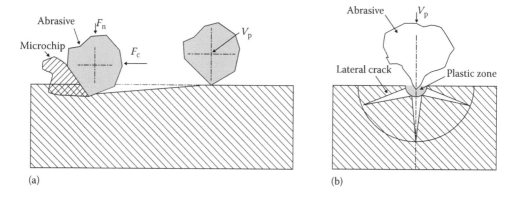

FIGURE 1.5
Mechanism of material removal in (a) ductile (From Finnie, I., *Wear*, 3, 87–103, 1960) and (b) brittle material (From Bitter, J.G.A., *Wear*, 6, 5–21, 1963).

Beam-based micromachining processes utilize different forms of thermal energy such as an electron beam in the case of electron beam micromachining (EBMM) (Sidpara and Jain, 2010), a laser beam in the case of laser beam micromachining (LBMM) (Vasa, 2010), the heat energy of sparks in the case of electric discharge micromachining (EDMM) (Joshi, 2010), a plasma arc in the case of plasma arc micromachining (PAMM), and the kinetic energy of ions in the case of ion beam micromachining (IBMM) (Aravindan et al., 2010). The thermal energy (except in IBMM) is concentrated on a small area of workpiece, resulting in melting or vaporization or both. These processes are widely used for machining hard and tough materials. To have a feel of the capabilities of these processes, let us take some examples of applications of these processes. LBMM (excimer laser) can write alphabetic letters on the human hair while EBMM has been used to make filters having thousands of 6-μm-diameter holes on a metallic sheet (thickness <100 μm) with a hole density of 200,000 cm^{-2}. IBMM has the capability to fabricate real-life nanodevices, for example, springs and sensors of nanometer dimensions.

Chemical or electrochemical micromachining processes include mainly electrochemical micromachining (ECMM) and photochemical micromachining (PCMM). The ECMM process has a wide field of applications. It is a controlled anodic dissolution process that yields high material removal rate (MRR), and its performance does not depend on the physical and mechanical properties of the work material (Jain, 2002). However, the work material should be electrically conductive. In this process, there is no tool wear and no residual stresses developed in the machined workpiece. On the other hand, PCMM is a type of etching process in which the workpiece is selectively etched by masking the area where no etching/machining is required. The selection of an appropriate etchant depends on the properties of the workpiece material to be machined. Figure 1.6 shows a micropattern, micromixer, microtool, and micronozzle fabricated using the ECMM process (Jain et al., 2011).

To further enhance the capabilities of micromachining processes, two or more than two micromachining processes are combined to take advantage of the worthiness of the constituent processes. For example, a conventional grinding process produces a good surface finish and tight tolerances, but the machined parts are associated with burrs, heat affected zones (HAZs), and residual stresses. However, electrochemically machined components do not have such problems. Therefore, a hybrid process called *electrochemical grinding* (ECG) has been developed. Similarly, other hybrid processes such as electrochemical spark micromachining (ECSM), electrodischarge grinding (EDG), and electrolyte in-process dressing (ELID) have been developed. ECSM is a combination of both the electrochemical machining (ECM) and electric discharge machining (EDM) processes, which are applicable to only electrically conductive materials, but ECSM is applicable to both electrically conductive and electrically nonconductive materials (Kulkarni et al., 2002, 2011a). This process

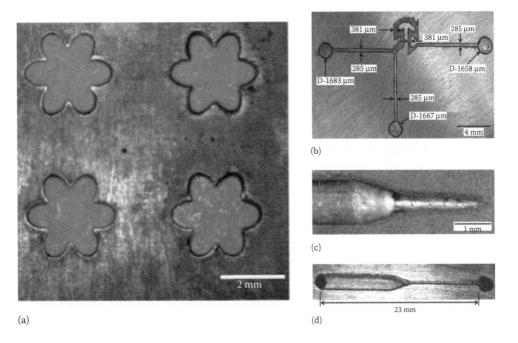

(a)

(b)

(c)

(d)

FIGURE 1.6
(a) Micropattern, (b) micromixer, (c) microtool, and (d) micronozzle fabricated by the ECMM process.

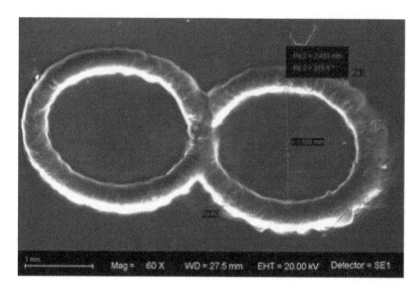

FIGURE 1.7
An 8-shaped microchannel produced on glass by the ECSM process (voltage = 50 V, electrolyte concentration = 16% NaOH).

has been used to create microchannels on an electrically nonconductive material such as glass. Figure 1.7 shows an 8-shaped microchannel produced on glass by the ECSM process (Priyadarshini, 2007; Kulkarni et al., 2011b).

Nowadays, the design requirements for components are stringent, such as, extraordinary properties of materials (high strength, high heat resistance, high hardness, and high corrosion resistance), complex 3D shapes [e.g., turbine blade, miniature features (filters for food processing and textile industries having a few tens of micrometers as hole diameter and thousands in numbers)], and nanolevel surface finish on complex geometries, which are not easy to achieve by many traditional machining and finishing methods.

1.5 Nanofinishing Processes

To finish surfaces in the nanometer range, it is required to remove material in the form of atoms or molecules individually or in groups. Figure 1.8 shows the classification of micro/nanofinishing processes into two groups: traditional and advanced. Traditional finishing processes are out of the scope of this book. Advanced finishing processes can be divided into two groups: the first one includes abrasive flow finishing (AFF) and chemomechanical polishing (CMP). In these processes, it is not possible to externally control the forces acting on a workpiece during the finishing process. The second category includes magnetorheological finishing (MRF), magnetic abrasive finishing (MAF), magnetorheological abrasive flow finishing (MRAFF), and magnetic float polishing (MFP). In these processes, it is possible to externally control the forces acting on a workpiece by varying the electric current flowing in the electromagnetic coil or by varying the working gap between the magnet and the workpiece, especially when using a permanent magnet.

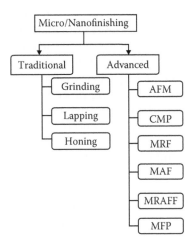

FIGURE 1.8
Classification of the nanofinishing processes.

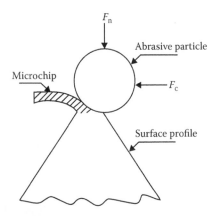

FIGURE 1.9
Forces acting on the workpiece by an abrasive particle.

In all these processes, two main types of forces act (Figure 1.9), a normal force (F_n), which is responsible for the penetration of an abrasive particle inside the workpiece, and a cutting force (F_c), which is responsible for the removal of the material in the form of micro/ nanochip.

The AFF process was originally identified for deburring and finishing critical hydraulic and fuel system aircraft components in aerospace industries. It can polish any surface on which air, liquid, or fuel flows. Rough, power robbing cast, machined, or EDM surfaces, such as the internal passageways of intake parts with complex geometry, are improved substantially regardless of their surface complexities. To enhance the capabilities of this process, the rotational abrasive flow finishing (R-AFF) process has been developed (Sankar et al., 2009). This process is capable of producing cross hatch patterns that help in retaining lubricating oil, and it is also able to produce a final surface roughness value less than 50 nm (Sankar et al., 2010).

CMP utilizes both mechanical wear and chemical etching to achieve surface roughness (Ra) in the nanometer range and a high level of planarization (Venkatesh et al., 1995a,b; Venkatesh et al., 2004; Zhong and Venkatesh, 1995). CMP is the most preferred process in semiconductor industries for the finishing and planarization of silicon wafers.

Except for MAF, which uses magnetic particles, abrasive particles, and small amount of oil, all other magnetic-field-assisted finishing processes use a magnetorheological (MR) fluid that consists of magnetic particles, abrasive particles, carrier liquid (water or oil), and additives (Singh et al., 2005, 2006). The concentration of different constituents in the MR fluid plays a significant role in the quality of the finished surface. Selection of different types of abrasive particles and the carrier fluid depends on the type of workpiece to be finished. An Ra value less than 1 nm has been achieved by some of these processes. Figure 1.10 shows the final Ra value of 0.48 nm achieved using the chemomechanical magnetorheological finishing (CMMRF) process (Jain et al., 2010), 8 nm on a silicon blank using the MRF process (Sidpara and Jain, 2011), and 14 nm on a stainless steel cylindrical workpiece using the R-MRAFF process (Das et al., 2010).

1.6 Additive Processes

In the additive manufacturing processes, a part is built by adding layers of material one upon the other. There are many processes that come under this category such as

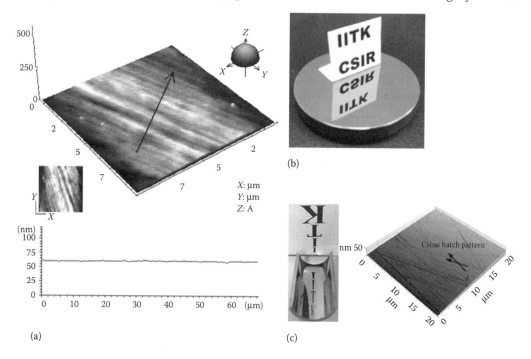

FIGURE 1.10
(a) Atomic force microscope (AFM) image and roughness plot along the arrow showing a silicon surface finished by CMMRF (Ra = 0.48 nm), (b) reflection of "IITK-CSIR" letters on a silicon surface finished by MRF, and (c) photograph and AFM micrograph of stainless steel tube finished by the R-MRAFF process.

chemical vapor deposition (CVD), physical vapor deposition (PVD), stereolithography, and *li*thographie, *g*alvanoformung, *a*bformung (LIGA is a German acronym for Lithography, Electroplating, and Molding that describes a fabrication technology used to create high-aspect-ratio microstructures).

1.6.1 Chemical Vapor Deposition

CVD is a generic name for a group of processes that involve depositing a solid material from a gaseous phase. Precursor gases (often diluted with carrier gases) are delivered into the reaction chamber at approximately ambient temperature. As they come into contact with a heated substrate, they react or decompose, forming a solid phase on the substrate. The substrate temperature is critical and it affects the reaction between the substrate and the material to be deposited. CVD coatings are usually only a few microns thick and are generally deposited at fairly low rates (few hundred microns per hour). CVD is an extremely versatile process that can be used to process almost any metallic or ceramic compound.

The main application of the CVD process is coating—for a variety of applications such as wear resistance, corrosion resistance, high temperature protection, and erosion protection. There are many other areas where CVD is used, such as in semiconductors and related devices, optical fibers for telecommunications, and composites.

1.6.2 Physical Vapor Deposition

PVD is a coating technique in which material is transferred at the atomic level. It is an alternative process to electroplating. The process is similar to CVD except that the raw material/precursor (the material that is to be deposited) starts out in solid form, whereas in CVD, the precursors are introduced to the reaction chamber in the gaseous state. PVD processes are carried out under vacuum conditions.

The process involves four steps: (1) *Evaporation*: the target to be deposited is bombarded by a high-energy source, such as a beam of electrons or ions. As a result, the atoms dislodge from the surface of the target. (2) *Transportation*: the atoms from the target move toward the substrate to be coated. (3) *Reaction*: the atoms of the metal react with the appropriate gas during the transport stage. (4) *Deposition*: a coating builds up on the surface of the substrate.

A PVD coating is applied to improve hardness and wear resistance, to reduce friction, and to improve oxidation resistance. The use of PVD coatings is aimed at improving efficiency through improved performance and longer component life. They may also allow coated components to work in the environment in which the uncoated component would not otherwise have been able to work.

1.6.3 Stereolithography

Stereolithography is a widely used rapid prototyping (RP) additive process. This process is also called *layered manufacturing*. Nowadays, lead time (time to market) is becoming an increasingly important factor for a product's success. A major contributor to the product development cycle is the time needed to produce a prototype. Furthermore, when customers demand individually tailored components, the cost and time to produce tools or molds may become considerable.

RP processes help to overcome these problems by building complex shapes through additive processes, producing components without the use of tools (Jacobs, 1996). The stereolithography apparatus (SLA) uses UV or laser light to selectively cure resins. The SLA produces plastic prototypes by photopolymerization of a liquid monomer. The 3D computer-based component design is divided into submillimeter-thick layers. The laser is then scanned onto the surface of the monomer, drawing the cross-section of the component. Once this resin is cured, the platform drops one more layer into the tub filled with the monomer. The liquid monomer flows over the layer and the second layer is scanned. The procedure is repeated until the part is finished.

1.6.4 LIGA

LIGA is a German acronym for *lithographie, galvanoformung, abformung* (lithography, electroplating, and molding) that describes a fabrication technology used to create high aspect ratio microstructures. LIGA is also an additive process that can make complex microstructures with very high aspect ratios. It can make structures that are composed of electroplated metals or injected plastics, using the LIGA-fabricated structure as a mold. LIGA consists of three main processing steps: lithography, electroplating, and molding (Madou, 2001). High aspect ratio structures are fabricated by x-ray LIGA, which uses x-rays produced by a synchrotron, and low aspect ratio structures are fabricated by ultraviolet (UV) LIGA, which uses ultraviolet light. The LIGA process can build microparts smaller than those built using conventional machining processes and also bigger than those through AMPs. In this manner, it fills a gap in the size range of microparts.

1.7 Mass Containing Processes

Since 1990, as the awareness of microtechnology grew, the production of microparts using turning and milling processes started. Then researchers in the metal-forming area started thinking of how to make these microparts by metal-forming processes. The researchers also concluded that the know-how of conventional metal-forming technologies cannot be easily scaled down to microscale owing to the problems related to the process, tooling, and machine tool, for which the solutions were not available.

1.7.1 Microforming

Forming is a plastic deformation process, without any addition or removal of material. Forming processes are particularly suited for mass production of metallic parts owing to their distinct advantages, such as high production rate, minimum or zero metal loss, excellent mechanical properties of the final product, and close tolerances, making them suitable for near net shape production (Geiger et al., 2001). Forming processes that have been successfully applied for fabrication of microparts are blanking/punching, deep drawing, bending/wire bending, extrusion, and embossing/coining.

The term *microforming* is defined as the production of parts or structures with at least two out of three dimensions in the submillimeter range (Geiger et al., 2001). Microforming is an appropriate technology to manufacture very small metal parts in large quantities at a high production rate, in particular for electronic components in microsystem technologies (MSTs) and MEMS. Some commonly produced microformed parts are leverages, connector pins, resistor caps, contact springs, chip lead frames, IC-carriers, fasteners, microscrews, and sockets (Engel and Eckstein, 2002). However, transformation of the products from macro to micro is challenging in the case of forming processes.

In the *sheet bending process*, the force increases when the sheet thickness is reduced to a value such that only one grain is located over the thickness (Engel and Eckstein, 2002; Kals and Eckstein, 2000; Eckstein and Engel, 2000). This is explained by the fact that each single grain is deformed according to the shape of the tool, regardless of its possibly unfavorable orientation (Alting et al., 2003). The normal mean anisotropy of sheets decreases with miniaturization, leading to deterioration of the forming characteristics because the thickness reduction increases in deep-drawing processes.

Bending is important for the production of connectors, contact springs, and lead frames of micron sizes. Micro wire bending is successfully used for the production of filaments and springs that are used in medical applications and electronic industries. The wire is bent by means of several shaping tools as it comes out of the die at a controlled speed. Finally, the finished part (drawn microwire) is cut off from the main wire at the end of the process.

Deep drawing is used for the production of cups for electron guns in color TV sets and shafts of small motors. The most commonly used material in deep-drawing processes at a microscale level is brass. It has properties suitable for drawing, such as good thermal and electrical conductivity and good malleability (Gau et al., 2007). Deep drawing is commonly used in the micromanufacturing of devices useful in the industries related to biomedicine (e.g., medical equipment and surgical implants) and electronics (e.g., connectors, sensors, and ion containers).

In the *extrusion process*, first, the wire has to be cut in order to produce a billet of well-defined length. Thus, a transfer from the cutoff station to the forming station and a precise positioning in the die are necessary. Hence, gripping and precise positioning of the low-weight billet are the major issues in microforming. Apart from the difficulties experienced in handling small parts, making the complex inner geometries of tools such as extrusion dies, the alignment of punch, die, and knockout pin, and the overall precision of the machine impose additional constraints on the microforming processes. Microstructures of 10 μm with radii of approximately 100 nm and a surface roughness of 20 nm in aluminum, copper, brass, and steel have been produced (Geiger et al., 2001).

The *microblanking* process is used for shearing of cutting blades for shavers and punching of holes in fuel nozzles for engines. Micropunching is a popular method for making microholes for mass production. However, this process has two serious issues to be resolved when it is applied to micromanufacturing: fabrication of micron-sized punch and die and maintenance of uniform clearance between micropunch and microdie.

The first problem can be solved by applying appropriate micromachining technologies discussed in Section II. The second is more difficult; therefore, the realization of micropunching depends on the development of a system that ensures easy setting of microtools. One possible solution is to introduce the concept of on-the-machine tool fabrication (Masuzawa, 2000). Production of miniature parts is gaining importance because of the trend of miniaturization in industries such as telecommunication, computers, and medicine, and from sensor technology to optoelectronics.

1.7.1.1 Challenges in Microforming Processes

The challenges in the conventional forming process, such as die design, die wear, and its appropriate treatment, are also the challenges in microforming. These problems are strongly coupled with miniaturization of the parts produced. Figure 1.11 shows the classification

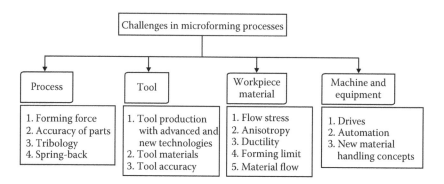

FIGURE 1.11
Classification of challenges in microforming processes.

of some of the challenges faced in the development of microforming processes. These challenges are now discussed briefly.

The main challenge in microforming is the fabrication of tools. However, new manufacturing methods have been developed in order to overcome these difficulties. Newer technologies such as electron beam lithography have been used to develop an embossing tool with dimensions down to 200 nm (Geiger et al., 2001; Saotome and Inoue, 2000). Focused ion beam (FIB) machining is another technology that has been used mainly to produce nanofeatures and nanodevices. The FIB process has also been used in the additive (deposition) mode to produce nanodevices.

As the material behavior becomes less manageable with miniaturization (e.g., low ductility and increased scatter), new strategies have to be developed for microforming. Warm forming is one such alternative that combines the benefits of cold forming, such as high surface quality and strain hardening, and the advantages of hot forming, such as lower process force and larger formability. Furthermore, in the case of production of microbillets from thin wire, special attention has to be paid to shearing failures such as edge draw-in, fractured area, and burr, which can affect the geometrical accuracy of the formed part.

Wire-EDM using wires down to 10 μm in diameter enables the production of very fine punching and extrusion dies (Masuzawa, 2000). The smallest punches of 60 μm diameter have been made by grinding. Structures with dimensions in the range of 10 μm have been fabricated by *laser ablation* (Erhardt et al., 1999; Hellrung and Scherr, 2000). By electron beam lithography, it has been possible to create structures even down to 200 nm in width (Saotome and Inoue, 2000).

1.7.1.2 Mechanical Properties of Microformed Workpiece

The behavior of the workpiece material during microforming changes because of the size effects while scaling down the process. That is, different material properties such as flow stress, anisotropy, ductility, and forming limit of the workpiece materials become important. Different process phenomena such as forming forces, spring-back, and friction, depend on the size changes. Hence, the size effects should be considered during a microforming process design because these phenomena affect the accuracy of the parts produced (Alting et al., 2003).

High speed manufacturing of high-precision components is a prime requirement. A mass production machine has to locate and position the billet in the die with an accuracy of a few microns in the least possible time. The surface where the part can be gripped is extremely small and the weight of the part is too low to overcome adhesive forces. As such, the parts do not separate from the gripper by themselves. Additionally, they have to be placed into a multistation die with tolerance of a few microns. The clearance or backlash between the die and the punch is not an issue in a conventional forming machine; however, it becomes a major issue when producing microparts, which requires strokes of a few hundred microns. This type of equipment can achieve the punching of microholes with an accuracy of around 1 μm. Additionally, adequate measurement technology is required in order to ensure product quality and process control. Finally, the process for making such small parts may require clean rooms, which is cost intensive and leads to new machine concepts.

Besides tool manufacturing, *handling of microparts* is another area of investigation. Particularly, part transfer in multistage formers demands fast and accurate positioning of very small parts despite the unfavorable effect of adhesion between the part and the gripper. Geiger et al. (2002) developed a prototype of a transfer system containing a vacuum gripper that enables the transport of 4.3 parts per second, with a transfer length of 25 mm and an accuracy of 5 μm.

Knowledge of material behavior when subjected to microdeformation is important from the point of view of designing microforming processes. There is a decrease in flow stress during upsetting for microforming that has been explained by the surface layer model (Geiger et al., 1995). On small scales, the material can no longer be considered as a homogeneous continuum (Messner, 1998). The decisive criterion is the ratio of the grain size to the billets' dimensions. The integral flow stress that is measured in upsetting or tensile tests is a function of the share of surface grains as compared to the grains surrounded by other grains. This share increases with grain size and with miniaturization.

Experimental investigations into the effect of miniaturization on friction have shown that friction increases with miniaturization in the case of lubrication with oil, whereas it is size independent in the case of no lubrication (Geiger et al., 1997). This frictional behavior can be explained by the model of open and closed lubricant pockets, also called *dynamic and static lubricant pockets* (Messner, 1998).

Downscaling the part dimensions results in an increase in the punch pressure because of an increase in friction. There is an increase in the bending forces during microbending operations (Eckstein and Engel, 2000). A different orientation of grains in the workpiece could be the reason for the increasing bending force.

1.7.2 Micromolding

Nowadays, a large number of microproducts are partly or fully made of plastics to reduce both their weight and cost. Many of these small-sized products are manufactured by injection molding. Micromolding is equally applicable for the parts made of metals as well. It is a near net shape process, and hence the need for machining operations is minimized or sometimes eliminated. Micromolding processes can be classified as injection molding, reaction injection molding (RIM), hot embossing, and injection compression molding. Some of these processes are briefly described in the following sections.

1.7.2.1 Injection Molding

In microinjection molding, the micromold cavity is created by joining two mold inserts. The microcavity is evacuated and heated above the glass transition temperature of the polymer. An injection unit heats the polymer and supplies the melted polymer into the mold.

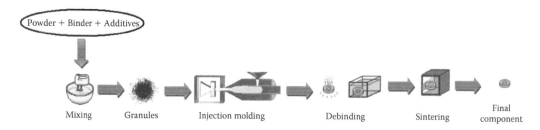

FIGURE 1.12
Process sequence of metal injection molding.

Once the microcavity is filled completely, the polymer is cooled down and demolded from the mold inserts. The metal injection molding process is especially suitable for the production of highly complex parts with small dimensional tolerances and good surface quality. In injection molding, metal or ceramic powder, binder (thermoplastic), and some additives are mixed together in a mixer (Figure 1.12). These granules are then processed in an injection molding machine to form the required shape. Debinding is a binder removal process to obtain the parts with an interconnected pore network without destroying the shape of the component. There is often some chance that the binder is present in the parts holding the metal powder particles together; hence, sintering is needed to evaporate the residual binder quickly. The sintering process evacuates most of the pore volume that was earlier occupied by the binder.

In these methods, the particle size of the powder must be much smaller than the size of the mold. The most important requirement is that the mold must be evacuated before molding because proper venting during molding is not possible because of microcavities. Some deformation usually occurs during the removal of the microparts from the mold. Apart from this, many other problems arise, such as the release of compressive stress, temperature change, chemical postreaction, and shrinkage by sintering.

Injection molding has the possibility of achieving highly complex parts made by plastic injection, which have strong mechanical properties similar to metal parts. Hence, it can obtain the advantages of both the injection molding process and the powder metallurgy methods simultaneously.[*] Micromolded components are the prime choice of implantable medical devices. The medical field demands parts/devices that are smaller and easy to fit in the body. By way of the micromolding process, it is possible to produce medical parts that are small and complex with tight tolerances comparable to machined parts. Some medical devices are made of a combination of plastic and metal, but now they can be micromolded entirely out of plastic.[†] The only time it does not make sense to switch to micromolding is when a component requires a mechanical, electrical, or thermal property that plastic cannot match.

1.7.2.2 Reaction Injection Molding

RIM is also used for fabrication of microparts made by thermosetting and elastomeric materials, while injection molding is used only for thermoplastic materials. In this process, two components are injected into the micromold cavity (Figure 1.13) instead of one type of plastic as in injection molding. The most difficult part of this process is to make a good

---

[*] http://www.micromanufacturing.net/didactico/Desarollo/microforming (Dr. Endika Gandarias, project coordinator, Micromanufacturing).

[†] http://www.micromanufacturing.net/tutorials/didactico/Desarollo/microtechnologies (Dr. Endika Gandarias, project coordinator, Micromanufacturing).

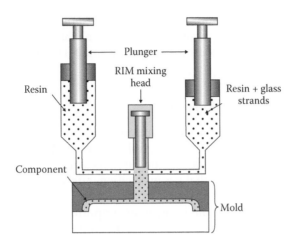

FIGURE 1.13
Schematic of the reaction injection molding process.

mixture of the two constituents in the microdomain. However, this problem can be solved by using ultraviolet curing for mixture preparation of the components.

The two liquid reactants—a resin mixture and a resin with strands—are held in separate containers at an elevated temperature with agitators. These liquids are fed through supply lines at high pressure to the mixing head. When the valves open in the mixing head, the liquids enter a chamber at high pressure and high speed. In the mixing chamber, they are mixed by high velocity impingement. The mixed liquid now flows into the mold at atmospheric pressure and undergoes an exothermic chemical reaction, forming a polymer in the mold.

1.7.2.3 Injection Compression Molding

Injection compression molding combines injection molding and hot embossing processes to overcome the problem of heating the polymer by the mold inserts. In this process, the polymer is injected by a screw into the semiclosed mold inserts and then pressed. This way, the problem of injection through a small gap is avoided when producing a microstructure of a thin carrier layer. Injection compression molding is widely used to produce CDs and DVDs where features of less than 1 μm and small aspect ratio are required.

1.7.2.4 Manufacturing of Micromold

The prime concern of the micromolding process is manufacturing of micromold inserts. The mold inserts should have smooth sidewalls to avoid friction. For manufacturing high aspect ratio features, narrow grooves, sharp corners, and mold inserts need a strong backing in order to provide the necessary mechanical stability against the forces. Therefore, the selection of the appropriate manufacturing technique is an important issue. The micromold can be prepared with micromachining processes such as EDMM, LBMM, mechanical micromachining equipped with computer numerically controlled (CNC) machines, USMM, lithography, and electroplating (Masuzawa, 2000).

Mold inserts with tapered walls and 3D curved surfaces are fabricated by mechanical micromachining. Wire-EDM and die-sinking EDM are also used for fabrication of mold inserts. However, the surface roughness of the mold inserts fabricated by EDM is higher than that made by mechanical micromilling. Lithographic processes such as LIGA, UV lithography, and x-ray lithography are suitable for fabrication of micromold inserts with minute features, high structural tolerances, small lateral dimensions, and high surface finish.

1.7.2.5 Some Issues Related to Micromolding

There are many issues that arise when conventional molding is scaled down to the micro-domain. The first important issue is to control a very small volume of material to be molded, which requires a special machine or auxiliary equipment. It is also necessary to redesign the system in such a way that gas evacuation, quick injection, and removal of the part can be performed easily.

Evacuation of the mold before molding is an important requirement because some deformation may occur during the removal of microparts from the mold. It may be because of the release of compressive stresses, temperature difference, postchemical reaction, and shrinkage after sintering (Piotter et al., 1999). Sometimes, the delicate features of the mold inserts may even be destroyed if the micromold is not designed properly or molding parameters are poorly chosen. Small orifices in the micromolding machine create extremely high shear stress in the melt, which changes the melt viscosity and material properties as well.

The design of the runner system is also important for micromolding. The runner has the thickest wall cross-section in the system and it consumes most of the cycle time of the process. A system with full round runner can be used to increase productivity. Cold-runner systems are preferable for micromolding because hot-runner systems have not been designed for small parts. Other challenges in the design of the system include planning for the gate location, ejection surfaces, slides, and parting lines. For micromolding, the presses need higher injection speeds and pressures to push the melt through tiny nozzles and flow channels. Higher injection pressure helps reduce the viscosity of the molten material by shear thinning and ensures complete filling of the features before it gets cooled.

The selection of proper material is also an important issue, especially when dealing with fine features, thin wall sections, and long aspect ratios. Liquid crystal polymer (LCP) produces complex features and holds tight tolerances. Materials such as polyetheretherketone (PEEK) also make good micromolded parts, but they do not fill the mold cavities. Other materials that are used for micromolding applications are polyamide-imide, polyetherimide, cyclo-olefin (COC) resins, acrylonitrile–butadiene–styrene (ABS), nylon, acetal, and thermoplastic polyester.

1.7.3 Microcasting

Microcasting is another process used for mass production of microcomponents. Microcasting is a competitive process for the production of metallic microparts with high mechanical strength and high aspect ratio. Microcasting processes can be classified into four different categories: capillary action, investment casting, vacuum pressure casting, centrifugal casting (Baumeister et al., 2005).

1.7.3.1 Capillary Action

In the capillary action process, the microcavities are filled by capillary action of the molten material (Moehwald et al., 1999). There are two different ways to fill the microcavities:

(1) the molten material is sucked into the mold by capillary pressure or (2) the material melts inside the mold and fills the microcavities through capillary force. The pressure is then applied to the mold to force out the excess molten material through the slit. The problem in this process is that the composition of the original material changes owing to absorption of coating material during solidification. Furthermore, the process is applicable only for those geometries that can be filled by capillary action.

1.7.3.2 Investment Casting

Investment casting produces near net shape parts with precise details and dimensional accuracy. Therefore, machining and other secondary operations are usually not required. Furthermore, by systematic design of the casting, multiple parts can be eliminated by creating one casting that will replace the assembly. These castings are also subjected to secondary operations. Molten casting alloys must exhibit a low viscosity in order to fill the microfeatures completely. For ceramic investment, the important factors are the ability for high-precision replication, expansion behavior, and low surface roughness. Preheating temperature and pressure influence the entire form filling process (Baumeister et al., 2002), and as a consequence, the achievable grain size and the resulting mechanical properties.

1.7.3.3 Vacuum Pressure Casting

In vacuum pressure casting, molten material is kept in an induction furnace (Figure 1.14). On top of the induction furnace, the mold is fixed upside down in a special type of fixture. The mold is kept partially submerged in the molten material. When vacuum is created, it reduces the air pressure inside the mold to about two-thirds of atmospheric pressure, which facilitates drawing of the molten material into the mold cavities through a gate at the bottom of the mold. A pressure in the range of 3.5–4 bar is sufficient to make a micron-sized casting (Baumeister et al., 2002). The molten metal in the furnace is usually above the melting temperature and it begins to solidify within a fraction of a second. After the mold is filled, it is withdrawn from the molten material.

1.7.3.4 Centrifugal Casting

The centrifugal casting process utilizes the inertial force produced by rotation to distribute the molten metal into the mold cavities. In this process, the melt fills the mold cavities by

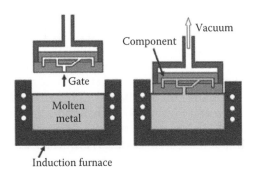

FIGURE 1.14
Schematic diagram of vacuum pressure casting.

centrifugal force. The molten metal is poured into a rotating mold and the axis of rotation may be horizontal or vertical. This process is good for hollow cylindrical parts with outer shapes such as square, polygon, and cylinder. The molds are generally made of steel, iron, or graphite, and coated with a refractory material to increase its life.

There are two ways of filling the mold: the molten metal is poured into the runner and then the system rotates or the molten metal is poured into the already rotating system. In the latter case, the centrifugal force is higher because the machine rotates at full speed while pouring the molten material. The rotational speed of centrifugal casting machines is generally in the range of 350–3000 rpm. Centrifugal casting machines have better cavity-filling pressure than vacuum pressure casting machines, which are beneficial for casting that requires small features. However, the high pressure may cause defects such as gas entrapment owing to high turbulence.

1.7.3.5 Issues Related to Microcasting

The materials used for microcasting should have high flowability, high form-filling ability, low shrinkage and contraction, low segregation, high surface quality (which depends on the microstructure and dimensional accuracy), low susceptibility to hot cracks, and good mechanical properties. The pattern should be designed so that all the patterns get filled homogeneously. In microcasting, the parts are generally smaller than the feeder and runner, and solidification must start in the microparts (not in the feeder and runner). Therefore, the cross-sectional area of the sprue should increase from the top to the bottom where it connects to the microparts (Baumeister et al., 2005).

The study of flow behavior in microcasting is also an important factor because the surface-to-volume ratio is comparatively higher than that of macrocasting. Hence, the effect of surface roughness and the turbulent flow of the molten material are necessary to consider. Another important issue is that the solidification in the small cavities occurs very quickly, which also obstructs the form-filling process of other features. This happens because of the very low temperature of the mold and insufficient overheating of the molten material. Surface shrinkage, holes, and surface pores are also caused by fast solidification and the very high temperature that is needed for casting. If the filling pressure is too low, the microcavities do not get filled because the surface tension of the molten material is so high that it is not able to enter the microcavities. High pressure gives better replication of the mold surface but produces high surface roughness. Low pressure produces low surface roughness.

The properties of the ceramic slurry also play an important role in the quality of casting. The ceramic slurry should have high flowability, be easy to process, have good mechanical and thermal stability at high temperatures, be chemically inert with the molten material, and its microcast part should have high mechanical strength, low surface roughness value, and low porosity (Baumeister et al., 2005). The selection of the ceramic slurry depends on the cast metal and the required strength of the mold. Phosphate bonded ceramic slurries are easy to handle and they have a higher strength as compared to plaster bonded ceramic slurries (Low and Swain, 2000). Therefore, they are widely used in dental applications for precious metals. Ceramic slurry with fine particles is required for producing a high-precision replica and accurate casting. However, excessively fine particles are also problematic because their surfaces require higher liquid content, and the part may develop drying cracks during hardening. The surface roughness of the cast part significantly depends on the ceramic slurry because the surface of the cast product is a replica of the mold. Other parameters that affect the surface roughness are mold temperature and casting pressure. Hence, a smooth surface of the mold is necessary to produce high-quality castings, and it also depends on the size of the slurry particles.

The cooling rate of the casting decreases with increase in mold temperature, which affects the microstructure of the cast parts. The minimum possible size of the casting depends on the aspect ratio of the features (the ratio of flow length to wall thickness), which, in turn, is determined by the preheating temperature of the mold and filling pressure.

1.7.3.6 Applications of Micromolding and Microcasting

Micromolding processes have been employed to fabricate a variety of polymer components. Most applications are in the field of micro-optics for components such as wave guides (Ulrich et al., 1972), optical switches, optical gratings (Knop, 1976), optical fiber components, and antireflective surfaces (David et al., 2002). Microfluidics (Gerlach et al., 2002) also employs these processes for components such as inkjets, flow sensors, channels, pumps (Schomburg et al., 1999), nozzles, valves (Goll et al., 1997), mixtures, pressure sensors, capillary analysis systems, and nebulizers. The most widely used micromolded products are CDs and DVDs for data storage. There are also some examples of micro-electrical and micromechanical devices such as electrical switches, acceleration sensors, spectrometers, photonic structures, and nanoimprinting (Lebib et al., 2002; Malaquin et al., 2002; Chou et al., 1995).

The medical field is a major growth market for micromolding. Today's demand for micro-mold products such as tiny catheter products, surgical instruments, implantable devices, tissue anchors, needle sheaths, absorbers, and stent blanks (Lee et al., 2001) has opened up a new frontier for medical products and device manufacturers.

Microcasting is also widely used in the medical field and for making finely detailed jewelry. Medical applications include instruments for minimally invasive surgery, dental devices, and instruments for biotechnology.

1.8 Microjoining

Macro- or microjoining (welding, soldering, brazing, and adhesive bonding) is a part of the manufacturing and assembly processes. Joining can be mechanical connection, electrical connection, and optical coupling. The continuing miniaturization of engineering devices brings new challenges to the technology of microjoining in terms of cost reduction, enhanced performance, and reliability. Microjoining deals with the joining of parts with characteristic dimensions less than a few hundred micrometers but more than hundreds of nanometers (Zhou, 2008).

Most microjoining processes were developed around the 1950s (Anderson et al., 1957; Cullison, 1996). The combination of thermocompression and ultrasonic wire bonding, called *thermosonic wire bonding*, was reported in 1970 (Harman, 1997). This technology of wire bonding is one of the key chip-level interconnection technologies in microelectronic products. Later, electron beam welding (EBW), which is suitable for macro-, micro- and even nanoscale joining because of the precise beam quality, fixturing, and controls integrated with the system, was developed (Stohr and Eriola, 1958).

Microjoining has been an integral part of micromanufacturing for many decades in the microelectronics, medicine, aerospace, and defense industries. Microjoining continues to

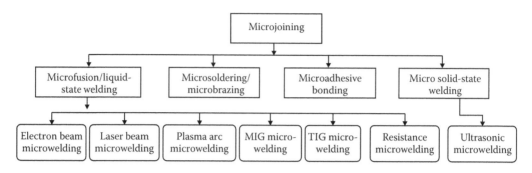

FIGURE 1.15
Classification of microjoining processes.

face challenges because of advances in miniaturization. The classification of microjoining processes is shown in Figure 1.15. These processes face challenges in interconnecting MEMS, wherein electrical, mechanical, fluidic, and optical components are connected as well as coupled to the macroscopic external environment (Menz et al., 2001). This microjoining of MEMS should overcome some problems; for example, it should yield undistorted physical and chemical output signals, and should also protect from corrosion as well as mechanical damage (Lancaster, 1984; Kou, 2003; Eagar, 1993; Norrish, 2006).

Many processes such as welding, soldering, and brazing have been used for joining similar as well as dissimilar materials. However, simply scaling down conventional joining machines for micromanufacturing is not a final solution. Apart from scaling down the process, optimization of the welding parameters, creating a stable welding/joining structure, and overcoming the size effect require attention. When selecting a specific joining process, the particular requirements and conditions involved must be examined. Some of the parameters of microjoining processes that can be considered are depth of penetration, joint preparation, cleaning, inert gas or vacuum environment, weld joint accessibility, proximity to heat-sensitive materials, productivity, and cost.

Micromanufacturing is bound to play an important role in the development of highly integrated microsystems. During the manufacturing of complex microsystems, many components require bonding or joining to form an assembly. Hence, microjoining processes are essential for micromanufacturing and assembly. As mechanical assembly (screws, gaskets, and clamps) causes time-related failure, welding is often the preferred joining process for both intermediate and final assemblies. Microjoining processes are needed for packaging and interconnecting of MEMS where individual electrical, mechanical, fluidic, and optical components need to be connected and coupled to the macroscopic external environment. Furthermore, it would be beneficial if no additional material is included (e.g., filler material), and pollution during processing is minimized (Guo, 2009).

Electron beam, laser beam, plasma arc, and gas tungsten arc are the dominant choices for microwelding processes. The precise narrow welds and low energy input prevent distortion, minimize HAZ, and produce welds of high metallurgical quality. For high-volume welding of electronic or precision assemblies, electron beam microwelding (EBMW) and laser beam microwelding (LBMW) are the best choices. For both LBMW and EBMW, beam size, beam characteristics, beam–material interaction, numerical simulation of the microwelding process, temperature measurement, and integration of image processing for control of the welding process are the central points of research (Guo, 2009). Most of these process parameters affect the cooling rate, stability of fluid flow, surface tension, and distortion, which ultimately affect the welding/joining quality.

1.8.1 Microwelding

Welding of thin and delicate materials by depositing very small lumps of welded metal using precisely controlled minimal heat energy is called *microwelding*. These processes are designed to maintain tight operational tolerances and not to interfere with the product function as well as quality. Miniature devices and applications for which components must be microwelded require equipment that is capable of locating the joining with the required precision. This should join minimal material in the exact spot where it is needed, with minimum interference with the surroundings. Microwelding is a key joining process utilized in the manufacture of miniature devices such as micro medical implants and MEMS. Commonly employed microwelding processes are shown in Figure 1.15. Some of them are discussed in the following sections.

1.8.2 Fusion Microwelding

1.8.2.1 Electron Beam Microwelding

EBW is a popular process for macro applications and it is characterized by low thermal load, precise energy input, and fine beam manipulation capabilities. However, this method is also gaining importance in micro applications because of its capability to focus the beams exactly within a diameter of a few microns. The most common industries that employ EBW are the aerospace, automotive, and nuclear industries, and manufacturers of electronics, consumer products, and medical devices.[‡] EBW is capable of joining any metal such as refractory metals (tungsten, molybdenum, and niobium) and chemically active metals (titanium, zirconium, and beryllium)[‡] as also dissimilar metals. EBMW is also able to join dissimilar metals (Ogawa et al., 2009; Dilthey and Dorfmuller, 2006; Reisgen and Dorfmuller, 2008; Smolka et al., 2004).

In this process, the KE of electrons is utilized for melting the targeted material. Electrons accelerated by an electric field are focused into a thin beam by the focusing coil. The deflection coil moves the electron beam along the direction of the weld (Figure 1.16). The heat energy fuses the edges of workpieces, which are joined together, forming a weld after solidification. The process is carried out in a vacuum chamber to prevent loss of electron energy in collisions with air molecules. The vacuum also protects the components against contamination and possible oxide formation.

This process allows almost inertia-free manipulation of the electron beams via electromagnetic coils. Another important feature of the process is the ability to magnetically deflect the electron beams to form different patterns or shapes. The beam deflection control unit is a useful tool for microjoining miniature details and complex configurations.[§] Pulsed electron beam energy, when coupled with programmable beam deflection, can offer the user versatility for microjoining applications. The heat intensity is determined by the electron beam's spot size; a smaller spot size has higher heat intensity. The positioning accuracy must be of the order of the heat source diameter. The process yields high flexibility and accuracy and is applicable where joint tolerances are less critical.

Modern digital electronic controls, coupled with the modern computer-controlled operator console, allow for continuous cycle-to-cycle repeatability. The EBMW process offers many merits such as the ability to produce welds with high depth-to-width ratio, eliminating multiple-pass welding. Owing to low heat input, the process produces minimal shrinkage and distortion, making it possible to weld in close proximity to heat-sensitive

---

[‡] http://www.substech.com/dokuwiki/doku.php?id=electron_beam_welding_ebw.
[§] http://www.joiningtech.com/downloads/Micro-JoiningEBEnergy.pdf.

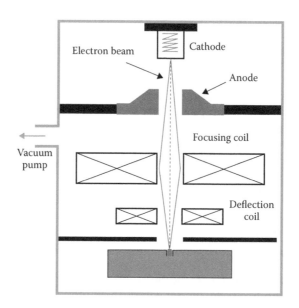

FIGURE 1.16
Schematic diagram of the EBW process.

components. Microwelding is performed under vacuum, and hence the microwelded parts have superior strength. The parts microwelded under vacuum can yield 95% strength as compared to that of base material. The vacuum environment eliminates impurities such as oxides and nitrides, and hence gives high-purity welded joints. The weld zone is narrow, that is, around 50 μm.

The EBMW process also has certain limitations, such as high equipment cost and limited work chamber size. X-rays produced during welding are harmful to the operator, and the rapid solidification rates can cause cracks in some materials.

1.8.2.2 Laser Beam Microwelding

In LBMW, the highly concentrated (10^8–10^{10} W/cm^2) laser beam forms a tiny weld pool in a few microseconds. Solidification of the weld pool surrounded by the cold metal occurs as fast as the melting. No shields (neutral gas, flux) are required because no contamination occurs owing to the very short exposure time of the molten metal to the atmosphere. Laser welding is a noncontact process that requires only single-sided access. Materials that can be microwelded using laser beam are stainless steel, inconel, aluminum, titanium, copper, and refractory alloys, and tool steels. LBMW is widely used for micromanufacturing in industries such as electronics, communication, aerospace, and medicine, and in those that produce scientific instruments.

In LBMW, exact control of energy in the joining area can be achieved, which increases precision and minimizes HAZ. Some of the special features of the process are selective joining, single-step process, short cycle time, low mechanical and thermal load on the component, good strength, high weld strength to weld size ratio, no weld debris generation, and a high level of flexibility. Even transparent materials can be welded without additional absorbers by adapting the laser wavelength according to the absorption behavior.[1]

---

[1] http://www.substech.com/dokuwiki/doku.php?id=laser_welding_lw.

Infrared (IR) lasers are used for microwelding. If any oxide layer exists on the material to be processed, the first step is to remove that thin oxide layer (Gabriel et al., 2007) because oxide absorbs more heat compared to metals, and it may disturb the wettability of the surface and produce welding of an inappropriate quality (Gabriel et al., 2007).[**] The welding energy depends strongly on the surface reflectivity of the metal and on the roughness of the surface. Seam widths below 200 μm can be achieved by continuous fiber lasers and pulsed Nd-YAG lasers.[††] Laser beam welding of thermoplastic polymers provides weld seams of high optical and mechanical quality.

Laser beam equipment consists of a cylindrical ruby crystal (ruby is aluminum oxide with chromium dispersed throughout) with both the ends absolutely parallel to each other. One face of the ruby crystal is highly silvered so that it reflects the incident light. The other face is partially silvered and contains a small hole (non reflecting small area) through which the laser beam is trapped. The ruby crystal is surrounded by a helical flash tube containing the inert gas "xenon" and this tube is connected to a pulsed high voltage source, so that xenon transforms the electrical energy into light energy. As the ruby is exposed to intense light flashes, the chromium atoms of the crystal are excited and pumped to a high energy level. These chromium atoms immediately drop to an intermediate energy level with the evolution of a discrete quantity of radiation in the form of red fluorescent light. As the red light emitted by one excited atom hits another excited atom, the second atom gives off a red light, which is in phase with the colliding red light wave. The effect is enhanced as the silvered ends of the ruby crystal cause the red light to reflect back and forth along the length of the crystal. The chain reaction collisions between the red light wave and the chromium atoms become so numerous that finally the total energy bursts and escapes through the tiny hole as a "laser beam." This laser beam is focused by an optical focusing lens onto the spot to be welded. Optical energy, as it impacts the workpiece, is converted into heat energy. Because of the heat generated, the material melts over a tiny area, and on cooling the material within becomes a homogeneous solid structure to make a strong joint. The laser beam is focused to a very small diameter and it can be used to weld foils as thin as 25 μm. Weld penetration can be up to 1.5 mm depth (Asibu, 2009; Chang et al., 2010). In the LBMW process, reflectivity of the workpiece is one of the major constraints. However, once some laser power is delivered to the workpieces at the joint and its temperature raised, the reflectivity decreases.

The alternative for microwelding other than the ruby crystal is low-power fiber lasers. These are very compact and robust lasers in terms of beam quality and wall plug efficiency of approximately 25% (Norman et al., 2004; Dominic et al., 1999). The single-mode fiber laser is an efficient, reliable, and compact solution for microwelding.

LBMW can be used to weld very small, intricate devices or in close proximity to heat-sensitive locations on the assembly, such as glass-to-metal seals, without disturbing the seal. In addition to aesthetic benefits, clean welds result in products that are easier to sterilize or fit into other assemblies. Laser provides high strength with a minimum number of welds.

1.8.2.2.1 Laser Bonding, Soldering, and Brazing

Laser bonding of silicon and glass is a nonmelting solid-state joining process. This is important for thermally sensitive components such as organic light-emitting diodes (OLEDs) or organic electronics. It is used for the hermetic sealing of electronic packages and MEMS

[**] Mohanan S. Laser assisted film deposition, advanced materials. http://www.uniulm.de/ilm/Advanced Materials/Presentation/MahananLaserassistedfilmdeposition.pdf.

[††] http://www.ilt.fraunhofer.de/eng/ilt/pdf/eng/products/Micro_Joining_with_Lasers.pdf.

components in glass, silicon, or ceramics. Laser forms oxygen bridges, and the selective laser irradiation of the joining area produces the bond (seam widths less than 200 μm).[††] Laser bonding generates minimum thermal load on the entire component. This process is particularly suited for bonding and encapsulation of microsystems with moving parts and thermally sensitive components. Laser soldering and brazing are two joining processes used in electronic, photovoltaic, and medical devices for components that are sensitive to temperature and mechanical impact. The operation can be completed in a few hundred milliseconds. A prominent feature of laser beam soldering is based on the ability to work on pitch dimensions ranging from 100 to 2000 μm by an appropriate selection of focusing lens and irradiation strategy.[††]

1.8.2.3 Resistance Microwelding

Resistance microwelding is mostly used to join nonferrous materials in the fabrication of precision components and devices (such as sensors, actuators, and medical devices). Resistance welding uses the high resistance of a weld interface to create heat by passing current through the parts to be joined. The circuit is created by electrodes that contact both parts either from the same side or from the opposite sides of the workpieces (Figure 1.17). Force is applied before, during, and after the application of current to prevent arcing. Unique microresistance welding techniques have been developed for attachment of wires/ foils, which range in thickness from 12.5 to 750 μm (Kuroda and Ikeuchi, 2006).

Materials that can be microwelded using resistance microwelding are nichrome, nickel, stainless steel, inconel, and copper alloys, and precious metals. The process requires comparatively low investment, and it is more environmentally friendly. However, it may not be recommended when the parts are mechanically delicate.

1.8.3 Solid-State Microwelding

Solid-state welding is a group of welding processes that produce temperatures essentially below the melting point of the base materials being joined, without the addition of a filler metal. In these processes, time, temperature, and pressure, individually or in combination, produce coalescence of the base metal without melting. Some of these processes offer certain advantages since the base metal does not melt and form a nugget (Takahashi and Maeda, 2011). The metals being joined retain their original properties. When dissimilar metals are joined, their thermal properties are of less importance for solid-state welding

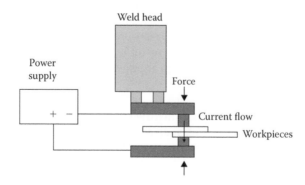

FIGURE 1.17
Schematic diagram of the resistance microwelding process.

than for the arc welding processes. However, for microwelding, ultrasonic microwelding (UMW) is mostly used. Other solid-state microwelding processes include microsoldering and microbrazing (Han et al., 2008), which have not been discussed here.

1.8.3.1 Ultrasonic Microwelding

The UMW process is used for semiconductor chip-level interconnections using wire diameters typically less than 25 μm but occasionally at a few hundred microns for power electronics applications. Ultrasonic bonding uses vibration energy at the joint interface to create the bond. Vibration energy is delivered to the interface by a horn that contacts the top part. The horn vibrates at frequencies of hundreds to thousands of times per second, with motion amplitude between 10 and 100 μm. The lower side of the part is supported with an "anvil," which can be either in the static mode or in the vibrating mode. The "scrubbing" action that occurs at the weld interface displaces metallic oxides and foreign materials to expose fresh clean metallic surfaces for bonding. The vibratory action under applied force causes plastic deformation of surface asperities at the weld interface, leading to highly intimate contact and diffusion of metallic atoms, resulting in a joint without any melting. There is a controlled deformation or thickness reduction of the parts. Contact of the horn with the part is maintained by the horn's friction, enhanced by a knurling pattern on the horn along with a force applied on the parts (Figure 1.18) (Yasuo and Masakatsu, 2011). Materials that are difficult to weld by other welding processes are commonly welded using UMW, for example, aluminum alloys, copper alloys, and gold. A basic drawback of the UMW process is that the horn is consumable, which requires timely inspection and replacement. Joint geometry is mostly limited to lap welding only.

1.8.4 Microjoining by Adhesive Bonding

For the assembly of microparts, the use of mechanical means (e.g., screws) is not a common practice, but microadhesive bonding is a feasible solution in such cases. The desirable properties of adhesives when used for such subassemblies can be totally different from those in classical macroscopic dimensions. The handling of adhesive materials is also different. Only a small amount of adhesive is needed. It has to be placed very precisely, and it may be swallowed by capillary action into the microstructure. The transport of a single drop with defined size must be provided (Böhm et al., 2006; Stefan et al., 2006).

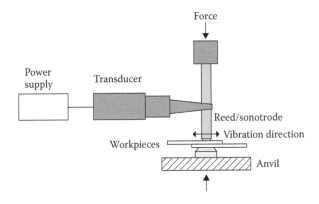

FIGURE 1.18
Schematic diagram of the ultrasonic microwelding process.

This process has many applications, such as in conductive joints in dynamically stressed systems such as electronics in automotive components, thermal management (high density packing of microelectronics), fiber-optic couplers, electro-optic transducers (telecommunications), and high definition sealing in microfluidics (biotechnology, life sciences). This process extends many advantages over other processes, for example, the joining of many dissimilar materials, low heat/cold joining, low mechanical stresses, uniform stress distribution, new avenues for design, and innovative technical solutions.

1.9 Miscellaneous Applications

Micromanufacturing processes coupled with automation lead to economical and high-yield mass production. Some of the application areas are discussed in this section. Typical micromechanical devices are used in three areas: sensors (i.e., pressure sensors and accelerometers), actuators (i.e., solenoids and valves), and microstructures (i.e., inkjet nozzles and connectors). According to a study by the European Network of Excellence in Multifunctional Microsystems (NEXUS) organization, the worldwide market for microsystem technologies is growing at an average rate of 11% per year.[‡‡] Microsensors and microactuators have created markets for read/write heads and inkjet heads, in addition to creating new opportunities in areas such as microphones, memories, micro energy sources, and chip coolers.[§§]

A major boost to the growing market is the consumer electronics segment. Experts see rear and front projection TVs for home theater, as well as HDDs serving the increasing storage requirements of digital equipment such as DVD recorders, digital cameras, camcorders, and portable MP3 players. A big driving force is the mobile phone, which already features motion sensors and is amenable to a variety of additional sensors and functions such as liquid lenses for camera zoom, fingerprint sensors, microfuel cell power sources, gas sensors, and weather barometers.

1.9.1 IT and Telecommunications

The push for smaller and faster electronics has led to enormous technical advances in miniaturization and integration of components. The IT field depends strongly on miniaturization capabilities. Micromanufacturing is a driving force for computer revolution. Micromanufacturing packs more and more devices into a single chip. All signs point toward a revolution that advances to the limits set by natural laws and the molecular graininess of matter. Trends in miniaturization point to remarkable results around 2015; device sizes will shrink to molecular dimensions and switching energies will reduce to the scale of molecular vibrations. With devices built on this scale, a million modern circuits could fit into a pocket.

1.9.2 Automotives

The automotive industry remains a driving force for microsystem innovation that offers high-volume applications, while aerospace applications are mostly related to sensors and

---

[‡‡] http://www.micromanufacturing.net/tutorials/didactico/Desarollo/microtechnologies.
[§§] NEXUS, www.nexus-mems.com.

instrumentation. The automotive microsystems supply chain shows that it can innovate at all levels (materials, components, systems, usage).§§ The automotive sector remains a major application field with several high-volume safety products including air bags and tire pressure monitoring systems. Some of the most demanding sensors positioned currently in cars are accelerometers (airbag release); micronozzle systems (injection system); pressure sensors (gas recirculation); gyroscopes (navigation); level, light, and temperature sensors (oil/gas indicators, turn-on lights, and outside/inside temperature measurement); parking sensors (collision avoidance); and air flow sensors (air/flow ratio control) (Groover, 2002).

1.9.3 Medicine

The biomedical and telecommunication sectors are also major drivers for current and future microsystem developments. In this field, for example, *in vitro* diagnostics (mostly biochips, bioarrays, and microplates) is expected to become a very strong large-volume market.§§ The medical and biomedical market has a large variety of applications of microproducts.

1.10 Remarks

In this chapter, the importance of micromanufacturing processes and microproducts is discussed in terms of their functionalities and physical dimensions. Micromanufacturing processes are classified into different groups on the basis of material interactions and some of them (such as microcutting, micromolding, microcasting, microwelding, and micro-forming) are discussed in brief in this chapter. Micromanufacturing is a fast-growing field, which has become an inevitable part of household life and such sectors as medi-cine, defense, electronics and information technology, telecommunication, television, and automotive. Advances in micromanufacturing have significantly improved the living standards and comfort of mankind. In the present era, medical and consumer markets are dominated by microproducts.

References

Adams, D.P., Vasile, M.J., Benavides, G., and Campbell, A.N. 2001. Micro milling of metal alloy with focused ion beam–fabricated tools. *Journal of International Societies of Precision Engineering and Nanotechnology* 25: 107–113.

Alting, L., Kimura, F., Hansen, H.N., and Bissaco, G. 2003. Micro engineering. *CIRP Annals—Manufacturing Technology* 52(2): 635–658.

Anderson, O.L., Christensen, H., and Andreatch, P. 1957. Technique for connecting electrical leads to semiconductors. *Journal of Applied Physics* 28: 923–924.

Aravindan, S., Rao, P.V., and Masahiko, Y. 2010. Focused ion beam machining. *Introduction to Micromachining*, ed. V.K. Jain. Oxford, UK: Alpha Science International Ltd. pp. 3.1–3.28.

Asibu, E.K. 2009. *Principles of Laser Materials Processing*. Hoboken, NJ: John Wiley & Sons, Inc.

Balasubramanium, R. and Suri, V.K. 2010. Diamond turn machining. *Introduction to Micromachining*, ed. V.K. Jain. Oxford, UK: Alpha Science International Ltd. pp. 3.1–3.28.

Baumeister, G., Haußelt, J., Rath, S., and Ruprecht, R. 2005. Microcasting. *Advanced Micro and Nanosystems— Microengineering of Metals and Ceramics*, eds H. Baltes, O. Brand, G.K. Fedder, C. Hierold, J. Korvink, and O. Tabata. Weinheim: Wiley-VCH Verlag GmbH & Co. KGaA.

Baumeister, G., Mueller, K., Ruprecht, R., and Hausselt, J. 2002. Production of metallic high aspect ratio microstructures by micro casting. *Microsystem Technologies* 8: 105–108.

Bitter, J.G.A. 1963. A study of erosion phenomena—I and II, *Wear* 6: 5–21.

Böhm, S., Hemken, G., Stammen, E., and Dilger, K. 2006. Micro bonding using hot melt adhesives. *Journal of Adhesion and Interface* 7(4): 28–31.

Brinksmeier, E., Riemer, O., Gessenharter, A., and Autschbach, L. 2004. Polishing of structured molds. *CIRP Annals—Manufacturing Technology* 53(1): 247–250.

Brinksmeier, E., Riemer, O., and Stern, R. 2002. Machining of precision parts and microstructures. *Initiatives of Precision Engineering at the Beginning of a Millennium* 1(3–11). DOI: 10.1007/0-306-47000-4_1.

Chang, B., Bai, S., Du, D., Zhang, H., and Zhou, Y. 2010. Studies on the micro-laser spot welding of an NdFeB permanent magnet with low carbon steel. *Journal of Materials Processing Technology* 210(6–7): 885–891.

Chou, S.Y., Krauss, P.R., and Renstrom, P.J. 1995. Imprint of sub-25 nm vias and trenches in polymers. *Applied Physics Letter* 67: 3114–3116.

Cullison, A. 1996. Welding: A heavyweight in a miniature world. *Welding Journal* 75: 29–34.

Das, M., Jain, V.K., and Ghoshdastidar, P.S. 2010. Nano-finishing of stainless-steel tubes using R-MRAFF Process. *Machining Science and Technology* 14: 365–389.

David, C., Haberling, P., Schnieper, M., Sochtig, J., and Zschokke, C. 2002. Nano-structured anti-reflective surfaces replicated by hot embossing. *Microelectronic Engineering* 61–62: 435–440.

Dilthey U. and Dorfmuller, T. 2006. Micro electron beam welding. *Microsystem Technologies* 12(7): 626–631.

Dominic, V., MacCormack, S., Waarts, R., Sanders, S., Bicknese, S., Dohle, R., Wolak, E., Yeh, P.S., and Zucker, E. 1999. 110 W fibre laser. *Electronics Letters* 35: CPD11/1–CPD11/2.

Eagar, T.W. 1993. Energy sources used for fusion welding. *ASM Handbook of Welding, Brazing, and Soldering*, Vol. 6. ASM International. Cambridge, pp. 3–6.

Eckstein R. and Engel U. 2000. Behaviour of the grain structure in micro sheet metal working. *Proceedings of the 8th International Conference on Metal Forming*, Krakow, Poland, pp. 453–459.

Engel, U. and Eckstein, R. 2002. Microforming—From basic research to its realization. *Journal of Materials Processing Technology* 125–126: 35–44.

Erhardt, R., Schepp, F., and Schmoeckel, D. 1999. Micro-forming with local part heating by laser irradiation in transparent tools. *Proceedings of the Seventh International Conference on Sheet Metal (SheMet'99)*, eds M. Geiger, H.J.J. Kals, B. Shirvani, U.P. Singh. Bamberg: Meisenbach. pp. 497–504.

Finnie, I. 1960. Erosion of surfaces by solid particles. *Wear* 3: 87–103.

Gabriel, M.P., Zsolt, I.V., Balint, B., Dumitru, U., and Antonie, C. 2007. Laser applications in the field of MEMS. Industrial laser applications (INDLAS 2007). *Proceedings of SPIE* 7007. DOI: 10.1117/12.801970.

Gau, J.T., Principe, C., and Yu, M. 2007. Springback behavior of brass in micro sheet forming. *Journal of Materials Processing Technology* 191(1–3): 7–10.

Geiger, M., Egerer, E., Engel, U. 2002. Cross transport in a multi-station former for microparts. *Production Engineering* 9(1): 101–104.

Geiger, M., Kleiner, M., Eckstein, R., Tiesler, N., and Engel, U. 2001. Microforming. *CIRP Annals—Manufacturing Technology* 50(2): 445–462.

Geiger, M., Messner, A., and Engel, U. 1997. Production of microparts—Size effects in bulk metal forming, similarity theory. *Production Engineering* 4(1): 55–58.

Geiger, M., Messner, A., Engel, U., Kals, R., and Vollertsen, F. 1995. Design of micro-forming processes: Fundamentals, material data and friction behavior. *Proceedings of the 9th International Cold Forging Congress Solihull, UK, May-1995*, pp. 155–164.

Gerlach, A., Knebel, G., Guber, A., Heckele, M., Herrmann, D., Muslija, A., and Schaller, T. 2002. Microfabrication of single-use plastic microfluidic devices for high-throughput screening and DNA analysis. *Microsystem Technologies* 7: 265–268.

Goll, C., Bacher, W., Büstgens, B., Maas, D., Ruprecht, R., and Schomburg, W.K. 1997. Electrostatically actuated polymer microvalve equipped with a movable membrane electrode. *Journal of Micromechanics and Microengineering* 7: 224–226.

Groover, M.P. 2002. *Fundamentals of Modern Manufacturing*. 2nd ed. John Wiley & Sons, Inc., New Jersey.

Guo, K.W. 2009. A review of micro/nano welding and its future developments. *Recent Patents on Nanotechnology* 3: 53–60.

Han, Z.-J., Xue, S.-B., Wang, J.-X., Zhang, X., Zhang, L., Yu, S.-L., and Wang, H. 2008. Mechanical properties of QFP micro-joints soldered with lead-free solders using diode laser soldering technology. *Transactions of Nonferrous Metals Society of China* 18(4): 814–818.

Harman, G.G. 1997. *Wire Bonding in Microelectronics—Materials, Processes, Reliability and Yield*. 2nd ed. New York, NY: McGraw-Hill.

Hellrung, D. and Scherr, S. 2000. Zuverlassige herstellung miniaturisierter pragewerkzeuge. *wt Werkstattstechnik* 90(11–12): 520–521.

Honegger, A., Langstaff, G., Phillip, A., VanRavenswaay, T., Kapoor, S.G., and DeVor, R.E. 2006a. Development of an automated microfactory: Part 1—Microfactory architecture and subsystems development. *Transactions of the North American Manufacturing Research Institution of SME* XXXIV: 333–340.

Honegger, A., Langstaff, G., Phillip, A., VanRavenswaay, T., Kapoor, S.G., and DeVor, R.E. 2006b. Development of an automated microfactory: Part 2—Experimentation and analysis. *Transactions of the North American Manufacturing Research Institution of SME* XXXIV: 341–348.

Jackson, M. Robinson, G.M., Brady, M.P., and Ahmed, W. 2006. Micromilling of bipolar fuel cell plates using titanium- and diamond-coated cutting tools. *International Journal of Nanomanufacturing* 1(2): 290–303.

Jacobs, P.F. 1996. *Stereolithography and Other RP&M Technologies: From Rapid Prototyping to Rapid Tooling*. New York, NY: Society of Manufacturing Engineers and the Rapid Prototyping Association.

Jadoun, R.S. 2010. Ultrasonic micromachining. *Introduction to Micromachining*, ed. V.K. Jain. Oxford, UK: Alpha Science International Ltd. pp. 3.1–3.28.

Jain, V.K. 2002. *Advanced Machining Processes*. New Delhi, India: Allied Publishers.

Jain, V.K. 2010. Micromachining: An introduction. *Introduction to Micromachining*, ed. V.K. Jain. Oxford, UK: Alpha Science International Ltd. pp. 3.1–3.28.

Jain, V.K., Kalia, S., and Sidpara, A. 2011. Some aspects of fabrication of micro devices by electrochemical micro machining and its finishing by magnetorheological fluid. *International Journal of Advanced Manufacturing Technology*. DOI: 10.1007/s00170-011-3563-4.

Jain, V.K., Ranjan, P., Suri, V.K., and Komanduri, R. 2010. Chemo-mechanical magneto-rheological finishing (CMMRF) of silicon for microelectronics applications. *CIRP Annals—Manufacturing Technology* 59: 323–328.

Joshi, S.S. 2010. Micro-electric discharge micromachining. *Introduction to Micromachining*, ed. V.K. Jain. Oxford, UK: Alpha Science International Ltd. pp. 3.1–3.28.

Kals, T.A. and Eckstein, R. 2000. Miniaturization in sheet metal working. *Journal of Materials Processing Technology* 103: 95–101.

Knop, K. 1976. Color pictures using the zero diffraction order of phase grating structures. *Optics Communications* 18: 298–303.

Ko, T.J. 2010. Abrasive jet micromachining. *Introduction to Micromachining*, ed. V.K. Jain. Oxford, UK: Alpha Science International Ltd. pp. 3.1–3.28.

Kou, S. 2003. *Welding Metallurgy*. Hoboken, NJ: John Wiley & Sons Inc.

Kulkarni, A., Sharan, R., and Lal, G.K. 2002. An experimental study of discharge mechanism in electrochemical discharge machining. *International Journal of Machine Tools and Manufacture* 42: 1121–1127.

Kulkarni, A., Jain, V.K., and Misra, K.A. 2011a. Electrochemical spark machining: Present scenario. *International Journal of Automation Technology* 5(1): 52–59.

Kulkarni, A.V., Jain, V.K., and Misra, K.A. 2011b. Electrochemical spark micromachining (microchannels and microholes) of metals and non-metals. *International Journal of Manufacturing Technology and Management* 22(2): 107–123.

Kuroda, T. and Ikeuchi, K. 2006. Micro flash resistance welding of super duplex stainless steel. *Journal of the Society of Materials Science Japan* 55(9): 855–859.

Lancaster, J. 1984. The physics of welding. *Physics in Technology* 15(2): 73–79.

Lebib, A., Chen, Y., Cambril, E., Youinou, P., Studer, V., Natali, M., Pepin, A., Janssen, H.M., and Sijbesma, R.P. 2002. Room-temperature and low-pressure nanoimprint lithography. *Microelectronic Engineering* 61–62: 371–377.

Lee, G.-B., Chen, S.-H., Huang, G.-R., Lin, Y.-H., Sung, W.-C., and Lin, Y.-H. 2001. Microfabricated plastic chips by hot embossing methods and their applications for DNA separation and detection. *Sensors Actuators* B 75: 142–148.

Low, D. and Swain, M.V. 2000. Mechanical properties of dental investment materials. *Journal of Materials Science: Materials in Medicine* 11(7): 399–405.

Madarkar, R.D. 2007. Investigations into magnetic abrasive deburring (MADe), M.Tech. Thesis, Indian institute of technology Kanpur, India.

Madou, M.J. 2001. *Fundamentals of Microfabrication.* Boca Raton: CRC Press.

Malaquin, L., Carcenac, F., Vieu, C., and Mauzac, M. 2002. Using polydimethylsiloxane as a thermo-curable resist for a soft imprint lithography process. *Microelectronic Engineering* 61–62: 379–384.

Masuzawa, T. 2000. State of the art of micromachining. *CIRP Annals—Manufacturing Technology* 49: 473–488.

Messner, A. 1998. Kaltmassivumformung metallischer Kleinstteile-Werk-stoffverhalten, Wirkflachenreibung, ProzeBauslegung. *Reihe Fertigungstechnik Erlangen.* Band 75, eds M. Geiger and K. Feldmann. Bamberg: Meisenbach.

Menz, W., Mohr, J., and Paul, O. 2001. *Microsystem Technology.* Weinheim, Germany: Wiley-VCH Verlag GmbH.

Miller, D.S. 2004. Micromachining with abrasive waterjets. *Journal of Materials Processing Technology* 149(1–3): 37–42.

Moehwald, K., Morsbach, C., Bach, F.-W., and Gatzen, H.-H. 1999. Investigations on capillary action microcasting of metals. *1st International Conference and General Meeting of the European Society for Precision Engineering and Nanotechnology,* Vol. 1. Bremen: Shaker Verlag. pp. 490–493.

Norman, S., Zervas, M.N., Appleyard, A., Durkin, M.K., Horley, R., Varanham, M., Nilsson, J., and Jeong, Y. 2004. Latest development of high-power fiber lasers in SPI. *Proceedings of SPIE Fiber Lasers: Technology, Systems, and Applications* 5335: 229–237.

Norrish, J. 2006. *Advanced Welding Processes.* Cambridge, UK: Woodhead Publishing Limited. pp. 144–148.

Ogawa, H., Yang, M., Matsumoto, Y., and Guo, W. 2009. Welding of metallic foil with electron beam. *Journal of Solid Mechanics and Materials Engineering* 3(4): 647–655.

Park, B.J., Choi, Y.J., and Chu, C.N. 2002. Prevention of exit crack in micro drilling of soda-lime glass. *CIRP Annals—Manufacturing Technology* 51(1): 347–350.

Piotter, V., Benzler, T., Hanemann, T., Ruprecht, R., Wollmer, H., and HauBett, J. 1999. Manufacturing of micro parts by micro molding techniques. *Proceedings of 1st EUSPEN Conference,* Vol. 1, Bremen, pp. 494–497.

Priyadarshini, D. 2007. Generation of micro channels in ceramics (quartz) using electrochemical spark machining, M.Tech. Thesis, Indian Institute of Technology Kanpur, India.

Reisgen, U. and Dorfmueller, T. 2008. Developments in micro-electron beam welding. *Microsystem Technologies—Special Issue on Colloquium on Micro-Production, Karlsruhe, Germany* 14(12): 1871–1877.

Sankar, M.R., Jain, V.K., and Ramkumar, J. 2009. Experimental investigations into rotating workpiece abrasive flow finishing. *Wear* 267(1–4): 43–51.

Sankar, M.R., Jain, V.K., and Ramkumar, J. 2010. Rotational abrasive flow finishing (R-AFF) process and its effects on finished surface topography. *International Journal of Machine Tools and Manufacture* 50(7): 637–650.

Saotome, Y. and Inoue, A. 2000. New amorphous alloys as micromaterials and the processing technology. *Proceedings of the IEEE 13th Annual International Conference on Micro Electro Mechanical Systems*, Miyazaki, pp. 288–292.

Schomburg, W.K., Ahrens, R., Bacher, W., Martin, J., and Saile, V. 1999. AMANDA—Surface micromachining, molding, and diaphragm transfer. *Sensors Actuators A* 76: 343–348.

Sidpara, A. and Jain, V.K. 2010. Electron beam micromachining. *Introduction to Micromachining*, ed. V.K. Jain. Oxford, UK: Alpha Science International Ltd. pp. 3.1–3.28.

Sidpara, A. and Jain, V.K. 2011. Nano level finishing of single crystal silicon blank using MRF process. *Tribology International*. DOI: 10.1016/j.triboint.2011.10.008.

Singh, D.K., Jain, V.K., and Raghuram, V. 2005. On the performance analysis of flexible magnetic abrasive brush. *Machining Science and Technology* 9: 601–619.

Singh, D.K., Jain, V.K., and Raghuram, V. 2006. Experimental investigations into forces acting during magnetic abrasive finishing. *International Journal of Advanced Manufacturing Technology* 30: 652–662.

Smith, S.T. 2000. Meso systems considerations. *M4 Workshop on Micro/Meso Mechanical Manufacturing*. Northwestern University, USA: Centre for Precision Metrology.

Smolka, G., Gillner, A., Bosse, L., and Lultzeler, R. 2004. Micro electron beam welding and laser machining-potentials of beam welding methods in the micro-system technology. *Microsystem Technologies* 10(3): 187–192.

Stefan, B., Gregor, H., Elisabeth, S., and Klaus, D. 2006. Micro bonding using hot melt adhesives. *Journal of Adhesion and Interface* 7(4): 28–31.

Stohr, J.A. and Eriola, J. 1958. Vacuum welding of metals. *Welding and Metal Fabrication*. Vol. 26, pp. 366.

Sze-Wei, G., Han-Seok, L., Rahman, M., and Watt, F. 2007. A fine tool servo system for global position error compensation for a miniature ultra-precision lathe. *International Journal of Machine Tools and Manufacture* 47(7–8): 1302–1310.

Takahashi, Y. and Maeda, M. 2011. Environment friendly low temperature solid state micro joining. *Transactions of Joining and Welding Research Institute of Osaka University, Japan* 40(1): 1–7.

Ulrich, R., Weber, H.P., Chandross, E.A., Tomlinson, W.J., and Franke, E.A. 1972. Embossed optical waveguides. *Applied Physics Letters* 20: 213–215.

Vasa, N.J. 2010. Laser micromachining techniques and their applications. *Introduction to Micromachining*, ed. V.K. Jain. Oxford, UK: Alpha Science International Ltd. pp. 3.1–3.28.

Venkatesh, V.C. and Izman, S. 2007. *Precision Engineering*. New Delhi: Tata McGraw Hill.

Venkatesh, V.C., Inasaki, I., Toenshoff, H.K., Nakagawa, Y., and Marinescu I.D. 1995a. Some observations on polishing and ultra-precision machining of semi-conductor substrate materials. *Annals of CIRP* 44(2): 611–618.

Venkatesh, V.C., Inasaki, I., Toenshof, H.K., Nakagawa, T., and Marinescu, I.D. 1995b. Polishing and ultraprecision machining of semiconductor substrate materials. *CIRP Annals—Manufacturing Technology* 44(2): 1–7.

Venkatesh, V.C., Izman, S., and Mahadevan, S.C. 2004. Electro-chemical mechanical polishing of copper and chemical mechanical polishing of glass. *Journal of Materials Processing Technology* 149: 493–498.

Yasuo, T. and Masakatsu, M. 2011. Environment friendly low temperature solid state micro joining. *Transactions of JWRI* 40(1): 1–7.

Zhong, Z.Z. and Venkatesh, V.C. 1995. Semi ductile grinding and polishing of ophthalmic and aspheric and spherics. *CIRP Annals—Manufacturing Technology* 44(1): 339–342.

Zhou, Y. 2008. *Microjoining and Nanojoining*. Cambridge, UK: Woodhead Publishing Ltd.

2

Challenges in Meso-, Micro-, and Nanomanufacturing

V. Radhakrishnan

Indian Institute of Space Science and Technology

CONTENTS

2.1 Introduction

Development of a product is based on many factors, some of which are only remotely connected to the engineering or sciences involved in it. The socioeconomic environment, consumer needs, consumer confidence, and usability are some such factors. Initially, when human society was evolving, most of the product needs were focused on ensuring food and nourishment and the creation of a habitat for safe living. Such objectives were met in a primitive manner through the innovative use of naturally available materials and objects; stone implements, vessels, bows, and arrows shaped from naturally available resources are some examples. As communities evolved and civilizations got established, significant attempts were made toward improvement of living standards through the development of energy sources and related equipment. Water wheel, windmill, and steam engine were such developments to remove the drudgery of daily chores. Meanwhile, fundamental knowledge established through creative thinking gave impetus to new inventions and discoveries leading to mechanization, allowing the development of a wider variety of products that commanded a good market. This marked the Industrial Revolution and the resultant social revolution.

While technology was the primary driving force behind these developments, slowly the idea of systems got established. Soon many valuable concepts toward meeting the aspirations of the consumer were brought in, leading to product availability and affordability. Some of the conceptual changes that came about included standardization and interchangeability, mass production, and statistical quality control. These changes started with mass production approaches, accelerated later on with the development in electronics and information processing and culminating in the now well-known concept of automated and flexible manufacturing. Most of the products covered had direct application in day-to-day activities. At present, these are designed and manufactured with the help of advanced technologies and are in great demand. The impact of information technology on global competition has made things really tough. New products were developed at a fast pace to keep the competition at bay. By the mid-1980s, it was indeed possible to have microelectronic devices that could be integrated with other systems to evolve into new product ranges. This triggered a rush for innovative products, and at present, this synergy has resulted in a number of value-added products of varied applications. This trend is at present very strong and has resulted in products and systems that are relatively small in size, but big in performance. As indicated earlier, microelectronics was the forerunner for this paradigm shift in the size and specifications of products. To realize such products at a reasonable cost, there is a need for understanding the manufacturing technologies to be adopted. Trends in product development clearly indicate the need to develop parts and components with remarkably small dimensions that are capable of integrating well with the microelectronic devices used in such advanced products (Hoummady and Fujita, 1999). This calls for size reduction and performance reliability for the parts involved. Manufacture of such parts and systems poses real challenges for the manufacturing engineer. This chapter deals with the understanding of some of the technologies that are developed toward this and the challenges posed.

2.2 Basic Manufacturing Concepts

Fundamentally, three different approaches are used for the realization of a product or a part from raw materials: moving, removing, or adding the material. The specific technique adopted has much to do with the size of the part as well as its cost and quality requirements. Parts and products that are size compatible with humans are in great demand and are manufactured in large numbers. Here, the product size varies from a few centimeters to a few meters. Figure 2.1 shows product variety based on their size (Radhakrishnan, 2006). Further, depending on the applications, their quality levels could vary. When the accuracy needs are coarse or medium, the parts can be realized through material-moving technologies. Moving of material is a basic and relatively cost-effective approach for shaping parts. The technologies include casting, forging, forming, and drawing. While the quality achieved by these processes has improved dramatically over the years, they still lag behind the machined parts in terms of precision. Contemporary quality needs are high and precision parts are in great demand. As the precision levels achieved in

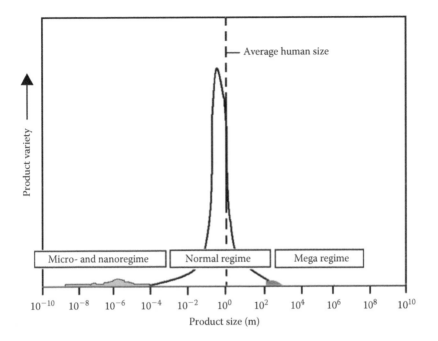

FIGURE 2.1
Product variety, based on size.

manufacturing have gone up dramatically, many of the common products have very close tolerances on their dimensions. Such products are realized through material removal processes. Present technologies available are excellent both from the quality and productivity perspectives and hence find wide acceptance in manufacturing. Relatively larger products are realized through assembly or by combining the parts; examples include welding of a ship hull, integrating aircraft wings onto a fuselage, or assembly of rocket segments. Although this cannot be termed as *additive manufacturing*, the concept is accepted as such. Additive manufacturing, as it is termed at present, is in vogue for relatively small and medium-sized parts, and electroforming is a typical example. Rapid prototyping (RP) or layered manufacturing technologies are examples of additive manufacturing that have found good acceptance. In brief, providing form and shape to a part through manufacturing is critical for its function, performance, quality, and cost. As of date, the material removal approach of manufacturing is considered to be the most sought after procedure to achieve close tolerances on the dimensions and features of parts.

2.3 Changing Product Scenario

Our forays in microelectronic manufacturing have allowed us to understand the possible avenues for creating parts and products that were not previously attempted. Over the years, this knowledge has led to the design and realization of a new range of products. Typically, these products are technologically suave and miniature in size. They could be classified as meso-, micro-, and nanoproducts. This classification is based on the physical

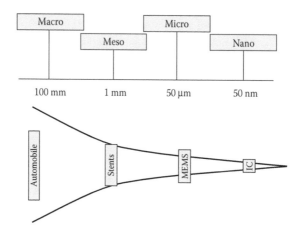

FIGURE 2.2
Scale of products from the macro to nanorange.

sizes involved. While the terms *micro* and *nano* are directly connected with the dimensions, *meso* only means middle. It is the size between macro and micro sizes. Figure 2.2 gives the range of these product segments together with some of the current product examples. Hence, meso-, micro-, and nanomanufacturing refer to the realization of features, components, or systems with dimensions most conveniently described in milli-, micro-, and nanometers, respectively.

The major products in the segment of micro- and nanomanufacturing have been microchips and other microelectronic devices. New products that use these chips and that are designed to meet many personal needs are now in great demand. These include personal computers, mobile phones, biomedical devices, entertainment systems, and other devices. This increases the demand for a variety of parts in the meso-, micro-, and nanorange. This demand is not only for parts of such dimensions but also for complex and precise features on such parts, which form the core of these new product lines. Many of these parts are to be integrated with microelectronic devices to realize such products. Currently, the health care, electronics, strategic, communication, and IT industries are the most active in this area. Unlike simple mechanical systems of the conventional type, most are integrated product systems covering microelectronics, micro-optics, and a good number of critical mechanical parts. Technology convergence is the catalyst for the development of many such products. Our understanding of the design and realization of such parts is still limited, and globally, there is great interest in developing cost-effective manufacturing technologies that could be adopted for such products and parts (Alting et al., 2003).

2.4 Issues of Importance

Before discussing the challenges in meso-, micro-, and nanomanufacturing, some of the issues involved must be understood. Micromanufacturing often refers to the realization of parts with dimensions conveniently expressed in micrometers. However, nanomanufacturing not only refers to dimensions but also to form and surface features. In other words, nanotechnology could mean achieving nanometer-level flatness on silicon wafers or achieving a finish of a few nanometers on a micromold as well as etching a line width of a few nanometers in an IC chip. Both micro- and nanomanufacturing have many features in common, which are different from conventional manufacturing. For mesoparts, the issues are not the same as those of conventional or micro- and nanomanufacturing. However, mesomanufacturing is a convenient stepping-stone to foray into micromanufacturing. Some of the common issues concerned with product realization in this area are the material behavior at this scale in manufacturing, tooling and setup, process selection, handling of parts, and their inspection and assembly. As these issues are quite different from those of conventional manufacturing, there are many challenges encountered.

2.5 Manufacturing Approaches

Even before the demand for micro- and nanoproducts grew, there was a trend toward achieving close tolerances in manufactured parts. This resulted in the development of manufacturing technologies capable of producing precision engineered parts. Technologies were developed on the basis of material removal processes that were capable of removing the material in very small quantities, thereby allowing very close tolerances to be achieved on parts. The same approaches were further developed to meet the challenges in the new regime of meso-, micro-, and nanomanufacturing. Ultrafine feeding of tools, vibration and chatter reduction in machining, high-speed machining, and machining using diamond tools were some of the technologically sound developments in this direction.

Scaling down the processes has been attempted, and conventional metal removal processes have been scaled down to meet the manufacturing needs of meso- and microparts. This covered the conventional and unconventional manufacturing processes such as micromilling, microdrilling, microelectric discharge machining (EDM), microelectrochemical machining (ECM), micro-ultrasonic machining USM, and laser micromachining. Scaling down of processes requires clear understanding of the mechanism of machining at the micro- and nanolevels, which could be totally different from normal manufacturing. Material-moving processes such as molding and forming have also been scaled down to meet the meso- and microproduct realization. However, these processes need tooling that is compatible with the sizes involved, which could be realized only through machining approaches. Processes such as laser forming offer a different approach for making microparts using thermal deformation.

The need for cost-effective manufacture of integrated chips leads to the advancement of photolithography—a material removal process achieved through a series of etching processes configured according to the design of the chip. As they involve planer processing of the raw material silicon with low aspect ratios, they are done in parallel and in batches. Common optical lithography is ideal for shallow etching. For deep etching with higher aspect ratios, x-ray lithography is used. By manipulating the process and the masks used, one can achieve different configurations, such as inclined side walls or stepped structures. The lithography process is quite different from normal manufacturing, in which case the operations are done serially. Parallel processing involved in IC manufacturing produces a batch of components, although the time taken is relatively long. To start with, lithography was used for static configurations. At present, with improved and innovative approaches, parts and systems with elements having limited movements are possible through lithography. To reduce the time for manufacturing microparts, micromolds are made through x-ray lithography and later used for micromolding of parts in large numbers. This is the well-known LIGA process. Forming tools are also produced this way, and thin microformed elements are made through this method.

Additive manufacturing is another approach for the realization of products, and RP technologies form the well-known approach for such manufacturing. By adapting sophisticated tools and procedures, it is possible to make parts of extremely small dimensions through the additive approach. The use of atomic force microscopy (AFM) for additive manufacturing has been demonstrated, wherein atoms could be moved and positioned to achieve the configurations needed. However, these are currently impractical because of the time taken, configurations, and costs involved.

As discussed, micromanufacturing had much to do with the development of microelectronics. In fact, without microelectronics in place, the need for microparts could have been limited. Likewise, the processes developed for the realization of integrated chips are used widely for micro- and nanopart fabrication. Currently, photolithography has established itself as a viable technology for the manufacture of microparts. It provides a unique opportunity to integrate mechanical systems with electronics to develop microelectromechanical systems (MEMS) *in situ*. This approach further eliminates the need for discrete component assembly, which is otherwise difficult at this scale.

2.6 Manufacturing Challenges

Unlike conventional products, the realization of meso-, micro-, and nanoscale products in an economical manner poses major challenges. To start with, let us consider the precision achieved in them. The tolerance limit defined for a part is an indication of the precision expected. However, the actual tolerance specified is related to the dimension of the part. Hence, it is prudent to take the ratio of diameter (D) to tolerance (T) as an indicator of precision. A D/T ratio above 10^4 is considered to indicate a precision part. As D scales down, T also has to scale down to maintain this ratio. This is a challenge that cannot be easily met. The trend in the minimum dimension achieved over time can be seen in Figure 2.3. With micro- and nanoscale products, the dimensions have shrunk dramatically in recent years. Meanwhile, conventional manufacturing approaches dealing with normal dimensions have improved in their precision dramatically. There are many reasons for this, including better understanding of the process mechanics, improved machine tools, and better control of process variables. Tolerances ranging from submicrometers to hundreds of nanometers are now common. This trend in tolerance is also plotted against time (in years) in Figure 2.3. The interesting aspect that can be noticed in Figure 2.3 is the merging of the tolerance and dimension graphs with time. D/T in this merging region could as well be 1. This is a far cry from our concept of precision where D/T has to be $> 10^4$.

While this argument may suggest that precision engineering and micro- and nanoscale manufacturing are two entirely different entities, it should be clear that these areas have considerable commonalities. It is obvious that if the dimensions are to be controlled precisely, material removal required to achieve this should also be precise. Therefore, the amount of material removed needs to be controlled to achieve the dimension. This should be obviously limited to extremely small amounts to achieve precise control over the dimensions; precision engineering requires this. Such procedures are also applicable to shape

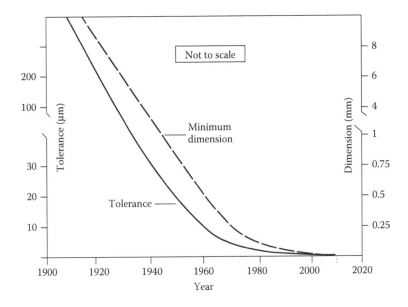

FIGURE 2.3
Trend in the minimum dimension achievable and tolerances given on dimensions.

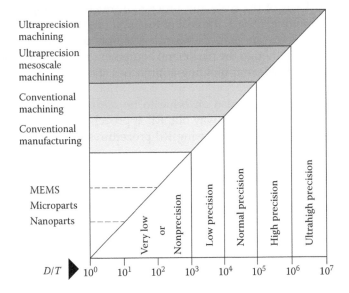

FIGURE 2.4
Precision levels achievable in present-day manufacturing.

micro- and nanoparts, as their dimensions are extremely small. Another commonality is in their dimensional inspection. Here again, certain procedures are common, which include surface measurement, form accuracy, and tool settings in the machine. Often, the optical instruments used for this purpose are generic and common. Figure 2.4 shows the precision levels achievable in manufacturing for different manufacturing regimens.

2.7 Process Mechanisms

Conventional manufacturing processes have been studied in depth and reasonable knowledge has been gained about their process mechanics, influence of the process variable on product quality, the role of machine tools, tooling and related equipment on process performance, and control strategies to achieve specific objectives in manufacturing. Material behavior at the micro- and nanoscales is quite different from that at the macrolevel. It has much to do with the microstructure of the material and its crystal structure. Hence, the mechanism of machining is to be interpreted in a different way altogether. One approach is to use molecular mechanics to understand the cutting mechanics. Likewise, other processes are to be understood in depth using different approaches. Heat transfer, solidification, plasticity, strain rate, melting, and vaporization at the micro- and nanoscales are to be studied in detail to understand and improve on the various processes applicable at this scale.

2.8 Tooling

Manufacturing at all scales needs tooling. The type and specification of the tooling will differ widely on the basis of the scale and process. In microlithography, this covers the masks and their positioning. In fact, the cost of tooling is significantly higher in this area compared to that in macroscale manufacturing. Attempts to avoid tooling of this kind provide a good solution, although the peripheral equipment needed could be costlier than the existing ones. Micromolds and molding approaches should again be examined from their yield and quality angles.

Deburring of molded parts is also an issue to be addressed. If parts can be made without any tooling, that would be a way out. RP approaches are a solution toward this. Microfluidic devices can be fabricated using RP procedures. Multilayer devices are also possible through lamination techniques.

2.9 Handling

It is quite evident that handling poses major challenges in this regime of manufacturing, and the obvious choice is to minimize, if not eliminate, handling as much as possible. This naturally puts some limitations on the designs as well as the processes selected for their realization. Planer processes such as masked etching are quite suitable for this. Similarly, parallel processing of parts keeps the micro/nanoparts under the macroregimen with reference to handling. Reduction of moving parts and the use of simple flexures also contributes to minimal handling, thus avoiding assembly. Current microsystems have significant design complexity and material heterogeneity. Assembly and packaging of such devices account for over 60%–70% of their cost; here, handling is a significant contributor to this cost. Multilayer manufacturing using RP technologies can lead to assembled manufacturing; in this case, the entire assembled system is made through a layered manufacturing procedure.

2.10 Micromachining

Many microparts are realized through direct machining or through molding, for which the mold is made through machining or by other processes such as LIGA. The meso- and microparts are conveniently obtained through direct micromachining. The challenges in such an approach are many, which cover thermal stability of the spindles of the machine tool, submicron feedback and control system, work and tool-holding systems, machine dynamics and vibration control that can give a repeatability of <1 μm, and in-process inspection. *In situ* tool making attempted in some machining operations such as micro-EDM is also a solution toward reducing the errors in tool positioning. As in conventional machining, here also one has to address tool wear, tool geometry, tool change, vibration in mechanical machining, and surface finish. Specific cutting energy increases dramatically with reduction in the depth of cut, as shown in Figure 2.5. This could result in rapid tool wear. It is of interest to note that the cutting speed that can be achieved in microdrilling is very much below the recommended values for machining owing to the very small diameter of the drill. Similarly, in micromachining using a mechanical tool, the rake angle encountered is negative to the tune of 40°–50° owing to the presence of edge radius of the tool (Figure 2.6). Furthermore, some tools are of micrometer dimensions and have to be ground or machined and inspected before use.

2.11 Toolless Machining

To avoid the difficulties encountered in mechanical tooling, toolless machining using laser and energy beams is used widely in micro- and nanoscale manufacturing. While a defined tool gives better consistency, toolless machining depends considerably on the beam source parameters in addition to the process parameters. However, at present, laser and ion beam machining approaches are used widely in micro- and nanoscale machining with success. Recently, micro water jet cutting has found applications in this area. RP or layered manufacturing is also a toolless procedure. However, this has considerable limitations with

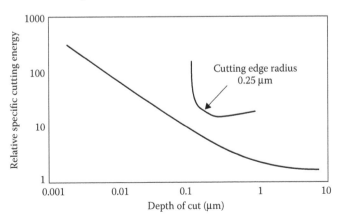

FIGURE 2.5
Relationship between relative specific cutting energy and depth of cut.

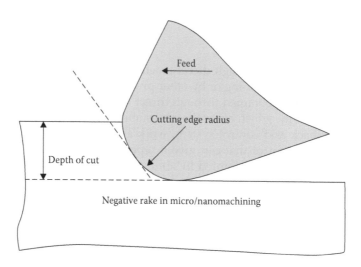

FIGURE 2.6
Role of edge radius on the rake angle in micro/nanoscale machining.

regard to the consistency. The challenge lies in selecting the right toolless machining and controlling the source as well as process parameters to achieve the goal.

2.12 Micromolding

Apart from the realization of the micromold, there are many issues connected with the molding of parts, for example, selection of the apprpriate molding machine, design of the mold with runners and slender core pins, ejection of parts, and handling method. This is all the more important in cases in which the material to be molded is costly, which is the case with many biomedical applications. While the molded part could be 1 or 2 µg, the peripherals that are needed should not result in wastage of material. Here, wasting of the material in the runner is an important aspect to be addressed. Shrinkage at the microscale is different from that seen at the macrolevel, especially for complex shapes. Balancing of injection pressures is essential to avoid displacement of slender core pins. While mold flow analysis software often renders a helping hand in the design of molds, there are issues to be addressed to achieve quality products. Part handling poses major challenges, and rotating molds are an option available for this.

2.13 Microforming

There are many parts that are made through microforming. These include large-scale production of microformed parts from microwires and thin sheets, as well as through profile forming. Wires are drawn to very small diameters of a few micrometers and then cut to the required sizes and formed further by rolling into microscrews. Parts obtained through wire drawing can also be extruded to get the required shape. Other approaches for forming different profiles include laser forming, incremental forming, and laser-assisted mechanical forming. The major challenge in the forming operation at the microscale is the tribological understanding of the process mechanics. This is because of the high surface-to-volume ratio of microparts. In addition to this, adhesion forces, surface tension, and material grain sizes play a major role in microforming. Thus, scaling down directly from macro- to microscale is not that simple.

Another forming approach is through microstamping, in which micropresses and micropunches/microdies are needed. Schematic representations of microstamping and a press are given in Figures 2.7 and 2.8. Microstamping is found to be suitable for mesoparts with microgrooves and other formable features on them. The knowledge base on these operations is still limited.

2.14 Surface Patterning

While the dimensions of the parts decrease, their surface features play an important role in their function (De Chiffre et al., 2003). This is all the more critical in biomedical

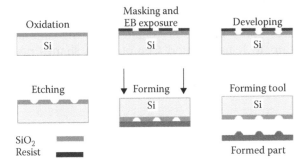

FIGURE 2.7
Microstamping sequence.

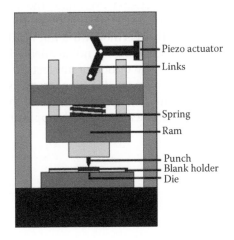

FIGURE 2.8
Micropress schematics.

applications, where the interaction with the cells depends on the surface features provided on the part. While we are aware of this requirement, there are still many gaps in our knowledge regarding the surface features that are ideal for specific applications. Providing the proper features on the surface of micro/nanoparts is tough. This is termed as *surface patterning*. Understanding and developing the processes suitable for producing the required surface patterns is a major challenge in this area. At present, it is a known fact that even in macroscale objects, surface patterning can provide considerable functional advantages and that such patterning will come under micro/nanomanufacturing.

2.15 Micro- and Nanometrology

Metrology is a key element in micro- and nanotechnology. It is the measurement of the parts that classifies them into these groups. However, measuring the dimension at these scales is not easy. While there are many instruments available for this, comparison among them, as far as the measurement reliability is concerned, is a challenge, as they work on different principles. Measurement using an electron beam depends on the position of the sensor with reference to the direction of emission. Interference and AFM measurement work on totally different principles. It will really be a challenge to monitor the uncertainty of measurements at this scale. Yet another challenge is the calibration of these instruments that measure up to a nanometer! Complex 3D shapes need micro-coordinate measuring machines (micro-CMMs) for measurement, and the probing system is not easy to implement. Many approaches are proposed, and the right one is yet to emerge.

2.16 Other Challenges

At the scale of micro- and nanometers, electrostatic and chemical forces dominate. Forces such as van der Waals and surface tension dominate the interaction between parts. Clustering of parts is thus a serious problem. Handling of microparts and presenting them is a tough proposition, and orienting them is more difficult. While macroparts are assisted by gravity in these operations, such advantage is not available for micro- and nanoparts. Designing the parts for self-assembly is an option of interest.

As with nanoscale products, health hazards due to macroparts are yet to be fully evaluated and regulations yet to be brought in. At present, most of them are encapsulated and pose less threat to the environment and health. As their applications grow, this will be an issue that has to be addressed.

2.17 Conclusion

There is a need to understand the basics of these processes, as they are totally different from bulk material manufacture at the macroscale. There is the need to study the role of materials in micromachining and forming. Tribological aspects of such processes and their related thermal aspects are to be understood well for improving the processes, for controlling them, and for making them economical. Additive micromanufacturing is another emerging process with wide applications. All these technologies need fundamental understanding of the processes involved, which is different from what has been achieved until now.

Measurement and condition monitoring procedures for such parts and processes need detailed research and development. Part defect identification, burr removal, part handling, and micropart assemblies are bound to be challenging propositions for researchers in this field. Self-assembly of micro- and nanoparts is possible if parts are designed with self-assembly in view. It is quite understandable that soon we may also have to plan for customer-oriented flexible manufacturing in this area. More avenues are opening up for R&D, which is basic to this field. As we move to the new regime of meso-, micro-, and nanomanufacturing, there is a need for fundamental studies that are applicable to this area so that we are able to progress in this field as we have done in conventional manufacturing. Such a study has to cover all areas that have been explored at the macroregimen.

References

Alting, L., Kimura, F., Hansen, H.N., and Bissacco, G. 2003. Micro engineering. *Annals of CIRP* 52(2): 635.

De Chiffre, L., Kunzmann, H., Peggs, G.N., and Lucca, D.A. 2003. Surfaces in precision engineering, microengineering and nanotechnology. *Annals of CIRP* 52(2): 561.

Hoummady, M. and Fujita, H. 1999. Micromachines for nanoscale science and technology. *Nanotechnology* 10: 29.

Radhakrishnan, V. 2006. Challenges in meso, micro and nano manufacturing. *International Conference on Advances in Mechanical Engineering SRM Institute of Science and Technology Deemed University*, Chennai, India.

Section II

Micromachining

Traditional Micromachining

3

Microturning

R. Balasubramaniam and Vinod Kumar Suri

Bhabha Atomic Research Centre

CONTENTS

3.1 Introduction

Turning is a basic material removal process that is practiced quite extensively in manufacturing. Turning involves removing material from a workpiece that rotates about a spindle axis by way of a single-point tool mounted on a tool post. The tool is moved relative to the workpiece in order to generate axisymmetric shapes such as cylinder, cone, and sphere. Manually operated lathes were most prevalent for large and medium-sized components, and with the industrial revolution, cam-operated automations became more popular for mass manufacturing of smaller components. With the development of computerized numerical control (CNC), processing of both larger and smaller components became much faster with improved flexibility, precision, and accuracy. Furthermore, researchers and companies are continuously motivated to manufacture smaller components for newer applications with higher quality and improved performance at lower cost. Micromanufacturing processes such as micromachining, microforming, microjoining, and microassembly play a vital role in the miniaturization of components and systems. Of the many micromachining processes, microturning is one of the most basic technologies for the production of miniaturized parts and components. Working in the area of miniaturization in general and microturning in particular needs a thorough understanding of the microturning process, including the basic differences between microturning and macroturning processes, critical requirements of the microturning machines, underlying mechanisms of the material removal process, various factors that affect the accuracy of the component, tools for microturning, and the effects of various machining parameters on the process performance. This chapter provides an overview of microturning to researchers, teachers, and students.

3.2 Macro-, Meso-, and Microturning

Turning of small parts is not new, but challenges arise when the size of features reduce to tens or hundreds of microns, and the precision requirements of the miniaturized parts become less than a few microns. Even though microturning is a miniaturized version of macro/mesoturning, there exist a few fundamental differences between them. Some of them are as follows:

- The main feature that distinguishes macro/mesoturning from microturning is the *size attribute* of the component, specifically the diameter being machined. Macro/mesoturning is applicable to component diameters of ≥1 mm, whereas microturning is applicable when the diameter of the component is <1 mm. Although there are different definitions for the upper limit of the diameter in the case of microturning, machining of <1 mm diameter is generally accepted by many.

- The second feature that distinguishes microturning from macro/mesoturning is the *cutting mechanism*. Cutting in macro/mesoturning is the shearing of bulk material along the average shear plane with high defect density; machining parameters lead to larger chip cross section encompassing a large number of grains, whereas in microturning the material is removed with very small chip cross section, generally, less than the average grain size. Under this condition, the material is not removed by shear along the grain boundaries but across the grains. The size effect becomes a dominant phenomenon, and it tremendously increases the specific cutting energy.

- The third factor that distinguishes them is the *lack of rigidity* of the component in microturning. While machining, the workpiece deflects and causes a large variation in shape and size. The magnitude of the cutting force in microturning deflects the component not only along the direction of the thrust force but also in the tangential direction, thus changing the center height of the tool and leading to a damaged workpiece. Hence, control of the reaction forces in microturning is a major issue.

- The fourth factor is the *attainable tolerance* on the component. With microturning, a ratio of "tolerance to the size of the component" similar to that achieved in macro/mesoturning cannot be attained. For example, when the tolerance on a 10-mm-diameter pin is 0.014 mm, the ratio is 0.0014. When the same ratio is applied for a 0.1-mm-diameter pin, the applicable tolerance becomes 0.00014 mm (= 0.14 nm), which is not realistic. Hence, the tolerance applicable to microturned components becomes very vague; however, the micromachine needs to be ultra-precise to meet the improved tolerance to certain extent.

- The fifth factor is the *size and accuracy of the machine*. Macro/mesomachines have comparatively lower spindle speeds and coarser feed and depth of cut values, whereas in microturning the reverse is the case. The footprint of the machine is much smaller owing to the smaller size of various subsystems. Smaller and short distance moving elements are important to achieve better machine tool accuracy and stable machining parameter values. One of the most important requirements of a microturning machine is that it needs to be an ultraprecision machine; for example, if the positional accuracy of the microturning machine is 2 μm, depths of cut less than 2 μm cannot be ensured. Hence, higher accuracy requirements in terms of positional tolerance and repeatability need to be addressed. To achieve higher accuracy levels, micromachines are designed with high loop stiffness, high thermal and mechanical stability, and high-precision axis motion and axis control. Table 3.1 summarizes the major differences between macro/meso- and microturning, and the details are further discussed in appropriate sections in the chapter.

3.3 Applications of Microturning

Microturning has many areas of applications in industry. The initial users of microturning were the watch industry; microturning was extensively employed for the manufacture of various miniature components for mechanical-type watches over many decades. Product-specific special-purpose turning machines were extensively developed and used by them. Of late, the demand for miniaturized systems for various applications in bio-engineering, telecommunications, and automobile industries is increasing. Various types of micro/nanosensor pins and probes are extensively used in the electronic industries for testing ICs, and many of the components for these sensors are manufactured by the microturning process. Most of the hybrid mechanical machines use a single workstation for manufacturing microcomponents. Various processes are employed in sequence to

TABLE 3.1

Major Differences between Macro/Meso- and Microturning Processes

Feature	Macro/Mesoturning	Microturning
Unit removal of material	Few cubic millimeters to cubic centimeters per revolution	Few tens to few hundred cubic microns per revolution
Specific cutting energy (J/cm^3)	10^2–10^3	10^4–10^5
Mechanism of material removal	Shear along grain boundaries for ductile material. Brittle fracture for brittle material	Shear across grains
Size effect	Less dominant	More dominant
Ratio of surface finish to tolerance	Small	High
L/D ratio	Less to medium	Medium to large

TABLE 3.2

Application Areas of Microturned Components

Area of Application	Name of the Microturned Component
Automobile	Shafts for micromotors, shafts for radial rotors
Electronics	Sensor pins, probes
Watch industry	Cylindrical pins, curb pins, screws, and studs
Medical industry	Neurosurgical implants, maxillofacial microscrews
Micro EDM and micro ECM tool	Generation of holes and arrays of pins

realize the final product. One major use of microturning is to manufacture electrodes for micro electric discharge machining (EDM) and micro electrochemical machining (ECM). Applications of microturning in many other areas are still emerging. Table 3.2 summarizes some of the important application areas and some actual products.

3.4 Machines for Microturning

Microturning is carried out with different types of machines. Figure 3.1 shows the most commonly used configurations of the machines. The most popular among them are tool based; they are generally used for turning soft materials, whereas nonconventional-type machines, namely, electrical discharge machines and wire electrical discharge grinding, are used for turning hard materials and very small diameter electrodes that are difficult to machine by other means. For the mass manufacture of microturned components, machines with rotating tools are used when the raw material is in coil form, and machines with rotating workpieces and stationary multiple tools are used when the raw material is a bar stock. This type of machine is extensively used for making components with a uniform diameter and without any steps. Figure 3.2a and b shows the schematic arrangement of

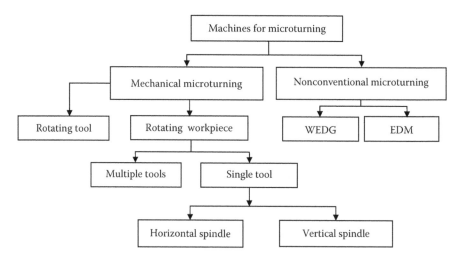

FIGURE 3.1
Various configurations of microturning machines.

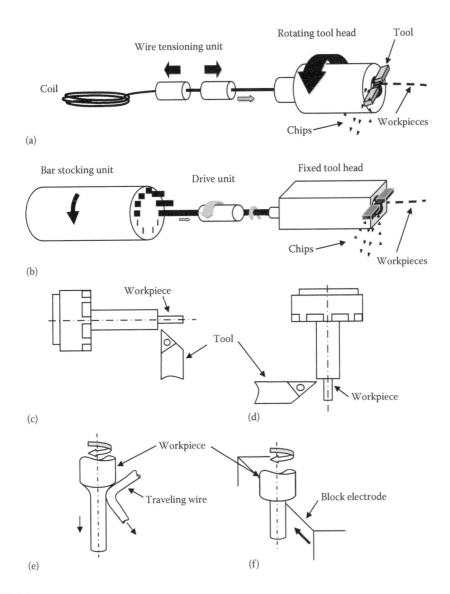

FIGURE 3.2
Schematic diagram of microturning machines: (a) rotating-tool-type microturning machine, (b) fixed-tool-type microturning machine, (c) horizontal spindle lathe, (d) vertical spindle lathe, (e) wire EDG microturning, and (f) reverse polarity EDM for microturning.

such machines. These machines are fitted with multiple tools in the same plane and positioned in such a way that they cancel the resultant thrust force acting on the workpiece. Another division of microlathes, which are most popular, is the miniaturized version of macro/mesolathes with either horizontal or vertical spindle configuration as shown in Figure 3.2c and d. Yet another family of machines used for microturning is electrical discharge grinding with traveling wire electrodes or electrical discharge machines with block electrode and reverse polarity. Figure 3.2e and f shows the schematic diagram of these machines. Generally, these machines are used for manufacturing components from hard materials such as tungsten or for preparing very small diameter electrodes for the micro EDM process. More emphasis is given on mechanical micromachines in the subsequent sections, and the relevant details are discussed.

3.4.1 Design Requirements of Microturning Machines

The accuracy of microturned components primarily depends on the following:

- Accuracy of the machine tool
- Control of the reaction forces on the component

Accuracy of the machine tool depends on the integration of many subsystems such as the mechanical structure, spindle and drive system, control system, and measurement and feedback system. However, product accuracy also depends on the integration of these subsystems with the tooling system, processing sequence, and the method of controlling the reaction forces on the component. Some of the design requirements are discussed in this section, and the influence of other factors on the accuracy of the component is discussed in subsequent sections.

Some of the critical design features that affect the accuracy of the microturning machine are as follows:

- Structural loop stiffness
- Damping property
- Thermal stability
- High natural frequency and low vibration
- Machine base stiffness and stability
- Smooth and accurate spindle motion
- Smooth, precise, and repeatable slide motion
- Tool quality and accuracy

In order to avoid deformation at the tool–workpiece interface, the loop stiffness of the structural path that comprises the workpiece, fixture, spindle, bearing, housing, machine frame, drive system, slide ways, and the cutting tool should be very high. Unlike other micromachining processes, microturning suffers from the lack of rigidity of the workpiece. Cutting forces cause larger deflection of the workpiece, and the loop stiffness is greatly reduced (Lu and Yoneyama, 1999). Selection of a suitable material for the machine bed and slides meets the damping requirements; the use of hydrostatic bearings provides better damping to the moving tables. Minimizing the spatial thermal gradient in the machine system and making the system quickly reach and maintain stable equilibrium are important criteria. Reduction of thermal deformation by a proper design rather than by providing correction techniques should be the design principle. The machine sliding systems are mounted on a granite base to provide high stiffness and stability and to isolate them from external vibrations.

Spindle smoothness largely affects the surface finish value as well as the geometrical accuracy of the component. As the microturning machines are ultraprecision machines that generally employ a few micrometers of feed and depth of cut, any variation in the spindle run out affects the uncut chip thickness significantly.

As microturned products demand stringent surface finish and dimensional tolerances, the required smooth, precise, and repeatable motions are achieved through hydrostatic slides or linear motion guideways. In order to overcome resonance at the operational speed range of the spindle, proper material selection and design of the machine components are necessary to have high natural frequency as well as less vibration during the operation of the machine.

3.4.2 Sources of Errors in Microturning

In microturning, which is a high-precision machining process, the margin of error must be extremely small. As discussed in the previous section, not only inaccuracies of the machine tool but also other factors such as the properties of the work material and the reaction forces against it significantly affect component accuracy. In order to maintain a high level of precision in microturning, it is important to know the factors responsible for errors during the process of generating the shape. Identification of these errors and minimizing their influence are essential to reach the level of the defined precision. Figure 3.3 shows the major sources of errors in microturning. The geometrical inaccuracies of the machine tool are reflected proportionately, whereas positional and repeatability errors are reflected directly on the component. They affect the size and shape of the component significantly. Similarly, any deformation caused by mechanical or thermal sources contributes to component error.

Rigidity of the work material becomes a major source of error. Unlike in the micromilling process, where the cutters have high rigidity and are subjected to less deflection, the rotating workpiece in microturning is less rigid and undergoes larger deflection during the material removal process. Components with higher rigidity in terms of material property, geometry, and L/D ratio are subjected to less deflection and fewer inaccuracies. Similarly, variation in the grain size and its orientation cause fluctuation in cutting forces and result in increased inaccuracy of the component.

The magnitude of the reaction force acting on the component is significantly affected by tool attributes such as cutting edge sharpness (cutting edge radius), tool nose

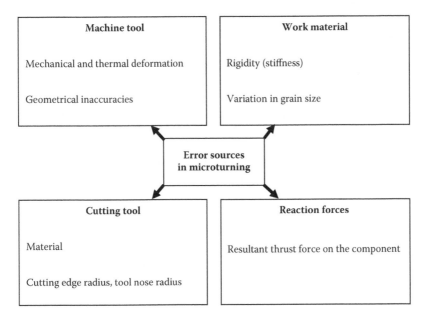

FIGURE 3.3
Error sources in microturning.

radius (TNR), tool material, and tool setting accuracy. The size effect and plowing effect become predominant when the cutting edge sharpness is less, and this increases the thrust force and deflection of the component. Hence, it is important to control the reaction forces during microturning for minimizing the inaccuracies of the component.

3.5 Mechanism of Material Removal in Microturning

When the components are miniaturized, the sizes of the features to be machined become smaller and smaller. The entire volume of the material to generate the size and shape of the microfeature cannot be removed as a single unit but can only be removed as smaller units. The unit volume should be such that it enables maintaining the tolerance on the size. The smallest volume of material removable from the workpiece by any process is termed *minimum unit removal* of the process, and the smaller this unit, the better is the control on the process. Processes such as EDM are capable of removing a cluster of atoms with a single spark (Masuzawa and Tonshoff, 1997). By controlling the energy and size of the spark, smaller units of material removal can be achieved. However, the smallest unit of material removable, that is, an atom, is not achievable by mechanical micromachining processes owing to various limitations.

In microturning, the minimum unit removal of material corresponds to the volume of the material removed in one revolution, namely, the chip cross section multiplied by the length of the chip, and the unit material removal becomes smaller with decreasing dimensions for a given chip cross section. Further reduction in unit material removal is achievable by reducing the chip cross section. However, the limits are dictated by the minimum feasible uncut chip thickness and feed. Among other factors that influence the uncut chip thickness, the most significant one is the cutting edge radius.

The ratio between the uncut chip thickness and the tool cutting edge radius plays a dominant role in microturning. Figure 3.4 shows the effect of uncut chip thickness on the mode of material removal for a given TNR. Table 3.3 lists the resulting conditions for various combinations of cutting edge radius and uncut chip thickness.

In microturning, as the tool has a finite cutting edge radius, depending on the size of the uncut chip thickness, the tool either removes the material as chips or rubs or burnishes the work surface, as shown in Figure 3.4. This can be explained by the elastic–plastic behavior of the work material. When the uncut chip thickness (t_c) is greater than the critical chip thickness ($t_{critical}$), shearing by plastic deformation becomes predominant and a chip is formed. When t_c is between $t_{critical}$ and the thickness that causes pure elastic deformation ($t_{elastic}$), elastic deformation becomes predominant and burnishing action takes place. When t_c is less

TABLE 3.3

Resulting Conditions for Various Combinations of Cutting Edge Radius and Uncut Chip Thickness

		Decreasing Uncut Chip Thickness ⟶		
		t_1	t_2	t_3
Increasing Cutting Edge Radius ↓	R_1	Cutting	Burnishing	Plowing
	R_2	Cutting	Burnishing	—
	R_3	Cutting	—	—

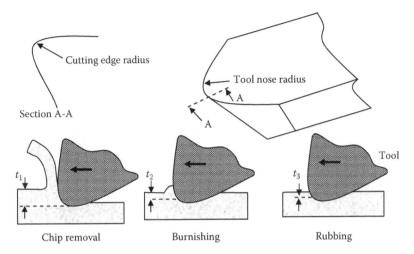

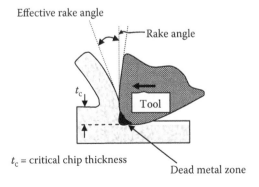

R = cutting edge radius; t_c = critical uncut chip thickness;

t_1, t_2, t_3 = uncut chip thickness

FIGURE 3.4
The effect of uncut chip thickness on the mode of material removal during microturning.

than $t_{elastic}$, there is rubbing of the surface. In the case of macroturning, the typical cutting edge radius is of the order of 5–10 μm and the uncut chip thickness varies from a few tens of micrometers to a few thousands of micrometers. But in microturning, the tool cutting edge radius is of the order of a few hundreds of nanometers (as in the case of single-crystal diamond tools) to 1 or 2 μm [as in the case of polycrystalline diamond (PCD) and cubic boron nitride (CBN) tools] and it is comparable in size to the uncut chip thickness (Woon et al., 2008). When the ratio of the uncut chip thickness to the cutting edge radius becomes much smaller, the effective rake angle becomes negative and the cutting edge radius creates a dead metal zone or a built-up edge as shown in Figure 3.5, which significantly increases the cutting energy as well as the cutting force generated in the material removal process.

In microturning, the size effect, which is predominant at lower uncut chip thicknesses, is typically characterized by a nonlinear increase in the specific cutting energy or specific cutting force with decreasing uncut chip thickness. The specific cutting energy becomes very significant at lower uncut chip thicknesses and causes undesirable effects such as

FIGURE 3.5
Negative rake angle and dead metal cap.

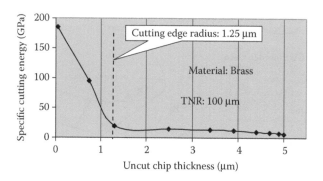

FIGURE 3.6
Specific cutting energy in microturning.

increased tool wear and thrust force. Figure 3.6 shows the effect of uncut chip thickness on specific cutting energy when the uncut chip thickness is reduced from a few micrometers to submicrometers on a brass workpiece with a PCD tool of 100 μm TNR and 1.25 μm cutting edge radius. In this case, the specific cutting energy increases tremendously when the uncut chip thickness becomes less than the cutting edge radius.

During the ductile material processing of any tool-based macro/mesomachining, as the tool advances in the feed direction with an uncut chip thickness much larger than the cutting edge radius, the material ahead of the tool is pushed and sheared along the average slip plane that has the highest defect density. The slip is initiated and propagated along the grain boundaries, and the material is removed as a cluster of grains by plastic deformation. Figure 3.7a schematically shows a typical crack initiation and material separation path. In the case of brittle materials, the failure is initiated again at grain boundaries, but the failure mode is brittle fracture (Shimada et al., 1995).

In microturning, when the actual uncut chip thickness is more than the critical chip thickness, the material is sheared along the dislocations within the grain. The range of uncut chip thickness under this condition is 0.1–10 μm. Figure 3.7b schematically shows a typical crack initiation point in the grain and the material separation path across the grain.

3.6 Forces in Microturning

Components of the cutting force, namely, tangential, thrust, and feed forces, act on the workpiece in three orthogonal directions, as shown in Figure 3.8, and each one of them causes certain undesirable effects on the component. The magnitude of the cutting force depends on the chip cross section, uncut chip thickness, cutting edge radius, and work material properties. Cutting force is proportional to the chip cross section; with decreasing chip cross section, as in macroturning, the cutting force decreases. However, the ratio of the uncut chip thickness to the cutting edge radius dictates the specific cutting force, and hence at lower uncut chip thickness, the magnitude of the cutting force increases to a very high value. The lack of rigidity and a lower strength of the workpiece, especially when the diameter becomes smaller, aggravate the undesirable effects of the cutting force. Properties of the workpiece material such as ductility and grain size also affect the magnitude of the cutting force.

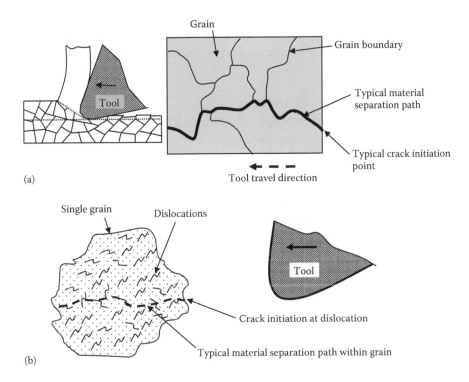

FIGURE 3.7
Material removal in (a) macro- (uncut chip thickness > 10 μm) and (b) microturning (uncut chip thickness range 0.1–10 μm).

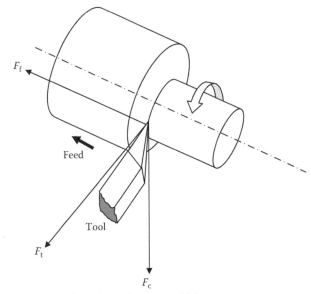

F_f = feed force, F_t = thrust force, F_c = tangential force

FIGURE 3.8
Components of the cutting force.

Thrust component and friction at the tool–workpiece interface cause deflection of the workpiece, and feed and thrust force components increase the length of the workpiece. Deflection of the workpiece during microturning causes destructive effects such as the following:

- Permanent deformation of the workpiece
- Wobbling or bending of the workpiece leading to breakage while machining
- Lengthening or stretching of the workpiece

The thrust component of the cutting force deflects the workpiece in the radial direction, thus causing variation in the uncut chip thickness, with larger variation at the free end and smaller variation at the fixed end of the workpiece. Figures 3.9 and 3.10 schematically show the deflection of the workpiece under the cutting condition and the variation in chip thickness, depicted as diameter, as a result of the deflection of a microturned shaft. It is

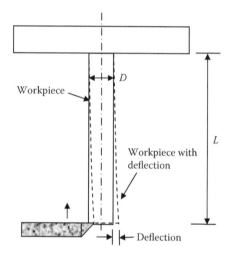

FIGURE 3.9
Radial deflection of a microturned shaft.

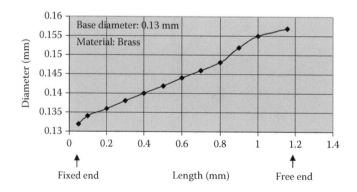

FIGURE 3.10
Variation in the diameter of a microturned shaft ($S = 2000$ rpm, $F = 4$ μm/rev, doc = 5 μm, cutting edge radius = 1.25 μm, material: brass).

seen that the diameter increases with increasing length of the workpiece, indicating that the uncut chip thickness is decreasing more at the free end and less at the fixed end as a result of deflection.

For a given uncut chip thickness, increasing value of the cutting edge radius leads to an increased frictional force and a tendency toward rubbing action at the tool–workpiece interface. This causes the resultant cutting force to shift toward the direction of thrust force and subsequently increases the magnitude of the thrust force. Hence, a sharper cutting edge is one of the prime requirements of the microturning process.

The friction at the interface of the tool flank and the work surface also results in deflection of the workpiece. Figure 3.11a shows the condition of the workpiece and forces acting on it when the workpiece is not deflected. Owing to the reaction forces, the center of the workpiece gets shifted away from the center of the spindle as well as from the rake face, which causes the following effects:

- Owing to the friction at the interface of the tool–workpiece, thrust force components act as shown in Figure 3.11b. The F_{tx} component deflects the workpiece by δx and the F_{ty} component deflects it by δy. The deflection δx significantly changes the uncut chip thickness, and δy causes the effective rake angle to become more positive. δy tends to increase the possibility of the workpiece riding over the rake face. As the value of δ keeps changing because of the dynamics of the cutting process, the resulting plastic deformation at the fixed end of the component leads to the breakage of the workpiece.

- Deflection of the workpiece results in reduction of the uncut chip thickness at the free end and gradually increases it to the programmed thickness at the fixed end of the workpiece. This reduces the process precision in terms of size and shape of the workpiece.

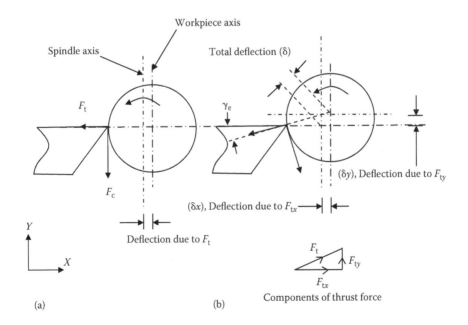

FIGURE 3.11
Deflection of the workpiece owing to cutting forces.

3.6.1 Deflection and Bending Stress

Deflection and the bending stress developed during microturning at any section of the workpiece can be represented as follows:

$$\text{Deflection, } \delta = \frac{Fl^3}{3EI} = \frac{64Fl^3}{3\pi Ed^4}$$

(3.1)

$$\text{Bending stress, } \sigma = \frac{32Fl}{\pi d^3}$$

(3.2)

where

F = thrust force,
d = diameter of the shaft,
l = length of the shaft,
σ = bending stress in the workpiece.

By measuring the thrust force at a particular workpiece length, deflection and maximum stress can be estimated. Maximum deflection will be at the free end, and maximum bending stress will be at the fixed end of the workpiece. The maximum stress that emerges should be restrained below the level that causes plastic deformation.

3.6.2 Cutting Force Ratio

As the depth of cut decreases, the ratio of the thrust force to the tangential force increases, and the thrust force becomes larger than the main cutting force at small values of depth of cut, indicating a transition of the material removal mechanism from cutting to plowing (Dornfeld and Takeuchi, 2006). Under this condition, the resultant force rotates toward the thrust direction. Hence, the cutting force ratio gives good indication of the effectiveness of the material removal in microturning.

3.6.3 Effect of Speed and Feed on Cutting Force

In any machining process, as the cutting speed increases, the generated cutting force decreases. Hence, microturning is generally carried out with very high cutting speeds in order to reduce the magnitude of the generated cutting force, and subsequently, the deflection of the workpiece. As the cutting force generated is proportional to the chip cross section, minimizing the chip cross section will enable reduction of the magnitude of the generated cutting force. As the cutting edge radius has a direct relevance on the uncut chip thickness, the feed cannot be reduced below a certain threshold value. At lower feed rates, instability of feed is observed and is found to affect the material removal process.

3.6.4 Effect of Grain Size on Force

As explained in the previous section, the material is removed across the grain in microturning. When the grain size of the work material changes, the dislocation density and its distribution within the grain also change; this being a significant factor in the material

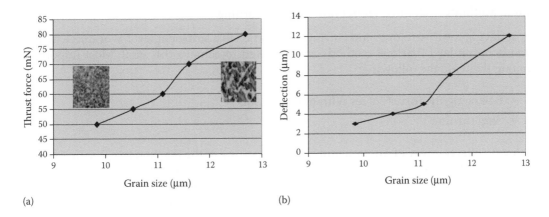

(a)   (b)

FIGURE 3.12
The effect of work material grain size in microturning: (a) thrust force (material: brass, diameter = 0.2 mm) and (b) deflection (tool: PCD with 0.125 mm TNR).

removal mechanism, it affects the magnitude of the generated cutting force. Subsequently, the thrust force and deflection of the workpiece are affected. Figure 3.12a and b shows the effect of grain size on the thrust force and the workpiece deflection, respectively. Both of them are found to increase with increasing grain size. One reason for this might be that in larger grain size machining, larger elastic recovery is encountered and rubbing of the tool with the surface results in a larger thrust force.

3.6.5 Increasing Length of the Workpiece

In microturning, yet another phenomenon observed is the increasing length of the workpiece. When the material is removed from the workpiece, the diameter decreases, and at the same time, some amount of the material flows in the axial direction of the workpiece, causing an increase in length. Figure 3.13 shows the increase in the length of the workpiece at various diameters when machining brass work material for a length of 1 mm. It is observed that when the diameter of the workpiece decreases, lengthening/stretching of the shaft increases.

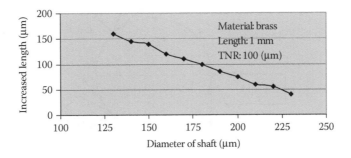

FIGURE 3.13
Increasing length in microturning.

3.6.6 Methods to Reduce Deflection in Microturning

By reducing the thrust force and cutting edge radius to a sufficient level, workpiece deflection is minimized. Various methods are practiced to achieve this goal. Most important among them are the following:

- Balancing the thrust forces with multiple tools
- Step turning

A very effective way of microturning is the balancing of the reaction forces acting on the workpiece. Reactive forces created by the tools should be such that the resultant reaction that causes deflection of the workpiece becomes zero. A schematic arrangement as shown in Figure 3.14 with three tools positioned at regular spacing balances the thrust force and overcomes the deflection.

According to the deflection equation, for a given diameter, the deflection is proportional to the cube of the length. Hence, instead of turning the entire length in one go, if it is turned in shorter step lengths, the deflection can be reduced. The step size (l), for which the shaft will not deflect plastically, can be determined by applying material strength equations. However, a certain amount of error because of mismatch between the two steps is inevitable, and its magnitude increases with an increase in the step length (Azuddin et al., 2009; Azizur Rahman et al., 2005). When the steps are machined with a uniformly decreasing diameter, from stock to finished diameter, a sudden change is effected in the section modulus, and thereby the magnitude of the stress at the fixed end of the workpiece can be minimized.

3.7 Surface Finish in Microturning

As the tolerance values of the microturned components are very small, in order to maintain the required ratio of the surface finish to the tolerance value, a high level of surface finish needs to be maintained. Being an ultraprecision machining process, microturning

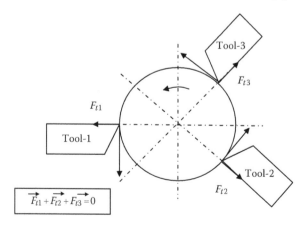

FIGURE 3.14
Balancing of thrust forces on the workpiece with multiple tools.

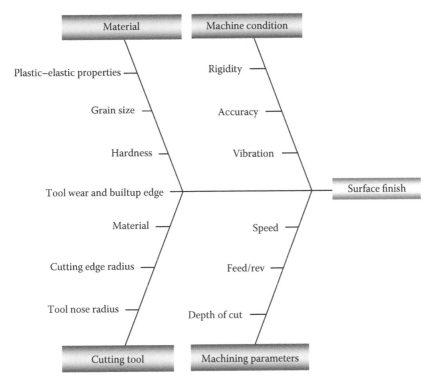

FIGURE 3.15
Factors that affect surface finish in microturning.

is capable of generating high levels of surface finish on the components. However, the surface finish value is affected by many factors, including machine tool condition, work material properties, tool geometry, and cutting parameters. Some of the important factors that affect the surface finish value in microturning are shown in Figure 3.15.

When very fine feeds are employed in microturning, significant variation is observed in the surface finish value produced and the theoretical surface finish value estimated using tool geometry and feed. It is well known by the equation $Ra = f^2/(8R)$ that the factors that influence the most are TNR and the feed rate. However, in the case of microturning, the equation gets modified because of various other factors (Feng and Wang, 2002; Dai and Chiang, 1992). When errors on the cutting edge are reproduced on the surface being machined, it alters the surface finish value. Similarly, the plastic flow of the material and the machine condition affect the surface finish value. Hence, the modified surface finish can be represented by the equation

$$Ra_{Actual} = Ra_{Theoretical} + Ra_{Plastic\ flow} + Ra_{Tool\ error} + Ra_{Machine\ condition} \qquad (3.3)$$

Figure 3.16 shows the 3D topography of microturned brass surfaces with TNR values of 50 μm and 100 μm, respectively. It can be seen that the surface finish value produced at larger radius, in this case 62 nm Ra with TNR of 100 μm, is better than that produced at a smaller TNR, that is, 115 nm Ra with 50 μm TNR. Many investigators (Dai and Chiang, 1992; Liu and Malkote, 2006) have reported that the surface finish decreases with an increase in the feed, reaches a minimum, and then increases with further reduction in

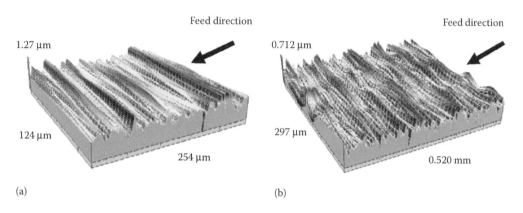

FIGURE 3.16
The effect of tool nose radius on surface finish: (a) Ra = 115 nm, tool nose radius: 50 μm and (b) Ra = 62 nm, tool nose radius = 100 μm (f = 2 μm/rev, doc = 3 μm, S = 5000 rpm, material: brass).

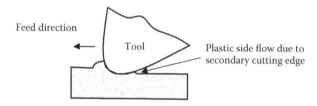

FIGURE 3.17
Schematic of plastic side flow in microturning.

the feed. At lower feed values, the plastic side flow, as shown in Figure 3.17, at the secondary cutting edge of the tool becomes noticeable, and it is found to be partly responsible for the increased surface finish value after reaching a minimum for feeds less than a certain threshold value. Another significant factor that affects the surface finish value is the cutting edge radius. As the radius of the cutting edge increases, the rake angle becomes increasingly negative and the plowing effect becomes more predominant, resulting in a deteriorated finish.

The machine tool rigidity, accuracy, and loop stiffness also affect the surface finish. Of all the components in the system, the one that is least rigid is the workpiece. Hence, the work material and its properties play a dominant role in achieving the required level of surface finish. With a lack of rigidity, chatter becomes predominant, especially at the free end of the workpiece.

3.8 Materials for Microturning

Most of the engineering materials that can be machined by tool-based macromachining can be machined by microturning. However, brittle materials pose larger problems and are more prone to breakage during machining. Materials with larger modulus of elasticity values are less prone to deflect and are more suitable for microturning. Hard materials such as tungsten can be microturned by wire electrical discharge grinding (WEDG) or by EDM. When very small diameter components with larger *L/D* ratios are to be micro-turned, nonconventional microturning is the most suitable method.

3.9 Microturning versus Microcutting

Microturning differs from microcutting in many ways. Microcutting aims at generating optical finish on metals, nonmetals, ceramics, semiconductors, and gems irrespective of the size of the workpiece, whereas microturning aims at machining smaller diameters without optical grade finishing. In microcutting, the material is removed from the work-piece by a very sharp tool with smaller uncut chip thicknesses, without any traces of rub-bing of the tool and without any microcracks on the finished surface; diamond turning, lapping, chemomechanical finishing, and magnetic rheological finishing are some of the microcutting processes (Balasubramaniam and Suri, 2010). Ductile regime machining of brittle materials, vibration-assisted microcutting, and molecular dynamic simulation are some of the research areas of microcutting that are in the forefront.

3.10 Concluding Remarks

Microturning is an ultraprecision machining process that generates features of smaller diameters with subgrain levels of uncut chip thickness. Some of the essential requirements of the microturning process are homogeneous work material, sharp cutting tools, and an ultraprecision machine tool. The major concern in microturning is the balancing of the reaction forces that otherwise cause deflection of the workpiece. Deflection is a major con-cern in microturning, and steps such as the use of multiple tools for balancing the reaction forces and the use of tool path compensation to overcome the effects of the deflection are extensively used to overcome this problem. As most of the microcomponents incorporate features that need to be manufactured by different processes, microturning is integrated with other processes in a single machine to eliminate setting errors.

References

Azizur Rahman, M., Rahman, M., Sennthil Kumar, A., and Lim, H.S. 2005. CNC microturning: An application to miniaturization. *International Journal of Machine Tools and Manufacture* 45: 631–639.

Azuddin, M., Afif, M., and Rosli, M. 2009. Development and analysis of taper tool path for micro turning operation. *Modern Applied Science* 3(1): 176–186.

Balasubramaniam, R. and Suri, V.K. 2010. Diamond turn machining, Chapter 3. *Introduction to Micro Machining*, ed. V.K. Jain. New Delhi: Narosa Publication.

Dai, Y.Z. and Chiang, F.P. 1992. On the mechanism of plastic deformation induced surface roughness. *Transactions of the ASME* 114: 432–438.

Dornfeld, M.S. and Takeuchi, Y. 2006. Recent advances in mechanical micro machining. *Annals of the CIRP* 55(2): 745–768.

Feng, C.X. and Wang, X. 2002. Development of empirical models for surface roughness prediction in finish turning. *International Journal of Advanced Manufacturing Technology* 20(5): 348–356.

Liu, K. and Malkote, S.N. 2006. Effect of plastic side flow on surface roughness in micro turning process. *International Journal of Machine Tools and Manufacture* 46: 1778–1785.

Lu, Z. and Yoneyama, T. 1999. Micro cutting in the micro lathe turning system. *International Journal of Machine Tools and Manufacture* 39: 1171–1183.

Masuzawa, T. and Tonshoff, H.K. 1997. Three dimensional micro machining by machine tools. *Annals of CIRP* 46(2): 621–628.

Shimada, S., Ikawa, N., Inamura, T., Takezawa, N., Ohmori, H., and Sata, T. 1995. Brittle–ductile transition phenomena in microindentation and micro machining. *Annals of CIRP* 44(1): 523–526.

Woon, K.S., Rahman, M., Neo, K.S., and Liu, K. 2008. The effect of tool edge radius on the contact phenomenon of tool-based micro machining. *International Journal of Machine Tools and Manufacture* 48: 1395–1407.

4

Microgrinding

P.V. Rao and S. Ghosh

Indian Institute of Technology Delhi

CONTENTS

4.1 Introduction

Grinding is basically a material removal process, where, instead of a single or few uniformly spaced and oriented cutting edges of identical and well-defined geometry as in any conventional machining, a very large number of randomly distributed very hard and stable abrasives of widely varying size, shape, and geometry accomplish material removal in the form of tiny chips as schematically shown in Figure 4.1. Generally, machining is done for bulk material removal, and grinding is done for finishing, aiming at high dimensional accuracy, good surface finish, and direct finishing of components of as such hard,

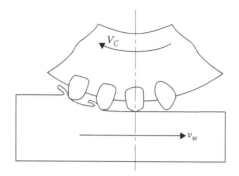

FIGURE 4.1
Schematic representation of chip formation during grinding.

or hardened, and exotic materials, which cannot be done by any conventional machining. Some major applications of grinding are listed below:

- Finishing to high dimensional accuracy and surface finish essentially required for better performance and longer service life of engineering components
- Simultaneous stock removal and finishing (abrasive milling)
- Sharpening and resharpening of cutting tools
- Slitting and parting of critical objects
- Deburring, descaling, and fettling
- Sample preparation for metallurgical studies

Nowadays, grinding is a major material removal process, which accounts for about 25% of the total cost of machining of engineering components. Grinding is generally performed by a grinding wheel, also known as a *bonded abrasive tool*, whose performance depends on the following factors:

- The type of abrasive particle used
- The size of the grit
- The hardness of the abrasive used
- The bond material
- The wheel's structure

4.2 Types of Grinding Wheels

Grinding wheels are usually made up of different types of abrasive grits and bonds. "Conventional" grinding wheels mostly use alumina or silicon carbide abrasives, and the bond may be vitreous or resinoid. "Superabrasive" wheels use diamond and cBN as abrasive particles, and the bond may be vitreous, resinoid, or metallic [1,2]. Conventional grinding wheels are mostly used for grinding ferrous and nonferrous engineering alloys, whereas superabrasives are mostly used for grinding hard and difficult-to-grind materials such as ceramics. The prime requirement of the abrasive used in grinding wheels is that it must be harder than the material it is going to abrade. The hardness of the abrasive is normally measured in terms of its static hardness as given by the Knoop hardness test or Vicker's hardness test. Another important property of the abrasive is its dynamic strength or toughness. Friability of an abrasive is very closely related to its toughness. The higher the toughness of the abrasive grit, the lesser is its tendency to fracture or fragment on being impacted upon the workpiece during the grinding process. Higher friability (lesser toughness), on the other hand, implies that the grits fragment when they impact the workpiece. Such fragmentation leads to the formation of new cutting edges and leads to self-sharpening of the wheels. Depending on the manufacturing process and the amount of additives used, hardness and friability of the abrasives can be changed. The super-abrasives, namely, diamond and cubic boron nitride (cBN), are the hardest materials available on earth. Diamond has the highest hardness and is followed by cBN. Diamond used for grinding purposes may be of natural or synthetically prepared form. The present trend is however to use the synthetic variety of diamond because of its higher friability and comparatively lower cost of manufacturing. cBN is a synthetically prepared material. The cBN grits produced these days are mostly monocrystalline in nature, although polycrystalline grains with submicron crystal size are also being manufactured. The important advantage of cBN over diamond is its better thermal stability. Because of such higher thermal stability, the cBN wheels can be manufactured with vitreous bonds, and such vitrified, bonded cBN wheels are widely used for precision grinding purposes.

Microgrinding has originated from the normal grinding process, and before discussing the microgrinding process it is imperative to look into the important aspects of normal grinding, especially the mechanics of grinding, and understand the various mechanisms associated with the grinding process. The following sections highlight the major aspects of the grinding process, which are equally important for microgrinding processes.

4.3 Machining and Grinding—A Comparison

Grinding is basically a metal cutting operation, but there are certain characteristics that are specific to the grinding process. The distinguishing characteristics of grinding over common machining are as follows:

- Grinding closely resembles the abrasive milling operation where hundreds of abrasive particles are dispersed in a matrix consisting of vitrified resin, rubber, and metals or embedded on the surfaces of metallic discs and accomplish material removal by their small sharp tips and edges.
- The size, shape, spacing, and geometry of the grinding abrasives randomly and widely vary unlike that of milling cutters [3].
- The cutting velocity during grinding is kept at least 30–50 times higher than that of machining. Such high speeds are used in grinding mostly for the following purposes:
  - To reduce the grinding forces generated.
  - To reduce chip load and forces per grit to achieve better surface finish and longer life of the working abrasives.

- Unlike in cutting tools, damage or dislodgement of a few abrasives or grits out of hundreds practically does not hamper the performance of the grinding wheel. Rather, in some cases, such dislodgement of grits is desired because that leads to the autosharpening phenomenon of the grinding wheel.
- However, a major drawback of the grinding process over the machining process lies in the requirement of higher specific energy. Grinding of a given material requires more (at least 10–20 times) specific energy than that required to remove unit volume of work material owing to the unfavorable geometry (e.g., large negative rake, on average 40°–60°) of the grit tips and additional rubbing action [4].

4.4 Grindability

Like machinability, there is also a term called *grindability*, which refers to the ease of grinding and is judged mainly by:

- Magnitude of the grinding forces and specific energy requirement
- Grinding temperature, which affects product quality and wheel life
- Surface integrity, including surface finish, residual stresses, and surface and subsurface microcracks
- Wheel life or grinding ratio
- Type of chips or mode of chip formation

Productivity, product quality, and the overall economy of grinding may be enhanced substantially by improving the grindability of a material through reduction, as far as possible, of grinding forces and specific energy requirement, grinding temperature, surface roughness, wheel damage, and chip formation mode. Depending on the material and the grinding process parameters chosen, the types of chip formation modes in grinding also vary. The following types of chips are formed for ductile materials:

- Long, continuous thread-like chips, produced ideally by shearing action
- Wider leafy chips, produced by shearing and plowing
- Spherical chips, produced under excessive temperature
- Undesirable highly strained short chips

For grinding of brittle materials such as ceramics, the chips are produced mostly by fracture and they are powdery in nature.

Grindability of any material can be improved by proper selection of the grinding wheel and grinding process parameters and by proper application of grinding fluids. Also, it is desirable that the grinding machines to be used must be rigid and robust in design with high stiffness.

Many improvements and innovative developments have taken place in the last few decades to meet the growing demand for higher productivity and quality in grinding. The major developments include the following:

- Grinding wheels, that is, abrasives, bond materials, and manufacturing process

- Design and construction of high capacity and precision grinding machines
- Preparation, selection, and application techniques of cutting fluids
- Automated condition monitoring and adaptive control

Even then, grinding still suffers from some major problems such as:

- Very high specific energy requirement
- Very high grinding zone temperature, which not only affects the effectiveness and life of the grinding wheel but also impairs the surface integrity of the products by oxidation, corrosion, and burning and also by inducing tensile residual stresses and developing microcracks at the surfaces and subsurfaces
- Wheel loading or glazing, which causes abrupt rise in grinding forces and grinding zone temperature, adversely affecting the surface integrity of the ground components
- Cutting fluids become less effective as they fail to penetrate the stiff air barrier film formed around the rotating grinding wheel
- Environment pollution and health hazards due to the use of cutting fluids, many of which are not environment friendly [5]

To overcome these problems and for better understanding of the grinding process, a careful examination of grinding mechanisms is essential.

4.5 Grinding Mechanisms

Material removal during grinding occurs because of the interaction between the hard abrasive particles and the workpiece. To understand the mechanism of material removal during grinding, some strategies can be adopted. The first and the simplest of them is a thorough study of the grinding chips or debris or swarfs. The chips produced during grinding are microscopic in nature, and hence they can be viewed only under a powerful optical microscope or, for a better and magnified view, under a scanning electron microscope (SEM) [6]. Figure 4.2 shows the SEM views of some typical grinding chips. Long shear-type chips directly support the shear mechanism of chip formation in grinding just as in machining, owing to the cutting action of microscopic abrasive grits. Leafy type chips and blocky fractured chips indicate the presence of some other phenomena, namely, plowing and sliding, which happen during the grinding process.

The second approach is to measure the force or the power requirement during the grinding process, and from those measurements calculate the specific grinding energy or the energy required for unit volume of material removed by the grinding process. For more in-depth analysis of the grinding mechanics, single grit experiments may be performed. In such single grit experiments, a single abrasive grit is used to cut through the work material. As the grit performs the cutting operation, a groove is produced. Figure 4.3 shows such grooves produced by a single grit on a ductile work material at different depths of cut under a constant grinding speed and table feed condition. Such grooves can be measured by a 3D Talysurf profilometer [7].

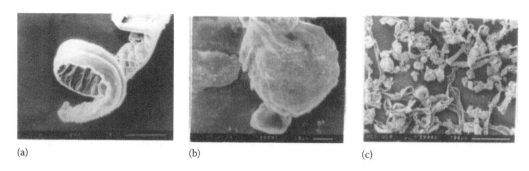

(a)   (b)   (c)

FIGURE 4.2
Microscopic views of (a) leafy, (b) spherical, and (c) blocky fractured chips. (From Paul, S., Improvements in grinding some commonly used steels by cryogenic cooling, PhD Thesis, IIT Kharagpur, India, 1994.)

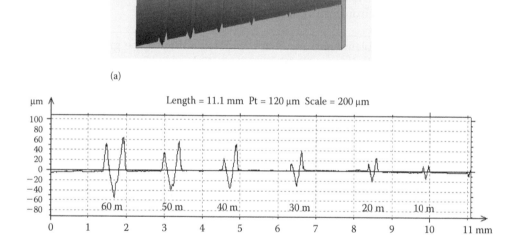

(a)

(b)

FIGURE 4.3
Pattern and sectional profiles of the grooves formed during single grit experiment as observed in (a) 3D scanning and (b) 2D scanning in a 3D profilometer. (From Ghosh, S., High efficiency deep grinding of bearing steel and modeling for specific energy requirement, PhD Thesis, IIT Kharagpur, India, 2007.)

From the profilogram of the grooves formed, it is clear that each groove has a pile-up of material on both sides. This piled-up material signifies that the grit has performed a plowing operation. This operation consumes much energy but without significant material removal. Other than this plowing action there is also a rubbing phenomenon that takes place between the grit tip and the workpiece. The specific energy requirement in grinding is very high because of both the plowing and rubbing phenomena. The energy consumers in grinding may be listed as:

- Chip formation due to shearing by the microcutting action of the abrasive grits

- Primary rubbing—rubbing between the grit tip and the workpiece material
- Secondary rubbing—rubbing along the cutting edge between abrasive grits and the work material over the total cutting path
- Plowing—displacement of the work material by highly negative rake abrasive grits of ill-defined shapes
- Microfracturing of the work material just ahead of the grit tip, which is especially of relevance in the case of brittle materials [8]
- Wear flat rubbing—friction between the wear flat developed on the abrasive grit and the workpiece material
- Friction between the loaded chip particles lodged at the inter-grit space and the work material
- Friction between the bond material and the workpiece material
- Removal of thermomechanical microwelding between the loaded chip particles and the workpiece, and between the grit tip and the workpiece
- Removal of redeposited chip particles from the ground surface

The above *energy consumers* can be divided into two broad subgroups. Chip formation, primary rubbing, secondary rubbing, and plowing can be categorized as *primary energy consumers* in grinding [9]. They are very relevant in any grinding operation irrespective of the cycle time. The other consumers are called *secondary energy consumers* and they are relevant in typically long-duration grinding, which may introduce a wheel wear and wheel loading. Microfracturing as a mode of chip formation mainly occurs while grinding rather brittle materials such as ceramics that retain their strength even at high grinding temperature.

4.5.1 Chip Formation Energy

The contact length in flat surface grinding is theoretically equal to $\sqrt{aD}$ [5], where a denotes the infeed or the depth of cut and D is the diameter of the wheel. However, researchers have observed that the actual contact length may be 1.5–2 times higher than the theoretical contact length during the grinding operation.

Malkin [5] suggested that the energy required for chip formation is equal to the amount of energy required for adiabatic melting of the same volume of work material, although it was asserted that melting does not occur in grinding. The timescale of chip formation, which is in microseconds, indirectly supports this hypothesis of Malkin.

The adiabatic energy may be calculated by using the equation

$$U_{\text{adiabatic}} = \rho \int_0^T C_p \, dt + L\rho$$

(4.1)

where L is the latent heat transfer, ρ is the density of the work material, C_p is the specific heat at constant pressure, and T is the melting point temperature. Malkin indicated the chip formation energy for ferrous alloys to be around 13.6 J/mm^3 but adiabatic melting as calculated from the above equation comes to about 8.27 J/mm^3, which is obviously less than chip formation energy, U_{ch}. The difference is around 30%. Adiabatic specific melting energy only accounts for the shearing energy; the energy expended owing to chip grit interfacial friction accounts for the difference between $U_{\text{adiabatic}}$ and U_{ch} [7].

However, the specific energy required for chip formation can also be estimated using classical theories of machining as follows:

$$U = \frac{P_z V_C}{t s_o V_C}$$

$$= \frac{t s_o \tau_s \left(\zeta - \tan \gamma_o + 1 \right) V_C}{t s_o V_C}$$

$$U = \tau_s \left[e^{\mu \left(\frac{\pi}{2} - \gamma_o \right)} - \tan \gamma_o + 1 \right]$$

(4.2)

Results obtained during simulation experiments using a highly negative rake turning tool indicated that the values of ζ and τ_s increase when the tool rake angle is large (negatively). Under normal machining conditions with a tool rake angle of −6°, ζ can be around 2.00, which makes the apparent coefficient of friction μ to be equal to 0.4 according to the following equation [3].

$$\zeta \cong e^{\mu \left(\frac{\pi}{2} - \gamma \right)}$$

(4.3)

During grinding the average rake angle is around −60°, which incorporates a higher change in momentum of the chip, leading to a more specific normal force at the chip–grit interface. In machining and grinding, the chip tool interfacial friction is non-Coulombian as the contact stress is as high as the yield strength of the work material. Thus, in grinding, the value of μ would be higher owing to higher contact stress (higher contact stress leads to more microwelding between the backside of the chip and grit leading to higher μ). An upper bound of $\mu = 1.0$ may be safely assumed for calculations [7].

Dynamic yield shear strength of the work material, as has been observed during the simulation experiments, was found to be affected by tool rake angle, cutting velocity, and strain rate. In the simulation experiment itself, τ_s has been observed to be 800 MPa. During grinding, the strain rate is relatively higher than that in machining, and a representative value of τ_s of 1000 MPa can be reasonably assumed. Using this value of τ_s in Equation 4.2 yields a simulated chip formation specific energy of 13.7 J/mm³. The very close match between the simulated specific energy and U_{ch} despite the assumptions shows that chip formation energy in grinding is quite high owing to high strain rate, leading to higher τ_s and high chip reduction coefficient due to high negative rake of the irregular grinding grits. Further, this upper-bound solution of specific energy for chip formation in the grinding of ferrous alloys remains more or less constant throughout most of the grinding domain [7].

4.5.2 Primary Rubbing

The phenomenon of primary rubbing in grinding has already been identified and defined in previous sections. Figure 4.4 shows the primary rubbing phenomenon. Loop stiffness of the machine is assumed to be k. Assuming all other contacts and elements to be infinitely rigid, loop stiffness has been shown between the grit and the wheel in Figure 4.4 [7].

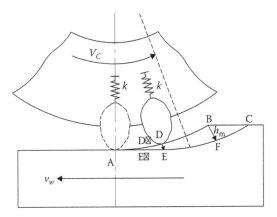

FIGURE 4.4
Schematic view of grits acting as a spring. (From Ghosh, S., High efficiency deep grinding of bearing steel and modeling for specific energy requirement, PhD Thesis, IIT Kharagpur, India, 2007.)

The grit tip is not infinitely sharp and has an edge radius of r_e. At any intermediate position, the grit tip is at D' as penetration has not occurred where it should have been, at E'. Thus, the grit would be experiencing a normal force, f_n, which may be expressed as

$$f_n = k\delta \tag{4.4}$$

where

δ = linear displacement,
k = stiffness of the machine.

The Hertzian contact stress can be determined, taking into account the contact between two rollers. But in the present case, the radius of curvature of the cutting path and that of the grit tip are almost 3–4 orders different. Thus the contact area has been assumed to be the square of the edge radius (r_e) and the contact stress has been taken as σ_n, which is expressed as

$$\sigma_n = \frac{f_n}{\text{contact area}}$$
$$= \frac{f_n}{r_e^2}$$
$$= \frac{k\delta}{r_e^2} \tag{4.5}$$

Penetration would occur only when the contact stress (σ_n) is more than or equal to the flow strength or *in situ* hardness (σ_H) of the material, that is,

$$\sigma_n = \frac{k\delta}{r_e^2} \geq \sigma_H$$
$$\therefore \sigma_n = \frac{f_n}{r_e^2} = \sigma_H \tag{4.6}$$

Thus, when the grit penetrates the workpiece at D, the normal force would be given by

$$f_n = \sigma_H r_e^2$$

$$(4.7)$$

The tangential force is typically half the normal force while grinding ferrous alloys with conventional and superabrasive wheels [5,6]. The same ratio could be as low as one-third to one-fifth while grinding harder materials such as carbides and ceramics.

The specific energy due to primary rubbing can thus be calculated as the ratio of average power requirement to material removal rate [7]:

$$
\begin{aligned}
U_{pr\_rub} &= \frac{(\text{avg power})_{pr\_rub}}{\text{MRR}} \\[2mm]
&= \frac{f_t^{avg} \times \text{no. of grits} \times V_C}{v_w ab} \\[2mm]
&= \frac{f_t^{avg} \times (\text{grit density} \times \text{contact area}) \times V_C}{v_w ab} \\[2mm]
&= \frac{f_t^{avg} \times C \times \sqrt{a_i \times d} \times b \times V_C}{v_w ab}
\end{aligned}
$$

$$(4.8)$$

a_i is the instantaneous infeed (at D) where penetration occurs and b is the width of the grinding wheel or the width of the job whichever is smaller. a_i is related to h_i (instantaneous grit depth of cut at D) as

$$h_i = \left\{ \frac{4}{\lambda C} \times \frac{v_w}{V_C} \times \sqrt{\frac{a_i}{d}} \right\}^{1/2}$$

$$(4.9)$$

Again, h_i can be related to δ. Thus,

$$h_i = \delta = \frac{\sigma_H r_e^2}{k}$$

$$(4.10)$$

The loop stiffness k can be determined from the growth of force data with number of passes. Typically, a fraction of the infeed is removed in grinding owing to finite stiffness of the grinding wheel. Figure 4.5 shows a typical growth of f_t with number of passes due to the finite stiffness of the machine.

In the very first pass only $a(1)$ amount of material is removed instead of the given infeed of a_g, giving rise to a normal force of $f_n(1)$ owing to deflection. Thus,

$$a_g = a(1) + \text{deflection}$$

$$a_g = a(1) + \frac{f_n(1)}{k}$$

$$(4.11)$$

Hence, the stiffness k can be determined experimentally as

$$k = \frac{f_n(1)}{a_g - a(1)}$$

$$(4.12)$$

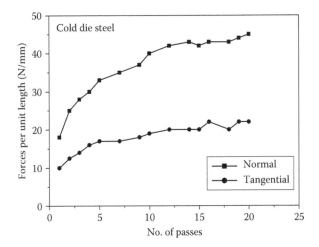

FIGURE 4.5
Growth of forces with number of passes while grinding cold die steel with a white alumina wheel. (From Paul, S., Improvements in grinding some commonly used steels by cryogenic cooling, PhD Thesis, IIT Kharagpur, India, 1994.)

$a(1)$, that is, the actual material removed or infeed in the first pass can be determined by actual measurement of or reduction in thickness or height of the workpiece by a precision dial gauge or height gauge.

4.5.3 Secondary Rubbing

Secondary rubbing occurs mainly along the entire contact profile of the scratch produced by the grinding grit. The grit may rub against the piled-up material, but that is neglected as the plowed material would not have so much contact stiffness as the parent material or the workpiece. Figure 4.6 shows a schematic view of the tip of a single grit along with the plowed material.

The tangential force (f_t) acting along the entire length of the grinding scratch is usually about half of the average normal force. Maximum normal force would occur when the grit is about to leave the cut, that is, when the grit depth of cut is maximum (i.e., equal to h_m). This maximum normal force per grit (f_n^{max}) can be estimated as

$$f_n^{max} = \text{contact stress} \times \text{contact area} \tag{4.13}$$

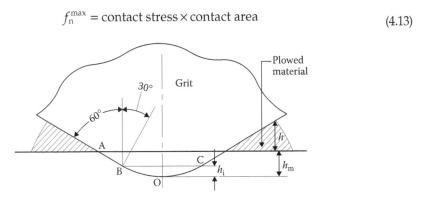

FIGURE 4.6
Schematic view of a rounded grit with plowed material.

The nature of contact during secondary rubbing could be elastic and the contact stress would be similar to the one during machining or turning at the wear land. The contact stress can be safely assumed as half of the elastic limit of the work material. Thus the average tangential force per grit has been estimated as

$$f_t^{avg} = f_n^{avg} = 0.5 \times 0.5 \times f_n^{max}$$

$$f_t^{avg} = 0.25 \times 2 \times \left[\frac{\left\{ h_m - r_e \left(1 - \sin \frac{\pi}{3} \right) \right\}}{\cos \frac{\pi}{3}} + r_e \times \frac{\pi}{6} \right] \times r_e \times \sigma_e \times 0.5$$

$$= 0.25 \times r_e \times \sigma_e \left(2 \times h_e + 0.255 \times r_e \right) \tag{4.14}$$

Now, the specific secondary rubbing energy can be estimated as

$$U_{sec\,rub} = \frac{f_t^{avg} \times (\text{no. of grits}) \times V_C}{\text{MRR}}$$

$$= \frac{f_t^{avg} \times \left(C \times \sqrt{ad} \times b \right) \times V_C}{v_w \times a \times b}$$

$$= \frac{f_t^{avg} \times \left(C \times \sqrt{ad} \right) \times V_C}{v_w \times a}$$

$$= \frac{f_t^{avg} \times C \times V_C}{v_w} \sqrt{\frac{d}{a}} \tag{4.15}$$

4.5.4 Specific Plowing Energy

In the previous sections and figures, plowing has been defined and described as the lateral displacement of the work material by the grinding grits as piled-up material along the length of the scratch.

De Vathaire et al. [10] proposed an upper-bound model of plowing by a pyramidal indenter. They conducted the experiments by plowing a low-carbon steel workpiece with pyramidal indenters. The term *scratch hardness* was defined as the resistance of the work material toward scratching by an indenter. Hardness is evaluated as the ratio of normal force introducing plastic deformation in the work material to the bearing area of the indentation. Similarly, scratch hardness was defined as the ratio of normal force during scratching to bearing area of the scratch. Bearing area can be estimated as $e \tan \alpha (e + h) \tan \alpha$, where α is the semi-angle of the pyramidal grit. Thus the normal force due to plowing, $f_{n,p}$, can be written as

$$f_{n,p} = H_s e^2 \tan^2 \alpha \left(1 + \frac{h}{e} \right)$$

$$= H_s e^2 \tan^2 \alpha (1 + H) \tag{4.16}$$

In grinding, e is comparable to the maximum grit depth of cut h_m and h are the height of the piled-up material as may be determined from Figure 4.6. De Vathaire et al. [10] have carried out a series of experiments to correlate scratch hardness to workpiece hardness, and typically they observed the following relationship:

$$H_s = g(\alpha)H_C \tag{4.17}$$

where the coefficient g that relates H_s and H_C is a strong function of the apex angle of the pyramidal grit α [11].

The tangential force during plowing, $f_{t,p}$, would be related to the normal force $f_{n,p}$. Typically, in the grinding of ferrous alloys, f_t/f_n is 0.5.

Now, the specific energy in plowing may be determined as

$$U_{pl} = \frac{f_t^p \times V_C}{MRR} \text{ at grit level} \tag{4.18}$$

MRR = Area of the scratch formed by the grit $\times$ velocity

$$= h_m^2 \tan \alpha \, V_C \tag{4.19}$$

$$\therefore U_{pl} = \frac{f_t^p \times V_C}{h_m^2 \tan \alpha \, V_C} \tag{4.20}$$

$$U_{pl} = \frac{f_{n,p} \times \left(\dfrac{f_{t,p}}{f_{n,p}} \right)}{h_m^2 \tan \alpha}$$

$$= \frac{g H_C h_m^2 \tan^2 \alpha (1+H)}{h_m^2 \tan \alpha} \times \left(\frac{f_{t,p}}{f_{n,p}} \right) \tag{4.21}$$

Now, U_{pl} can be determined using the expressions of g and H and the property of the work material hardness (H_C). H is the ratio of the height of piled-up material to the grit depth of cut, and it is also a function of the semi-apex-angle of a pyramidal grit [10,11].

The above sections throw light on how to individually estimate the specific energies involved in the shearing, rubbing, and plowing phenomena associated with the grinding process. The models have been made primarily for the primary energy processes but similar models may be obtained for secondary energy consumption processes as well.

In grinding, multiple grits are in action simultaneously to remove the work material. To understand the mechanism of so many grits working simultaneously, a quick stop apparatus has been developed, fabricated, and used to capture the transient grinding zone. The following section deals with design, construction, and use of such a quick stop apparatus in grinding processes.

4.5.5 A Quick Stop Apparatus to Study Grinding Mechanism

Total forces and the overall specific energy required in grinding are governed by the combined actions of several grits in the grinding (contact) zone. Unlike single grit action, combined action of so many grits at the grinding zone is more difficult

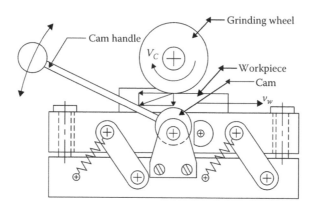

FIGURE 4.7
Schematic view of the quick stop apparatus. (From Ghosh, S., High efficiency deep grinding of bearing steel and modeling for specific energy requirement, PhD Thesis, IIT Kharagpur, India, 2007.)

to observe. In conventional machining, the actual chip–tool interaction is studied by quickly freezing the machining action by using a suitable quick stop or drop tool apparatus. Such a device is also needed for a close study of the actual wheel (abrasives)–work interaction. For this purpose a simple device has been designed and built as shown in Figure 4.7. While surface grinding a test specimen under pendulum mode, the work specimen is suddenly and very rapidly dropped manually, striking the lever. The workpiece gets detached from the wheel, leaving the transient grinding zone intact. Then the topography of that region is observed and recorded in SEM for a close study of the combined action of the grits.

The SEM views of the transient grinding zone obtained by using the quick stop apparatus device while down-grinding a hardened (R_C 60) bearing steel specimen are shown in Figure 4.8 [7].

Figure 4.8 shows that the streaks are clean, distinct, and almost free from plowing and redeposition in the grinding zone of the bearing steel, possibly for higher hardness and less stickiness of bearing steel. The irregularities in the borders of engagement and disengagement in and from the transient zone can be attributed to random variation in shape, orientation, and protrusion of the grits. It is also seen in Figure 4.8 that the width and depth of the grooves or streaks gradually decrease before disengagement, which is expected in the case of down-grinding. The present quick stop device needs to be improved further for more robust and speedy action. Close examination of the frozen grinding zone under SEM and Talysurf is expected to reveal a lot of interesting and important features, which can be used for understanding and modeling the grinding process [12].

4.6 Microgrinding—Its Definition and Applications

Recently, the electronics, computer, and biomedical industries demand the manufacturing of miniature-sized components and their parts with high precision. Consequently, many new processing concepts, procedures, and machines have evolved to fulfill these requirements and expectations [13]. Among the various micromanufacturing processes, mechanical micromachining has become very popular for various material removal

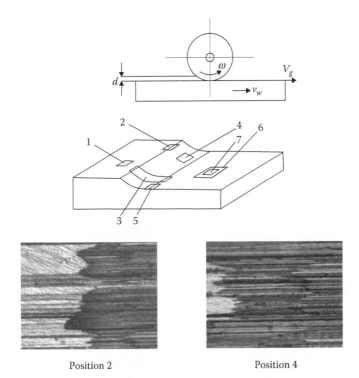

Position 2 Position 4

FIGURE 4.8
Microscopic views of the various surfaces observed after quick stop. (From Ghosh, S., High efficiency deep grinding of bearing steel and modeling for specific energy requirement, PhD Thesis, IIT Kharagpur, India, 2007.)

processes. Within these processes, microgrinding is usually the final processing step and similar to conventional grinding it provides a competitive edge over other processes in manufacturing microscale products such as microsensors, microactuators, and micromachine parts. The microgrinding process resembles a conventional grinding process, and is distinctive only because of the size effect phenomenon, whereby mechanical and thermal interactions between a single grit and a workpiece are related to the phenomena observed in micromachining. The quality of the parts produced by this process is influenced by process conditions, properties of microgrinding wheel, and microstructure of the work materials. As the diameter of the grinding wheel decreases, the effects of higher plowing forces and grinding wheel deformation become more predominant in microgrinding operations. Currently, there have been numerous studies to understand mechanical micromachining by investigating the size effect in the specific cutting energy and crystallographic effects. However, there is no general consensus on the fundamental understanding of this mechanism. Efficient and reliable microgrinding tools with diameters down to 13 μm are manufactured by using electroplated diamond grits on carbide microshanks [14]. The machine and the equipment for manufacturing such small-sized tools require the usage of high precision spindles, proper grinding devices, and coating system. Microgrinding tools are capable of ultraprecision grinding. Optical quality microstructured surfaces may be obtained by using microgrinding tools. This grinding technique may be safely employed in numerous microtechnology applications, for example, precision dies. With reduced tool diameter, very complicated shapes can be manufactured in hard and brittle materials. Normally, dressing and truing of the grinding tools are not necessary, and the chip accommodation space is also relatively large, thereby preventing the wheel loading phenomenon considerably [14]. However, there are various problems associated with miniaturization of the grinding tools. When tools are miniaturized, their surface microstructure and material texture begin to affect their mechanical properties and performances [15,16], thus undesirably increasing the structural sensitivity. In particular, microtools used for micropunching and micro-cutting require sufficient mechanical strength to withstand the load during machining. If the microtool is poor in surface integrity, a rough surface may act as a fracture origin, thereby degrading the strength of the tool. Hence, extreme care should be taken while manufacturing such microtools. Normally, these microtools are fabricated by using a superfine abrasive wheel in ultraprecision grinding machines. Such a method can produce a cylindrically shaped microtool with a tip diameter of less than 1 μm. It has been observed that the surface of a processed microtool may be strengthened by allowing the penetration and diffusion of oxygen atoms into the material, thereby allowing oxidation to take place.

4.6.1 Microgrinding Machine Tools

To successfully implement the microgrinding operation, there is a requirement for improved accuracy of machine tools. While designing such microgrinding machine tools, it has been found by Bryan [17] that thermal effects in grinding reduce machining accuracy. Weck and Schiefer [18] suggested that there is a need for highly dynamic loop stiffness microgrinding machines. Also, during the machine tool design machine tool spindle error motions need to be considered as well [19]. In industrial settings, a number of early ultraprecision grinding machines were set up by the adaptation of grinding spindles to diamond turning machines. The driving force behind these developments was the growing demand for rapid manufacture of aspheric glass optics. Ultraprecision grinding was expected to be a machining process by which optical surface quality could be achieved so that polishing time is significantly reduced. Diamond turning of aspheric metal optics and of crystalline materials (e.g., germanium) was already well established by the mid-1980s, with a roughness of 3–5 nm rms and a form accuracy of 100 nm on 100-mm-diameter parts being reliably achieved by diamond turning systems [20]. Addition of grinding spindles to diamond turning machines was carried out to extend ultraprecision machining of glass-type materials that could not be effectively machined by diamond turning [21,22]. During the same period, the microelectronics sector had a number of manufacturing demands whereby the reduction, or elimination, of polishing was considered a key enabler for improving manufacturing capabilities and product quality. Products, including silicon wafer substrates, read/write heads, and glass/ceramic memory disks [23] were identified as notable examples. All these micro-electronic applications demanded a grinding process that could produce ultraprecise surfaces with low surface and subsurface damages. To carry out a high-quality ultra-precision grinding operation, the following requirements are a must for the ultraprecision machine tool [24]:

- Precise, smooth, and backlash-free motions
- Low levels of synchronous and asynchronous spindle errors
- High static/dynamic loop stiffness
- Long thermal dimensional control

4.6.2 Microgrinding Tools

4.6.2.1 Abrasive Types

Diamond and cBN are the two most widely used abrasives in microgrinding operations. There are two types of diamond—naturally occurring and synthetically manufactured. Both have very high wear resistance, heat conductivity, hardness, and low coefficient of friction. But the major drawback of diamond abrasives is their high chemical reactivity with some metallic materials that results in the transformation of diamond into its graphite form. Consequently, diamond abrasives are used only for grinding of brittle nonferrous materials such as silicon, glass, and ceramics. Sometimes, diamond grains are coated with suitable materials that prevent oxidation and thereby improve the grinding ratio during the grinding process. Another approach to extend the life of the abrasive tool is to use blocky, coarse-grained diamond grits that have higher abrasion resistance and better thermal stability. Sometimes, ordinary grinding wheels are coated with polycrystalline diamond films obtained by the chemical vapor deposition (CVD) process. The sharp edges of the microsized diamond crystallites are used for micropencil grinding tools. Further, cBN has superior thermochemical stability compared to diamond. It can be used for grinding those ferrous materials that react chemically with diamond.

4.6.2.2 Bonding of Abrasive Tools

The major bond systems used in grinding wheels are (i) metallic, (ii) resinoid, and (iii) vitrified. Metal bonding can be separated into two different types: sintered metal bonding and electroplating. In ultraprecision grinding, the sintered metal bond system is used for thin wheels that cut brittle materials, for example, silicon wafers (slicing/dicing) [25], or for micropencil grinding tools [26]. Often, electroplated metal bonding is applied to single-layered grinding wheels with stochastically or well-defined positioning of the grains [27]. High heat conductivity and good wear resistance are the major benefits of metal-bonded grinding tools.

Resin-bonded grinding wheels are normally used for rough grinding or abrasive machining purposes. For ultraprecision grinding, epoxy or polyester resins are used to generate high surface qualities by soft, smooth grinding or polishing, for example, in silicon wafer grinding.

Vitrified bonds have a glass-like structure and are fabricated at high temperatures from mineral fluxes such as feldspars, firing clays, ground glass frits, and chemical fluxes [28]. Vitrified bonds have higher strength and are easier to dress. The elastic modulus is almost four times higher than the resin bonding.

4.6.2.3 Geometry of the Grinding Wheels Used in Microgrinding Operations

Most grinding wheels applied in ultraprecision grinding are made with diverse design concepts, for example, with undefined or defined grain settings, grooved wheels, or cup wheels. Typically, the wheel diameter ranges from 50 to 400 mm and the grain sizes from fine grained (~0.125 μm) to coarse grained (~200 μm). Such grinding wheels are used for machining of materials such as silica, silicon, compound semiconductors, and materials used in electronics components.

Grinding wheels with undefined grain settings are the most widely used type, and they are applied with radial feed for surface grinding of optical products such as lenses and mirrors. Cup wheels are a special type of grinding wheel in which the grains, instead of being arranged in the radial direction, are bonded along the axial direction. This type of wheel finds application in wafer grinding, mirror grinding, and thinning. Sometimes, grooved grinding wheels are also used where deep grooves are made on the wheel, and such grooves are used to minimize heat transfer into the workpiece, thereby improving the cooling conditions.

4.6.2.4 Grinding Fluids

Grinding fluids are mostly used to reduce heat generation in the grinding zone. They also sometimes help in reducing friction between workpiece and grinding wheel, thereby minimizing thermal damages to the workpiece. In the microgrinding process, the contact length between the abrasives and the workpiece is small, and consequently the heat flow into the workpiece is also relatively low compared to that of the conventional grinding process. In such microgrinding operations, it has been observed that the use of coolants with high lubricity impairs the surface finish. Therefore, only water-based coolants should be used in microgrinding operations [24].

4.6.3 Microgrinding of Ceramic Materials

Grinding of hard and brittle materials such as ceramics induces microcracks, which deteriorate the surface quality of the ground product. In general, it has been observed that while grinding brittle materials, there exists a transition zone where the material removal mechanism transforms the behavior from brittle to ductile. A large number of research efforts have been made to identify this transition zone and to understand the removal mechanisms. This brittle to ductile regime transition in material removal is of paramount importance in microgrinding operations. It has been observed that the most essential parameter for transition from brittle to ductile behavior in chip removal is the stress conditions around the cutting edge. It has been established from plasticity theory and fracture mechanics that hydrostatic stress fields in the shear plane are necessary for the ductile cutting of hard and brittle materials. Ductile machining of the hard and brittle materials depends strongly on a complex interaction between tool geometry, process parameters, and material response. The relatively high negative rake angle during grinding results in the generation of high frictional heat in the work material, which favors higher hydrostatic compression in the cutting zone, transforming to the ductile material removal mechanism. The ductile mode of material removal also depends on the depth of cut used during the grinding process. The contact of the abrasives with the machined surface leads to elastic material response and, with increasing depth of cut, to plastic behavior, that is, "microgrooving" and "microplowing." This type of interaction also depends on the overall abrasive geometry and the grain tip radius. It has been established by various researchers that generating hydrostatic pressure and maintaining a critical chip thickness are prerequisites for crack-free ductile grinding of brittle materials. Such ductile regime grinding of brittle and hard materials requires a maximum chip thickness $h_{\max}$, not exceeding the critical material-specific chip thickness $h_{\max,cr}$ to avoid crack initiation. Marshall et al. [29] established the following equation for estimating the critical material-specific chip thickness:

$$h_{\max,cr} \sim \left(\frac{E}{\mathrm{HK}}\right)\left(\frac{K_c}{\mathrm{HK}}\right)^2$$

(4.22)

where E is the Young modulus of the work material, HK is the Knoop hardness, and K_c is the critical fracture toughness.

The researchers have carried out various grinding experiments with brittle materials and they have found that the proportionality factor in the above equation comes out to be approximately 0.15.

4.6.3.1 Surface Quality in Microgrinding Process

In brittle material grinding it has been found that the surface quality is best when the material is ground under ductile mode. It has also been found that the effect of change in grain size on ductile regime grinding is more profound than that of wheel speed and work speed in grinding. Generally, it has been observed that if wheels with an average grain size of less than 10 μm were used, the grinding of optical glass in ductile mode was obtained without any surface or subsurface crack formation. The use of suitable grinding coolants also influences ductile mode grinding. The effect of depth of cut in ductile regime grinding is found to be of least significance. If ductile mode grinding is carried out on the brittle materials, surface roughness of less than 0.2 nm rms and 2 nm R_{max} is achievable, especially during the grinding of optical glasses without any polishing process.

4.7 Remarks

Microgrinding is primarily used for the generation of very high quality surfaces and functional parts made from difficult-to-machine materials for both optical and nonoptical applications. The major application area of such grinding processes is in semiconductor industries where silicon wafers with a nanolevel surface finish are ground. This process is also widely used in hardened steel and ceramic mold manufacturing. The major challenge in microgrinding is to obtain an optical surface finish without surface or subsurface damage in the work materials. Research has proven that such a defect-free optical surface finish in grinding is achievable only when the hard and brittle materials are ground by the ductile regime grinding method. Tool wear in microgrinding operations, especially in ductile regime grinding, proved to be a major obstacle in obtaining a defect-free ground component. To reduce such tool wear, "engineered" grinding wheels with large truncated grains have been developed. It has been demonstrated that the use of such wheels resulted in excellent surfaces with no appreciable tool wear. Contour surface grinding by microtools depends mostly on the improvement of wheel preparation. Also, there is a need for developing new machine tools with very high loop stiffness and grinding spindles dedicated to ultraprecision grinding.

To improve the microgrinding process, one needs to have a closer look at the different grinding mechanisms that are responsible for material removal. Suitable analytical models need to be established so that effective utilization of the grinding process can take place. Sometimes, single grit experiments can give valuable insight into the grinding mechanisms, and by conducting such experiments, pseudoanalytical models of the grinding mechanics can be established [30].

Microgrinding is expected to become more reliable, faster, and versatile in the near future and will gradually conquer newer areas of applications.

Nomenclature

a	Infeed (μm)
a_1	Chip thickness before cut
a_2	Chip thickness after cut
b	Width of the grinding wheel
c	No. of effective grits per unit area
d	Diameter of the wheel
f	Form factor
f_n	Normal force during grinding (N)
f_t	Tangential force during grinding (N)
g	Ratio of scratch hardness to workpiece hardness
H	The ratio of the average height of the plowed material to the depth of the groove
H_C	Workpiece hardness
h_i	Instantaneous grit depth of cut
h_m	Maximum grit dept of cut
H_S	Scratch hardness of a material
k	Loop stiffness of the grinding machine
m	No. of active grits per unit length on the wheel periphery
m'	No. of cutting edges per unit length along the cutting periphery
P_X	Feed force (N)
P_Z	Tangential or main cutting force component (N)
r_e	Edge radius of a single grit
t	Depth of cut (mm)
t_{avg}	Average chip thickness per grit
t_m	Maximum chip thickness
U_c	Specific energy requirement for chip formation in grinding
U_{ch}	Chip formation energy
U_g	Specific grinding energy requirement (J/mm^3)
U_m	Specific energy requirement during machining
U_{pl}	Specific energy requirement for plowing
U_{pr_rub}	Specific energy requirement for primary rubbing
U_{secrub}	Specific energy requirement for secondary rubbing
V_C	Grinding velocity (m/s)
v_w	Table speed (mm/min)
ζ	Chip reduction coefficient
η	Angle of apparent friction at the chip tool interface
μ	Coefficient of friction
ϕ	Principal cutting edge angle
λ	The ratio of width of the groove to the depth of the groove
ρ	Specific gravity
δ	Linear displacement
σ_e	Elastic limit of the material (MPa)
σ_H	*In situ* hardness
σ_n	Contact stress
τ_s	Dynamic yield shearing strength of the work material (MPa)

References

1. Shaw, M.C. 1996. *Principles of Abrasive Processing*. Oxford, UK: Clarendon Press.
2. Metzer, J.L. 1986. *Superabrasive Grinding*. UK: Butterworth Publications.
3. Bhattacharya, A. 1984. *Metal Cutting Theory and Practice*. Kolkata, India: New Central Book Agency.
4. Malkin, S. 1979. Negative rake cutting to simulate chip formation in grinding, *Annals of the CIRP* 28(1): 209.
5. Malkin, S. 1989. *Grinding Technology: Theory and Applications of Machining with Abrasives*. New York, NY: John Wiley and Sons.
6. Paul, S. 1994. Improvements in grinding some commonly used steels by cryogenic cooling, PhD Thesis, IIT Kharagpur, India.
7. Ghosh, S. 2007. High efficiency deep grinding of bearing steel and modeling for specific energy requirement, PhD Thesis, IIT Kharagpur, India.
8. Chattopadhyay, A.B. 2002. Some recent developments in grinding technology. Keynote Address, *ICM-2002*. BUET, Dhaka, p. 1.
9. Paul, S. and Chattopadhyay, A.B. 1995. A study of effects of cryogenic cooling in grinding. *International Journal of Machine Tools and Manufacture* 35(1): 109.
10. De Vathaire, M., Delamare, F., and Felder, E. 1981. An upper bound model of plowing by a pyramidal indenter. *Wear* 66(1): 55.
11. Ghosh, S., Chattopadhyay, A.B., and Paul, S. 2008. Modelling of specific energy requirement during high efficiency deep grinding. *International Journal of Machine Tools and Manufacture* 48(11): 1242–1253.
12. Ghosh, S., Halder, B., and Chattopadhyay, A.B. 2009. Design and use of a quick stop device for close study of grinding mechanism. *14th National Conference on Machines and Mechanisms, NIT Durgapur, Dec. 17–18*, pp. 404–408.
13. Park, H.W. and Liang, S.Y. 2008. Force modelling of micro-grinding incorporating crystallographic effects. *International Journal of Machine Tools and Manufacture* 48: 1658–1667.
14. Aurich, J.C., Engmann, J., Scheuler, G.M., and Haberland, R. 2009. Microgrinding tool for manufacture of complex structure in brittle materials. *CIRP Annals—Manufacturing Technology* 58: 311–314.
15. Dornfeld, D., Min, S., and Takeuchi, Y. 2006. Recent advances in mechanical micromachining. *Annals of the CIRP* 55(2): 745–769.
16. Ohmori, H., Katahira, K., Uehara, Y., Watanabe, Y., and Lin, W. 2003. Improvement of mechanical strength of micro tools by controlling surface characteristics. *Annals of the CIRP* 52(1): 467–470.
17. Bryan, J. 1990. International status of thermal error research. *Annals of the CIRP* 39(2): 645–656.
18. Weck, M. and Schiefer, K.H. 1979. Interaction of the dynamic behaviour between machine tool and cutting process for grinding. *Annals of the CIRP* 28(1): 281–286.
19. Vanherck, P. and Peters, J. 1973. Digital axis of rotation measurements. *Annals of the CIRP* 22(1): 135–136.
20. Gerchman, M.C. 1986. Specifications and manufacturing considerations of diamond machined optical components. *Proceedings of the Society of Photo Optical Instrumentation Engineer's Conference*. Vol. 607. Los Angeles, pp. 36–45.
21. Nicholas, D.J. and Boon, J.E. 1981. The generation of high precision aspherical surfaces in glass by CNC machining. *Journal of Physics D Applied Physics* 14: 593–600.
22. Ruckman, J., Fess, E., and Li, Y. 1991. Contour mode deterministic micro-grinding. *Proceedings of the 14th Annual Meeting of the ASPE*. Vol. 20. Monterey, USA, pp. 542–546.
23. McKeown, P.A. 1986. Ultra-precision diamond machining of advanced technology components. *Journal of Materials and Manufacturing Processes* 1(1): 107–132.
24. Brinksmeir, E., Mutlugünes, Y., Klocke, F., Aurich, J.C., Shore, P., and Ohmori, H. 2010. Ultraprecision grinding. *CIRP Annals—Manufacturing Technology* 59: 652–671.

25. Anderson, J., Hollman P., and Jakobson S. 2001. Abrasive capacity of thin film diamond structures. *Key Engineering Materials* 196: 141–148.
26. Engmann, J., Schüler, G.M., Haberland, R., Walk, M., and Aurich, J.C. 2009. Efficient technique for 3rd micro structuring of carbide and brittle materials with diamond micro-shaft grinding tools. *Proceedings of the 9th Euspen International Conference*. Vol. 2. San Sebastian, Spain, pp. 97–100.
27. Rickens, K., Grimme, D., and Brinksmeier, E. 2006. Deterministic machining of brittle materials applying engineered diamond wheels. *Proceedings of the 6th Euspen International Conference*. Vol. 2. Baden bei Wien, Austria, pp. 216–219.
28. Jackson, M.J. and Mills, B. 2009 Vitrification heat treatment and dissolution of quartz in grinding wheel bonding systems. *British Ceramic Transactions* 100(1): 1–9.
29. Marshall, D.B., Lawn, B.R., and Cook, R.F. 1987. Microstructural effects on grinding of alumina and glass-ceramics. *Communications of the American Ceramic Society* 70(6): 139–140.
30. Singh, V. 2010. Grindability improvement of conductive ceramics using cryogenic coolant and modeling for specific grinding energy, PhD Thesis, IIT Delhi.

Advanced Micromachining

5

Biomachining—Acidithiobacillus-Genus-Based Metal Removal

Hong Hocheng, Jei-Heui Chang, and Umesh U. Jadhav

National Tsing Hua University

CONTENTS

5.1 Introduction

Machining is a process of removing a material from the bulk or surface and leaving the remaining material in the designed shape and dimensions. With the advent of microelectronic manufacturing, various micromachining methods have been pursued for making miniaturized devices (Rai-Choudhury, 1994; Roy et al., 2001; Chang et al., 2008). The various machining techniques presently used can be classified into two categories, namely, physical processing and chemical processing (Pandey and Shan, 1980; Uno et al., 1996), of which wet chemical etching is the widely used method for micromachining (Williams et al., 2003). These well-established processes require chemical or thermoelectric energy to be concentrated at the machining point. Such machining methods may create either a damaged layer or a heat-affected zone on the work surface and could cause potential damage to the metallurgical properties of the workpiece (Uno et al., 1996). Also, the use of hazardous materials (i.e., acids) is unavoidable. The use of biological techniques for material processing has become a promising alternative in the past few years. Recent advances in biotechnology have led to the widespread application of microorganisms in material processing. The innovative use of microbes for microscopic metal removal to achieve microfeatures is considered more environmentally friendly than other means (Zhang and Li, 1996, 1999). The inorganic bacterial pathways responsible for extensive corrosion, which are expensive, can be exploited for beneficial purposes (Xia et al., 2010). Biomachining can be defined as "a controlled microbiological process to selectively form microstructures on a metal workpiece by metal removal (or dissolution) using microorganisms" (Uno, 2002). As the size of the bacterium is of the order of microns, it appears to be an ideal tool for micromachining.

Moreover, as the metabolic function of the bacterium is utilized, no physical or chemical energy needs to be focused at the machining point, thereby avoiding the possibility of generating a damaged layer or a heat-affected zone in the machined surface (Zhang and Li, 1999). The use of chemolithotrophs also means that the nutrient requirement is very low (Brierley and Brierley, 2001). If it is possible for microorganisms to "biomachine" the desired part(s) of the metals to the required dimensions, they will become useful tools for micromachining (Ting et al., 2000).

5.2 Role of Chemolithotrophic Bacteria in Biomachining

Traditional approaches for removing or recovering metals, such as precipitation, oxidation/reduction, ion exchange, filtration, electrochemical processes, membrane separations, and evaporation, exhibit several disadvantages such as high cost, incomplete removal, low selectivity, high energy consumption, and generation of toxic slurries that are difficult to be eliminated. Therefore, much attention has been paid to the removal of metal ions by microorganisms owing to its potential applications in environmental protection and recovery of toxic or strategic heavy metals (Tsezos, 1985; Fourest and Roux, 1992; Chang and Hong, 1994; Puranik and Paknikar, 1999; Celaya et al., 2000; Liu et al., 2002). Microbiologists have discovered several bacterial species with applications in mining and recovery of radioactive waste, known as *bioleaching* and *bioremediation*, respectively (Lovley, 2003; Liu et al., 2004). Among these, chemolithotrophic bacteria play a vital role. Chemolithotrophs are those organisms that retrieve the energy needed for their growth from inorganic matter. Organisms that fit into this category have been the subject of intensive research for their unique ability to either oxidize or reduce certain inorganic compounds, especially heavy metals. The predominant metal-sulfide-dissolving microorganisms are extremely acidophilic bacteria that are capable of oxidizing inorganic sulfur compounds and/or ions. These bacteria belong to the genus *Acidithiobacillus* (Kelly and Wood, 2000). Among this group are the first isolates of the extremely acidophilic sulfur- and/or iron(II)-oxidizing bacteria, the mesophilic *At. thiooxidans* and *At. ferrooxidans*. These leaching bacteria, as also the moderately thermophilic *At. caldus*, are gram-negative γ-proteobacteria (Rohwerder et al., 2003; Lilova et al., 2007). Although there are a number of chemolithotrophic bacteria, most of the researchers used *At. ferrooxidans* for biomachining of metals (Uno et al., 1996; Zhang and Li, 1998; Ting et al., 2000; Yasuyuki et al., 2003).

To carry out the experiments with *At. ferrooxidans*, first the bacterium had to be obtained and cultured successfully. *At. ferrooxidans* (BCRC 13820) was obtained from the Food Industry Research and Development Institute (FIRDI). Several variations of the basic liquid media for this species were tried. Basal 510 medium was found to be both effective and easy to prepare. $FeSO_4$ was used as an energy source in the 510 medium. *At. ferrooxidans* 13820 has the ability to oxidize ferrous iron in an acidic solution, utilizing the energy thus derived to support carbon dioxide fixation and growth. The biological oxidation of ferrous iron is based on the following reaction (bacterial oxidation of ferrous ion):

$$2\ Fe^{2+} + 2\ H^+ + 0.5\ O_2 \rightarrow 2\ Fe^{3+} + H_2O \tag{5.1}$$

It is the production of ferric iron that makes this organism useful in the micromachining of metals. Ferric iron is capable of oxidizing pure, neutral copper into charged, soluble forms, thus being able to corrode a solid surface of the metal (Nemati et al., 1998; Ting et al., 2000; Lilova et al., 2007). The reaction for metal dissolution is summarized as follows:

$$Fe^{3+} + e^- \rightarrow Fe^{2+}$$
$$M_{(s)} - e^- \rightarrow M^+_{(aq)} \tag{5.2}$$

where $M_{(s)}$ is copper.

In this study, *At. ferrooxidans* 13820 was employed for the dissolution of several metals. The specific metal removal rates (SMRRs) of copper, nickel, and aluminum by *At. ferrooxidans* 13820 were determined. In the previous study, a formula was developed for the determination of SMRRs (Chang et al., 2008). In the present study, the same formula was used to determine the SMRRs. The metal pieces used were from separated 0.2 mm sheets of copper, nickel, and aluminum and were cut into squares of 2 cm length on each side. The metals used in this study were 99% pure. They were covered with 100 mL of 510 growth medium in a 500-mL shaking flask. The flasks were placed in a shaker at 30°C and 150 rpm for the leaching experiment. After exposure to the leaching solution, the metals were removed and gently rinsed with deionized water and placed in an oven at 50°C to remove the remaining moisture. Over the course of the bioleaching, the metal leaching rate was measured with a precise electronic balance machine (Precisa XS225A and $d = 0.0001$ g) at constant time intervals.

The SMRRs were calculated from the slopes at saturated concentration (Figure 5.1). The SMRRs of copper and nickel are shown in Figure 5.1a and b. The slopes, from the beginning of bioleaching until 24 h, at the saturated concentration (5.0×10^8 cells/mL) were equal to 3.125 mg/h cm^2 for copper and 3.64 mg/h cm^2 for nickel. The SMRR of aluminum

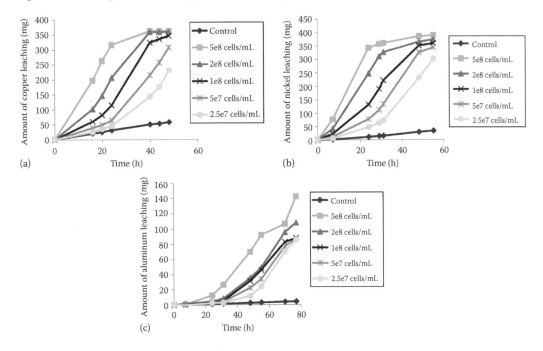

FIGURE 5.1
The amount of metal leaching of copper (a), nickel (b), and aluminum (c) by *At. ferrooxidans* at different cell concentration inoculums.

(0.46 mg/h cm^2) was lower than those of copper and nickel (Figure 5.1c). The SMRR of aluminum was calculated from the slope formed between 0 and 77 h of cell concentrations of 5.0×10^8 cells/mL.

The key process responsible for biomachining is the ferrous–ferric ion redox couple in acidic solution that causes metal dissolution. This could be based either on electron transfer from the metal workpieces (copper) to the microorganisms, in the case of physical contact between the microorganisms and the solids (i.e., direct bioleaching), or on the bacterial oxidation of Fe^{2+} to Fe^{3+}, where the ferric ion acts as the oxidizing agent and subsequently catalyzes metal dissolution (i.e., indirect bioleaching). The bacterium *At. ferrooxidans* derives energy from the oxidation of ferrous ions to ferric ions; the biogenic ferric iron then brings about the metal dissolution. The ferrous iron resulting from this reaction is reoxidized to ferric iron by the bacteria and as such can take part in the oxidation process again. These redox reactions are analogous to the direct and indirect bacterial leaching of mineral ores, where the solids are in the form of metal sulfides or oxides (Ting et al., 2000).

Although a previous study showed preliminary experimental results that demonstrated the possibility of metal removal by *At. ferrooxidans*, the mechanism for the biological action was not delineated (Uno et al., 1996). The authors stated that "it was reasonable to assume that a direct-leaching mechanism predominates for pure iron, while an indirect-leaching mechanism predominates for pure copper," but did not provide any evidence to support their assertion. Various workers have argued on the relative importance of direct and indirect bacterial actions in bioleaching (Sand et al., 1995; Ehrlichh, 1997). For instance, one study mentioned that the only role of the bacteria was the regeneration of ferric ions in solution and that there was no evidence of direct mechanisms for bacterial leaching (Fowler and Crundwellf, 1998).

The present study revealed one of the major functions of ferric iron in the potential metal oxidation. The unoxidized ferrous iron content was determined at various time intervals by taking 1 mL of the sample from each flask and titrating with potassium dichromate using *n*-phenyl anthranilic acid as the indicator (Vogel, 1978). The ferric iron concentration curves were plotted by subtracting the values of unoxidized ferrous iron from the initial ferrous iron content over a time course. Changes in ferric ion concentrations during bioleaching have the same trend as the metal removal curve. The number of secreted ferric ions indicated that inoculated cell concentrations below 5×10^8 for copper, nickel, and aluminum resulted in lagged bioleaching reactions (Figure 5.2). This lag persisted until ferric ions were significantly secreted and accumulated. Ferric ions accumulated with the progression of the bioleaching reactions and from the conversion of ferrous ions to ferric ions when the reaction was stopped by the consumption of metal. These results suggest that *At. ferrooxidans* can be successfully used for biomachining of metals. *At. ferrooxidans* is thus able to make the energy it needs to survive and simultaneously helps manufacturing engineers to produce a machining effect comparable to chemical etching.

5.3 The Process of Biomachining

For manufacturing engineers, the concept of bacterial machining opens up a new paradigm in micromachining. Bacteria, measured on the order of micrometers, can remove materials at the nanometer level. They exert negligible forces on the workpiece and produce no

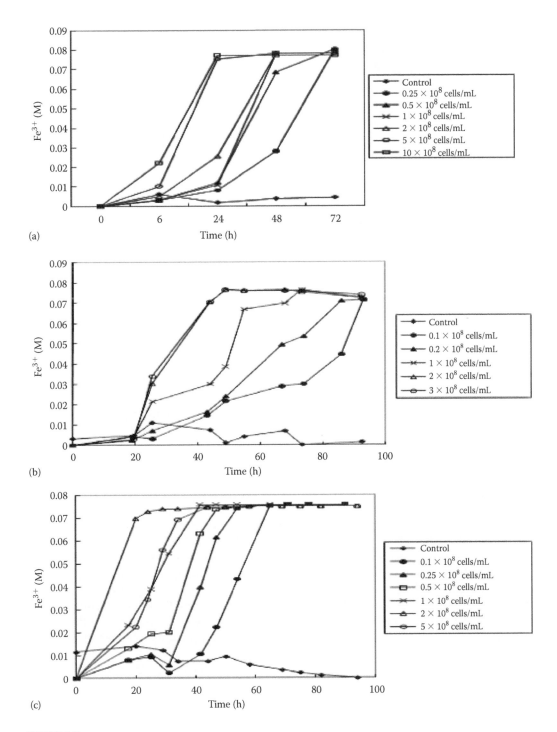

FIGURE 5.2
The change in ferric ion concentration during the biomachining of (a) copper, (b) nickel, and (c) aluminum by *At. ferrooxidans* with different cell concentration inocula.

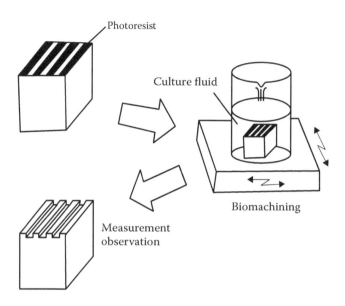

FIGURE 5.3
Schematic diagram of the experimental setup for the biomachining of metals. (From Uno, Y., et al., *JSME International Journal Series C*, 39, 837–842, 1996.)

thermal damage. All these characteristics suggest that bacteria could serve as a desirable tool in the machining of microscale features (Johnson et al., 2007).

Uno et al. (1996) carried out pioneering work in the biomachining of metals. They showed the potential of *At. ferrooxidans* in material processing. They employed a very basic experimental setup for the biomachining of metals (Figure. 5.3). For the biomachining experiment, pure copper and iron metals were employed. The mask pattern used to form the grooves was previously prepared on the workpiece by the photolithography process. The culture of *At. ferrooxidans* was grown in a 9 K medium at 28°C and 160 rpm. This culture solution was taken in an open beaker, and the pure copper and iron metals were added into the beaker separately to start the biomachining process. The researchers found that the groove depth for copper and iron increased linearly with machining time by using the culture medium with bacteria. On the other hand, the amount of metal removal by chemical etching was minimal. Also, there was some data on the effects of changing some of the parameters such as temperature and shaking conditions on biomachining of copper. In their work, they found that application of electric field during the biomachining process was effective. In electric-field-assisted biomachining, the metal removal rate at the anodic workpiece became much higher than that in normal biomachining, whereas the removal amount at the cathodic workpiece was minimal.

Zhang and Li performed an experiment very similar to the one by Uno et al. (1996). The sample was prepared once again using the existing photolithographic techniques. The prepared sample was then incubated with the bacteria for a period. Finally, the sample was removed periodically and measured both with a profilometer and a scanning electron microscope (Zhang and Li, 1996). Furthermore, they proposed a model of ion cycle in the biomachining processes. They also discussed measures for maintaining the stable equilibrium of the ion cycle.

Biomachining of copper by *At. ferrooxidans* is a complex thermodynamic process because Fe^{3+} is strongly hydrolyzed, resulting in the generation of H^+. In the experiments for machining pure copper, the rate of generation of H^+ owing to the hydrolysis of Fe^{3+} was lower than the rate of consumption of H^+ owing to the oxidation of Fe^{2+}, resulting in a gradual increase in the pH. If H^+ is not added at all times to adjust the pH to about 2, an increase in hydrolysis is observed, which ultimately decreases the growth of *At. ferrooxidans* (Zhang and Li, 1999). The necessary conditions for maintaining the stable equilibrium of the biomachining system are the supply of H^+ and the removal of Cu^{2+}. To overcome this problem, H_2S gas was bubbled into the solution. This gas supplied H^+ ions and generated CuS. The CuS was precipitated, and the Cu ions were removed from the solution. The results of Zhang and Li (1999) also indicate that the rate of biomachining depends directly on bacterial concentration.

The results of Yasuyuki et al. (2003) suggested the feasibility of material removal at microscale levels in the biomachining of controlled microstructures in low carbon steels using *At. ferrooxidans*. *At. ferrooxidans* converted Fe^{2+} into Fe^{3+}. The ferrite microstructures of steel were subjected to preferential corrosion, caused by the Fe^{3+} ions produced by *At. ferrooxidans*. The ferrite that existed in the submicrometer-scale gaps in the steel microstructure was dissolved selectively by Fe^{3+} ions. Kurosaki et al. (2003) also used the corrosive action of aerobes and anaerobes for the biomachining of metals.

Recently, a study was carried out to characterize the surface roughness and to quantify the material removal rate in biomachining (Johnson et al., 2007). Both parameters play important roles in obtaining precision products through the micromachining process. The quality of the surface produced is a very important aspect of the performance of the manufacturing process. Every manufacturing process, however, has limitations regarding this characteristic. It is therefore important to explore the surface finish characteristics of the biomachining process. Johnson et al. (2007) have included experiments on the surface roughness produced by biomachining. However, they did not explain the explicit relations and the trendlines of relations between surface roughness and machining time. Istiyanto et al. (2010) investigated the surface roughness and the material removal rate characteristics in the biomachining of copper for various machining times. In this study, the surface appearance of workpieces changed after biomachining for both 800- and 220-grit-polished workpieces. The biomachining process caused the arithmetic average of surface roughness (Ra) to increase at various rates for 6, 12, and 18 h of machining times. The researchers also found that the metal removal rate during biomachining was inversely proportional to the machining time but not simply linear.

In another study, several shims with a thickness of 0.1 mm, a diameter of only 2 mm, a thickness of 0.07 mm, and diameters of 15 mm and 16 mm were fabricated using this biomachining method (Yang et al., 2009). In this study, the change in surface roughness was also investigated. Surface roughness increased rapidly with time, from less than 0.2 µm to close to 2 µm. The surfaces of each slice of every material were damaged. They also found that in the first 2 h, roughness increased linearly with time. The surface quality deteriorated rapidly, and at the end of the 2 h, the surface roughness of the three kinds of materials reached the worst, and then became stable. Beyond the 2 h, it no longer increased significantly, but fluctuated within certain limits and then reached the stable stage eventually (Yang et al., 2009).

5.4 Micromachining of Metals by Culture Supernatant

In order to prove the mechanism of indirect biomachining, a study was carried out using a culture supernatant for the biomachining of copper, nickel, and aluminum, the equilibrium reactions of which are shown below. Equation 5.3 shows the oxidation of ferrous iron [Fe(II)] to ferric iron [Fe(III)] achieved by *At. ferrooxidans*, and Equation 5.4a–c shows the oxidation of copper, nickel, and aluminum, respectively, by ferric ions:

$$2Fe^{2+} + 1/2 O_2 + 2 H^+ \rightarrow 2 Fe^{3+} + H_2O \tag{5.3}$$

$$Cu^0 + 2 Fe^{3+} \rightarrow Cu^{2+} + 2 Fe^{2+} \tag{5.4a}$$

$$Ni^0 + 2 Fe^{3+} \rightarrow Ni^{2+} + 2 Fe^{2+} \tag{5.4b}$$

$$Al^0 + 3 Fe^{3+} \rightarrow Al^{3+} + 3Fe^{2+} \tag{5.4c}$$

It can be expected that the machining capacity can be calculated by the mole exchange of ferric ions present in the culture supernatant. As shown in Figure 5.4, the maximum amount of copper removed in 6 h was 132.7 mg. For nickel and aluminum, it was 106 and 43.4 mg, respectively, but the time required was more, that is, 10 and 47 h, respectively.

Industrial microdevices are mostly composed of simple patterns of linear structure, squares, and circles. To develop a new fabrication technology for micropattern, the aspect ratio, defined as the ratio between the machined depth and width, is significant. In the present study, a series of linear, circular, and square micropatterns with their feature

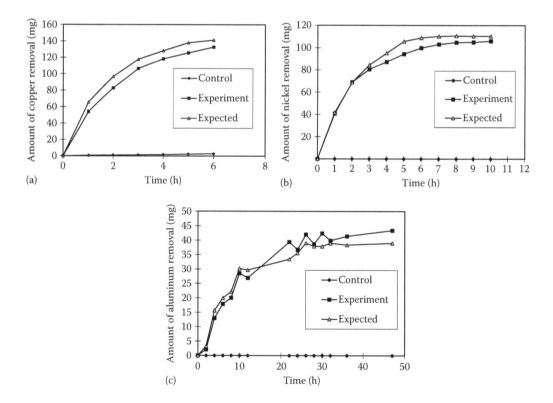

(a)

(b)

(c)

FIGURE 5.4
The expected metal loss was calculated by the change in the concentration of ferrous [Fe(II)] and ferric [Fe(III)] ions and the actual amounts determined by the balance machine during the biomachining of (a) copper, (b) nickel, and (c) aluminum using the culture supernatant that was incubated for 3 days (the culture was prepared by adding one part $FeSO_4$ to 510 optimal medium).

dimensions in width, diameter, and side, respectively, of 50, 100, 200, 300, 400, and 500 μm were made. In order to control the direction and extent of two-dimensional biomachining, a photoresist-patterned test piece was made. The pattern was drawn with Auto-CAD software on a computer. The graphic data were transferred to a laser plate-making device to make a photographic plate, called *photomask*, of the micrograph. The sample was covered with the photomask and coated with a layer of photoresist (JSR THB-151N), 20 μm for nickel and 40 μm for copper. The uncoated area was flushed out by the developer to expose the metallic surface, and the sample piece was completed. Micropattern machining experiments of copper and nickel were accomplished using a culture supernatant and incubating for 2 days. The aspect ratio was measured by SEM and WYKO machine. As shown in Figures 5.5 and 5.6, the aspect ratio of copper achieved was 1.2 at a line width of 65 μm, decreasing to 0.2 with an increase in the line width. For circular patterns, the aspect ratio achieved was 0.65 at a diameter of 65 μm, reducing to 0.2 with increase in size. For square patterns, the aspect ratio achieved was 0.9 for a 65-μm square, reducing to 0.2 when the pattern becomes bigger.

Similar results were obtained for nickel as well. The removal rate in depth remains the same, while larger mask openings (producing bigger patterns) make the aspect ratio lower.

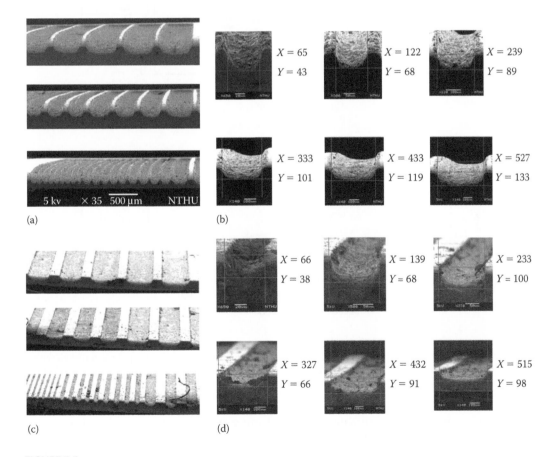

FIGURE 5.5
The SEM images of copper in the pattern of (a) line with a series of diameters from designed 50, 100, 200, 300, 400, and 500 μm, and (b) the aspect ratio calculated by actual measured width (X) and depth (Y) that were manufactured for 2 days by culture supernatant, respectively in nickel (c) and (d).

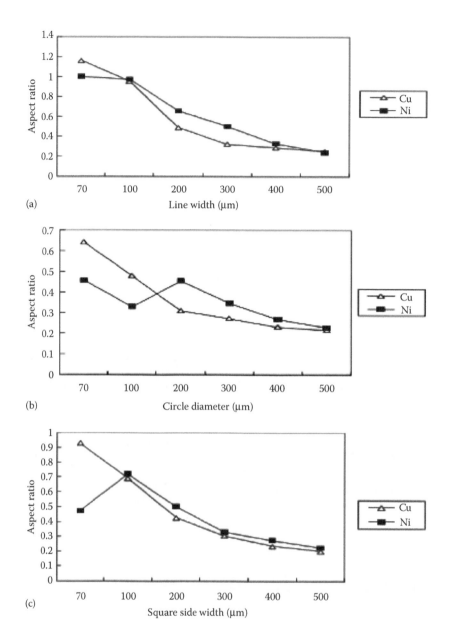

FIGURE 5.6
The WYKO-detected aspect ratio of (a) linear-, (b) circular-, and (c) square-patterned copper and nickel, with a series of diameters 50, 100, 200, 300, 400, 500 μm. The micropattern machining experiments were carried out using a culture supernatant and incubating for 2 days.

The behavior of aluminum during micromachining was unsatisfactory because it was often etched to large undercut and showed poor fidelity of the pattern prescribed by the mask. In addition, among the three micropatterns, circles and squares have lower aspect ratio than linear structures, possibly because of the limited space of circular and square patterns that allowed the culture supernatant to work into the opening, while the linear structure provided more open space (with a length of 1 cm) that allowed the culture supernatant to work more effectively. The aspect ratio of the linear structure was therefore higher.

5.5 The Use of *At. thiooxidans* in the Biomachining of Metals

While Uno, Zhang, and others were able to show that the use of an active population of *At. ferrooxidans* could remove a significant amount of material from a copper surface, they did not try any other bacterial species for the biomachining of metals. To expand on some of the work described above in the field of biomachining, a study was carried out using *At. thiooxidans* (BCRC 15616) for the biomachining of copper. Once the bacterium was successfully isolated and grown, an experiment was conducted for biomachining with a photoresist-patterned test piece of copper.

The workpiece was put into a flask in which the above-mentioned culture fluid of *At. thiooxidans* 15616 was present. The flask was kept in a shaking incubator. The exposed metal was removed by the two-dimensional machining process. The machining depth was measured with a dial indicator after 24 and 48 h of machining. The amount of undercut was very small, which influenced the degree of biomachining during the overall machining process. Figure 5.7 shows the SEM photograph of biomachined grooves (80 μm in depth and 45 μm in width) on a pure copper piece. The aspect ratio under the bioleaching condition of *At. thiooxidans* 15616 was found to be around 1.85 (Chang et al., 2008).

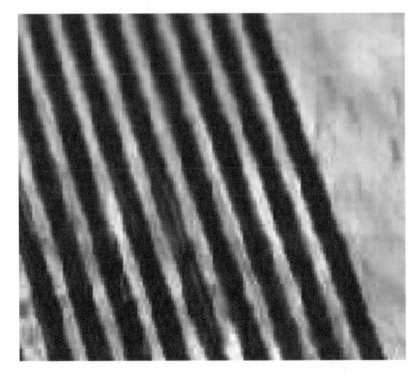

FIGURE 5.7
SEM photograph of biomachined grooves (80 μm in depth and 45 μm in width) on a pure copper piece. (From Chang, J., et al., *Journal of Materials Processing Technology,* 201, 560–564, 2008.)

5.6 Conclusion

An innovative metal removal process termed *biomachining* has been investigated using the *Acidithiobacillus* genus bacteria (*At. ferrooxidans* and *At. thiooxidans*). It has been demonstrated that several metals were successfully biomachined using these bacteria. Unlike conventional processes wherein chemical or thermoelectric energy is concentrated at the machining point, thereby creating either a damaged layer or a potential damage to metallurgical properties of the workpiece, biomachining does not pose such problems. More importantly, seen in the context of a future where industrial technologies must be more environmentally friendly and utilize "mild" process conditions, this metal processing technique appears promising. The focus of this research is now toward improving the rate of biomachining and the surface finish, and obtaining a better understanding of the mechanisms involved in biomachining.

References

Brierley, J. and Brierley, C. 2001. Present and future commercial applications of biohydrometallurgy. *Hydrometallurgy* 59: 233–239.

Celaya, R., Noriega, J., Yeomans, J., Ortega, L., and Ruiz-Manriquez, A. 2000. Biosorption of Zn(II) by *Thiobacillus ferrooxidans*. *Bioprocess Engineering* 22: 539–542.

Chang, J., Hocheng, H., Chang, H., and Shih, A. 2008. Metal removal rate of *Thiobacillus thiooxidans* without pre-secreted metabolite. *Journal of Materials Processing Technology* 201: 560–564.

Chang, J. and Hong, J. 1994. Biosorption of mercury by the inactivated cells of *Pseudomonas aeruginosa* PU21. *Biotechnology and Bioengineering* 44: 999–1006.

Ehrlichh, L. 1997. Microbes and metals. *Applied Microbiology Biotechnology* 48: 687–692.

Fourest, E. and Roux, J. 1992. Heavy metal biosorption by fungal mycelia by-product mechanisms and influence of pH. *Applied Microbiology Biotechnology* 37: 399–403.

Fowler, T. and Crundwellf, K. 1998. Leaching of zinc sulphide by *Thiobacillus ferrooxidans*: Experiments with a controlled redox potential indicate no direct bacterial mechanism. *Applied and Environmental Microbiology* 64: 3570–3575.

Istiyanto, J., Ko, T., and Yoon, I. 2010. A study on copper micromachining using micro-organisms. *International Journal of Precision Engineering and Manufacturing* 11: 659–664.

Johnson, D., Warner, R., and Shih, A. 2007. Surface roughness and material removal rate in machining using micro-organisms. *Journal of Manufacturing Science and Engineering* 129: 223–227.

Kelly, D. and Wood, A. 2000. Reclassification of some species of *Thiobacillus* to the newly designated genera *Acidithiobacillus* gen. nov., *Halothiobacillus* gen. nov., and *Thermithiobacillus* gen. nov. *International Journal of Systematic and Evolutionary Microbiology* 50: 511–516.

Kurosaki, Y., Matsui, M., Nakamura, Y., Murai, K., and Kimura, T. 2003. Material processing using micro-organisms. *JSME International Journal Series C* 46: 322–330.

Lilova, K., Karamanev, D., Flemming, R., and Karamaneva, T. 2007. Biological oxidation of metallic copper by *Acidithiobacillus ferrooxidans*. *Biotechnology and Bioengineering* 97: 308–316.

Liu, H., Chen, B., Lana, Y., and Cheng, Y. 2004. Biosorption of Zn(II) and Cu(II) by the indigenous *Thiobacillus thiooxidans*. *Chemical Engineering Journal* 97: 195–201.

Liu, C., Gorby, Y., Zachara, J., Fredrickson, J., and Brown, C. 2002. Reduction kinetics of Fe(III), Co(III), U(VI), Cr(VI), and Tc(VII) in cultures of dissimilatory metal-reducing bacteria. *Biotechnology and Bioengineering* 80: 637–649.

Lovley, D. 2003. Cleaning up with genomics: Applying molecular biology to bioremediation. *Nature Reviews Microbiology* 1: 35–44.

Nemati, M., Harrison, S., Hansford, G., and Webb, C. 1998. Biological oxidation of ferrous sulphate by *Thiobacillus ferrooxidans*: A review on the kinetic aspects. *Biochemical Engineering Journal* 1: 171–190.

Pandey, P. and Shan, H. 1980. *Modern Machining Processes*. New Delhi: Tata McGraw-Hill Publishers.

Puranik, P. and Paknikar, K. 1999. Biosorption of lead, cadmium, and zinc by *Citrobacter* strain MCM B-181: Characterization studies. *Biotechnology Progress* 15: 228–237.

Rai-Choudhury, P. ed. 1994. *Handbook of Microlithography, Micromachining and Microfabrication, Micromachining and Microfabrication*, Vol. 2. Bellingham: SPIE, Washington, London: IEE.

Rohwerder, T., Gehrke, T., Kinzler, K., and Sand, W. 2003. Bioleaching review part A: Progress in bioleaching: Fundamentals and mechanisms of bacterial metal sulfide oxidation. *Applied Microbiology Biotechnology* 63: 239–248.

Roy, S., Ferrara, L., Fleischman, A., and Benzel, E. 2001. Microelectromechanical systems and neurosurgery: A new era in a new millennium. *Neurosurgery* 49: 779–798.

Sand, W., Gehrke, T., Hallman, R., and Schippers, A. 1995. Sulphur chemistry, biofilm and the (in) direct attack mechanism—A critical evaluation of bacterial leaching. *Applied Microbiology Biotechnology* 43: 961–966.

Ting, Y., Kumar, S., Rahman, M., and Chia, B. 2000. Innovative use of *Thiobacillus ferrooxidans* for the biological machining of metals. *Acta Biotechnologica* 20: 87–96.

Tsezos, M. 1985. The selective extraction of metals from solutions by micro-organisms. A brief overview. *Canadian Metallurgical Quarterly* 24: 141–144.

Uno, Y. 2002. Micro machining technique of metals with bacteria: Biomachining. *Journal of the Institute of Electronics and Communication Engineers of Japan* 85: 129–131.

Uno, Y., Kaneeda, T., and Yokomizo, S. 1996. Fundamental study on biomachining. *JSME International Journal Series C* 39: 837–842.

Vogel, A. 1978. *A Text Book of Quantitative Inorganic Analysis Including Elementary Instrumental Analysis*. 4th ed. London: Elbs and Longman. pp. 317–321.

Williams, K., Gupta, K., and Wasilik, M. 2003. Etch rates for micromachining processing—Part II. *Journal of Microelectromechanical Systems* 12: 761–778.

Xia, L., Yin, C., Dai, S., Qiu, G., Chen, X., and Liu, J. 2010. Bioleaching of chalcopyrite concentrate using *Leptospirillum ferriphilum*, *Acidithiobacillus ferrooxidans* and *Acidithiobacillus thiooxidans* in a continuous bubble column reactor. *Journal of Industrial Microbiology Biotechnology* 37: 289–295.

Yang, Y., Wang, X., Liu, Y., Wang, S., and Wen, W. 2009. Techniques for micromachining using *Thiobacillus ferrooxidans* based on different culture medium. *Applied Mechanics and Materials* 16–19: 1053–1057.

Yasuyuki, M., Osamu, K., Chouanine, L., and Yasushi, K. 2003. Fundamental studies on biomachining of carbon steel by iron oxidizing bacteria. *Transactions of JWRI* 39: 239–242.

Zhang, D. and Li, Y. 1996. Possibility of biological micromachining used for metal removal. *Science in China* 41: 152–156.

Zhang, D. and Li, Y. 1999. Studies on kinetics and thermodynamics of biomachining pure copper. *Science in China* 42: 58–62.

6

Micro- and Nanomanufacturing by Focused Ion Beam

Vishwas N. Kulkarni, * **Neeraj Shukla, and Nitul S. Rajput**
Indian Institute of Technology Kanpur

CONTENTS

6.1 Introduction

Energetic ion beams of different elements have been used along with electron beam in different fields of science and engineering for modifying and analyzing target samples. Although broad beams can be used in most situations, certain advantages of the focused ion beam (FIB) such as maskless beam irradiation of a particular area of a sample have resulted in its increasing importance. FIBs can be used in implantation, modification of materials, selective modification of biological and soft materials, investigation of the mechanism of ion–matter interaction, and study of basic science phenomena. However, some

---

* Late (1954–2010).

properties of beams such as resolution and beam energy stability have been a challenge in the development of ion beam technology. FIB with high resolution, high brightness, optimum beam energy with high monochromaticity, and current density are required to characterize and micromachine materials at a small scale. These qualities are highly dependent on the type of ion beam source, the chamber pressure, the charged particle focusing system (lens), and the electronics. Owing to recent rapid developments, mainly in the quality of ion source and in the lens systems, FIB technology has evolved to make possible state-of-the-art machines that deal with materials in the nanoregime.

Initially, FIBs were mostly used for fabricating masks at a small scale. However, FIB technology has been redeveloped by various research groups in collaboration with industry to meet the challenges in its use as a manufacturing tool at the micro- and nanoscale levels. Nowadays, FIB machines have been ubiquitously used in different areas, for example, inspecting failures and editing electronic circuits, forming defects in a small predefined volume, fabricating complex 3D micro/nanostructures, preparing transmission electron microscopy (TEM) samples, and making arrays of nanoholes in metallic films (which has potential applications as photonic crystals) [1–3].

Owing to its rapidity, versatility, and the small number of steps involved (compared to various other techniques, e.g., lithography and etching), FIB has gained popularity as a micro/nanomachining tool.

6.2 Focused Ion Beam System (Dual Beam)

The ion beam, which has energy in the kiloelectron volts range, can be focused down to a few nanometers with the help of stable electrostatic lenses. With the use of such a focused beam, nanofabrication has been realized through maskless patterning (milling and deposition). FIBs are generally integrated with various tools to assist nanofabrication processes. It can be a complete package of several parts, namely, ion column, electron column, high-precision goniometer sample stage, gas injection system (GIS), needles for

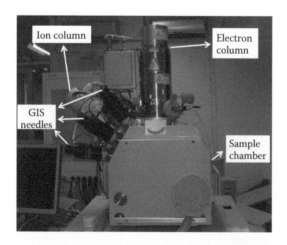

FIGURE 6.1
A dual-beam FIB system. (From Nova 600, Nano Lab manufactured by FEI Company.)

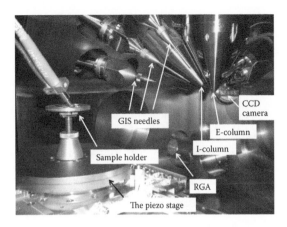

FIGURE 6.2
Different parts in the sample chamber.

pouring precursor gases, and various peripheral electronics. Some of these parts are shown in Figure 6.1, depicting a dual-beam system manufactured by FEI Company. Both the ion and the electron columns are aligned at an angle of 52° in the system. There is nothing sacrosanct about this number 52. It is just a limitation set by the population of various peripheral components inside the manufacturing chamber. The view of the sample chamber is shown and tagged in Figure 6.2, which shows the GIS needles, charge-coupled device (CCD) optical camera, sample stage, ion column, and electron column inside the sample chamber. The sample stage should be handled very carefully while operating the machine to ensure that it does not hit the peripheral components, with which the chamber is densely populated. Another important system is the vacuum unit. The chamber is evacuated by a combination of an oil-free rotary pump and a turbomolecular pump (TMP), which can provide a vacuum of the order of 10^{-6} mbar; the ion and electron columns are further pumped by ion getter pumps (IGPs).

Systems consisting of the ion beam and electron beam columns are generally referred to as *dual-beam systems*. In a dual-beam FIB system, the ion column is used for fabricating micro/nanostructures through bottom-up chemical vapor deposition (FIB-CVD) as well as top-down (FIB milling) processes. In the bottom-up procedure, growth of structures occurs through layer-by-layer deposition of predefined material on the substrate. The details of the processes are discussed in subsequent sections. The electron beam is generally used for imaging. Image processing can be done during the fabrication process as well as after patterning by the ion beam. The electron beam can also be used for fabricating structures using the bottom-up approach. This is known as the *electron-beam-induced chemical vapor deposition* (EB-CVD) process.

FIB systems can be integrated with other auxiliary parts or equipment such as a high-precision sample stage, an energy-dispersive spectroscopy (EDS) system, a nanomanipulator, and a residual gas analyzer (RGA). As integrating these parts requires sufficient space inside the chamber, one has to compromise on the chamber size when using these tools.

6.2.1 Ion Column

There are two types of sources mainly used to produce good-quality ion beams, which can fulfill the prerequisites to work at the nanoscale. They are gas-filled ion sources (GFIS)

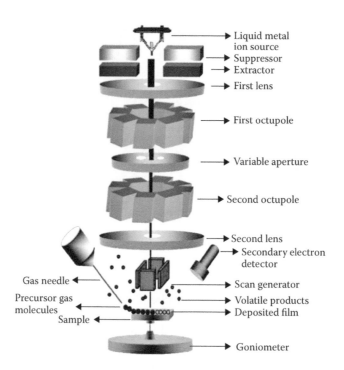

FIGURE 6.3
Various components of an ion column.

and liquid metal ion sources (LMIS). However, LMIS is preferable in FIB systems because of its suitable angular intensity, which is mainly related to its focusing capability. The main components of an ion column having LMIS are shown in Figure 6.3. The LMIS is placed at the top end. The suppressor is used to control the ions, which are then extracted by the extractor. The beam is focused well before it is passed through the variable aperture plate, which is used to control the number of particles crossing the aperture, thereby controlling the beam current. Then the beam is again passed through the focusing lenses (first lens, second lens, and octupole lenses) and the scan generator plates. The beam is finally allowed to fall on the sample, attached to a sample holder that is controlled by the goniometer stage.

6.2.2 High-Precision Five-Axes Goniometer Sample Stage

The sample stage usually has the ability to offer five-axes movement (X-, Y-, Z-, rotation, and tilt). All five of the stage motions may be motorized for automatic positioning. The movement of the stage in x-, y-, and z- directions is limited by the inner dimensions of the chamber. The stage can be tilted and rotated up to 60° and 360°, respectively. Depending on the chamber size, the stage can handle samples, typically, wafers of size ~10 cm. New FIB stages often have the potential for eucentric motion (center of the field of view is fixed with respect to the tilt) to avoid realignment of the sample every time the stage is moved so that when the stage is tilted, it takes center of the field of view as the origin. The high-precision stage works on the piezoelectric principle, and movements are controlled by optical encoders for each of the five axes. This enables precise stage movement in steps as small as 10 nm in the x-, y-, and z-axes. Also, the large stage must be very stable and should not get heated up from the mechanical action that is necessary for its movement. Thermal stability prevents specimen drift during FIB micro/nanofabrication. Automated stage navigation can be used for specific location of sites on large wafers and for device repairs that involve multiple layers of material.

6.2.3 Energy Dispersive Spectroscopy

A dual-beam system can be integrated with energy dispersive spectroscopy (EDS), by which the elemental composition of the target material can be analyzed. The electron beam is used for such characterization processes. The composition of the target material can be obtained using the EDS spectrum or by the mapscan and linescan models, which are used to investigate the variation of the elements spatially and along a line in the sample, respectively.

6.2.4 Nanomanipulator

Handling structures at the micro/nanoregime is important for circuit editing and integrating different parts of micro/nanoelectromechanical systems (MEMS/NEMS). Probes with nanodimension, which can be controlled precisely in x-, y-, z-, tilt, and rotational directions with micromotors, can do such a job. Recently, such systems, which can be integrated with the FIB or other microscopy systems, have been developed by a few companies. Zyvex and Kleindiek are the pioneers in this field that supply two-probe and four-probe systems along with microgripper facility and force measurement facility. Using such facilities, nanostructures can be moved from one position to another easily and precisely. Furthermore, such facilities can be used to study other physical properties of the structures such as conductivity, Young's modulus, and hardness.

6.2.5 Residual Gas Analyzer

The chemical/elemental composition of the chamber can sometimes affect the fabrication processes of the nanostructures or the measurements carried out in such a chamber. The variation in elemental composition inside the chamber can be monitored using an residual gas analyzer (RGA) system. This is a mass spectrometer of small size that can be connected to a vacuum system. During operation of the RGA, a small fraction of the gas molecules are ionized and the positive ions are separated, detected, and measured according to their molecular mass. The RGA is very effective in identifying different molecules present in the chamber within a couple of minutes.

6.3 Ion–Matter Interaction

In FIB systems, fabrication processes are based on the ion–matter interaction phenomenon, and hence it is important to understand the ion–matter interaction processes in order to properly understand the working principle of FIB. When a stream of ions enters a solid, the particles collide with the substrate atoms. The ion loses its energy, and several processes can occur depending on the incident beam parameters, target material, and other factors. Figure 6.4 shows a schematic view of the processes that can be observed during ion–matter interaction.

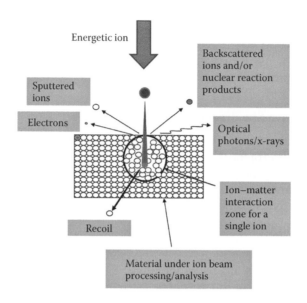

FIGURE 6.4
Schematic representation of ion–matter interaction.

Depending on the ion beam type, energy, dose, and type of substrate material, the following processes can dominate during the interaction [4,5]:

1. Ionization of the target atoms
2. Recoiling of the incident particles
3. Backscattering of the incident ions
4. Ion channeling
5. Generation of secondary electrons
6. Amorphization of the substrate
7. Implantation of the ions in the substrate
8. Production of optical photons and x-rays
9. Removal of the surface atoms from the substrate
10. Swelling of the substrate surface
11. Stripping, nuclear reaction, and generation of secondary particles

The zone within which the above processes occur is called the *interaction zone*. The size of the zone depends on the type of the incident ions, the energy of the beam, and the substrate material. The interaction constitutes a complicated, many-body problem. However, the interaction of the ions with the nucleus and the electrons can be dealt with separately. Furthermore, taking the binary collision approximation, Monte Carlo simulation-based programs were developed by Ziegler et al. [6] in MS-DOS systems, which were found to provide proper insight into the interaction processes. The software package was introduced in 1985 [6], and was initially called *the transport of ions in matter* (*TRIM*); it is now popularly known as *stopping and range of ions in matter* (*SRIM*) and is compatible with Windows systems. The software is freely available from the Web site www.srim.org. It can calculate 3D distribution of implanted ions in solids, damage of the substrate material,

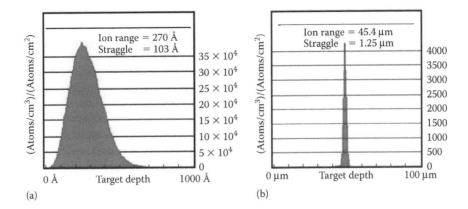

FIGURE 6.5
SRIM simulation showing the range of (a) 30-keV Ga ion beam and (b) 2-MeV proton beam in Si. (From SRIM 2008, http://www.srim.org.)

sputtering, phonon generation, and secondary ions. The target material can be single or composite. However, during the simulation it is assumed that the target material is amorphous. The other major assumptions in the simulation are as follows:

- Binary collision approximation (i.e., the influence of neighboring atoms is neglected).
- Inclusion of suitable screening potentials, for example, Thomas–Fermi potential.
- Recombination process of the interstitials with the vacancies is neglected.

The interaction cross section of ions in the kiloelectron volt energy is larger than that of the megaelectron volt energy range, hence the ions in the kiloelectron volt energy range are embedded more shallowly into the matter than their megaelectron volt energy counterparts.

In Figure 6.5, the distributions of a 30-keV Ga ion beam and a 2-MeV proton beam in Si material, simulated by SRIM are shown.

6.4 Working Principle of Focused Ion Beam

The following processes are involved in the FIB-assisted micro- and nanomanufacturing technique:

1. Ion beam milling
2. FIB-assisted CVD process
3. Ion-beam-assisted chemical etching process

6.4.1 Ion Beam Milling

Ion-beam-induced milling consists of (1) sputtering, (2) redeposition of the sputtered (ejected) material, and (3) amorphization. However, in general, the term *milling* in FIB processing means "removal of the target material."

6.4.1.1 Sputtering

The sputtering phenomenon is the removal of the substrate material by ion irradiation of sufficient energy. Sputtering rates are primarily characterized by the sputtering yield (Y), which is defined as

$$Y = \frac{\text{Mean number of emitted atoms}}{\text{Incident particle}}$$

(6.1)

The sputtering yield depends on the structure and composition of the target material, parameters of the incident ion beam, and the experimental geometry. The value of Y ranges from 10^{-5} to as high as 10^3 [7]. For the energy ranges used in FIB systems, this value varies between ~10^{-1} and 10^2 [5]. Figure 6.6 shows a schematic diagram of milling process performed by the ion beam.

When ions enter the substrate material, they lose their energy in the target atom matrix through two modes, namely, the elastic and inelastic processes [8]. In the elastic process, also called the *nuclear energy loss process*, energy is transferred to the target atoms through atomic collision by the incident ion. In the inelastic process, also known as *electronic energy loss process*, the transfer of energy occurs through electronic excitation (the outermost electrons of the target atom are excited by the incident ion) and ionization of the target atoms. Nuclear energy loss dominates at lower energy of the incident beam and ranges from a few kiloelectron volts to a few hundred kiloelectron volts, and electronic energy loss dominates at higher energy (typically in the range of megaelectron volts) of the incident beam, as shown in Figure 6.7 [8].

Therefore, when energy is transferred from the incident ions to the target material through elastic collisions (nuclear energy loss), a collision cascade is initiated within the surface layers (few monolayers). Because of these collisions, the energy of some of the surface atoms crosses the surface barrier energy (typically few electron volts), and these atoms get ejected from the surface. For this reason, in the case of micro/nanofabrication by FIB, emphasis is more on the nuclear energy loss of the incident ions. Hence, this forms the

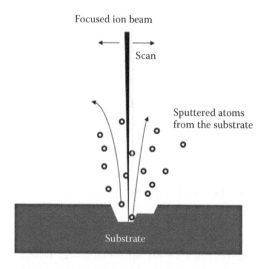

FIGURE 6.6
Schematic diagram showing the focused ion beam milling process.

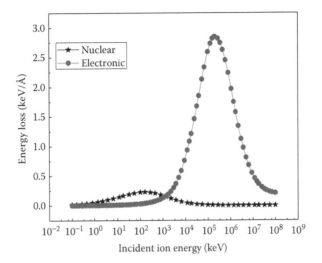

FIGURE 6.7
The graph shows the range of energy loss through elastic and inelastic energy loss mechanisms.

basis for selection of the optimum energy range for micro- and nanomanufacturing by FIB. This range lies between a few kiloelectron volts and a few tens of kiloelectron volts for Ga ions. In this range, sputtering yield of Ga ions on most of the targets is optimal. Apart from energy, sputtering yield depends on various other parameters, namely, target elements, angle of ion incidence, and temperature of the target material.

Sputtering yield measures the efficiency of the sputtering process, and hence understanding its mechanism is very important. Sigmund [9] gave a theory for the sputtering mechanism in 1969 in which the sputtering yield (Y) has been taken as a function of the material properties of the substrate (such as binding energy, atomic mass, and number density), density of transferred energy, and type, energy, and direction of the incident ion.

Sputtering yield should be proportional to the number of displaced or recoiled atoms. In the linear cascade region, this is applicable to medium mass ions, the number of recoils being proportional to the energy deposited per unit depth through nuclear energy loss. We can then express the sputtering yield Y for particles incident normal to the surface as [7,8]

$$Y = \Lambda F_D(E_0) \tag{6.2}$$

where Λ contains all the material properties such as surface binding energy, the range of the displaced target atoms, and the number of recoil atoms that overcome the surface barrier of the solid and escape, and $F_D(E_0)$ is the density of the transferred energy within the surface. This energy depends on the type, energy, and direction of the incident ion, and the target parameters Z_2, M_2, and N (atomic number, mass number, and atomic density of the target atoms, respectively).

The deposited energy at the surface can be expressed as

$$F_D(E_0) = \alpha N S_n(E_0) \tag{6.3}$$

where N is the atomic density of target atoms and $S_n(E_0)$ is the nuclear stopping cross section. In this equation, α is a correction factor, which ranges from 0.1 to 1.4, depending on the mass ratio (M_2/M_1) and angle of impact (θ). Here, M_1 and M_2 are the atomic masses of the incident ion and target atoms, respectively. A reasonable average value of α for normal incidence, sputtering with medium mass ions, is 0.25 [7]. The derivation of Λ involves a description of the number of recoil atoms that can overcome the surface barrier and escape from the solid. Sigmund has given the details of the derivation for the linear cascade regime [8,9]. The result is

$$\Lambda \cong \frac{0.042}{NU_0} \quad \text{Å/eV}$$

(6.4)

where N is the atomic density (Å^{-3}) and U_0 is the surface binding energy (eV). The value of U_0 can be estimated from the heat of sublimation and typically has a value between 2 and 4 eV. Using Equations 6.1 through 6.4, a simple form for the sputtering yield can be given as

$$Y_{tot}(E_0) = \frac{0.042 \times S_n}{U_0} \left(0.15 + 0.13 \frac{M_2}{M_1} \right)$$

(6.5)

For most of the ion sources used in FIB systems, the sputtering yield maximizes in the range of a few tens of kiloelectron volts, for most of the target materials that are used. For this reason, the energy limitation in most FIB systems is kept within a few tens of kiloelectron volts.

The sputtering yield depends highly on the incident beam angle (the angle between the incident ion direction and the surface normal of the target) and also on the ion beam energy. Figure 6.8 shows the simulation of sputtering yield versus ion beam incidence angle at different Ga ion energies. It can be seen that the yield increases sharply with the angle of incidence beyond ~50°. At higher incident angles, more shallow collision cascades create a greater density of displaced surface atoms, which create higher sputtering. The yield maximizes at an ion incident angle of around 80°. However, as the ion incident

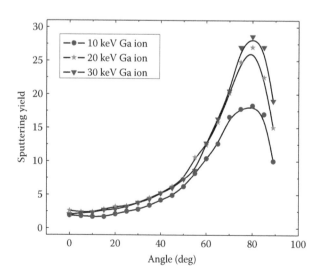

FIGURE 6.8
Theoretical SRIM simulation of sputtering yield versus ion incident angle for Ga$^+$ into Si.

angle approaches 90°, which is close to the grazing angle, the yield drops sharply because of the smaller cross-sectional area of the surface atoms covered by the impinging ion beam.

As the sputtering process occurs through the process of momentum and energy transfer in the target material, it depends on the density of the target material. It appears that in body-centered cubic (BCC) metals, sputtering yield is more in the [111] direction as compared to other directions. In face-centered cubic (FCC) metals, the yield is more in the [110] direction as compared to other directions. The density of atoms in those particular directions is greater when compared with that in other directions, and this affects the yield.

The energy of the sputtered particles is in the range of 2–7 eV only. In general, a cosine distribution of the sputtered particles has been observed when there is a normal irradiation of the incident beam.

6.4.1.2 Redeposition

After sputtering, the sputtered material forms a gas phase. Owing to their unstable thermodynamic state, the ejected particles tend to condense back to the solid state [10]. Sputtered material may therefore get redeposited in close proximity to the milling sites. Redeposition also occurs because of scattered ions from the target area [11–13]. This strongly affects the FIB milling process. Because of the redeposition process, it becomes difficult to make clean 3D structures, holes, rectangles, and trenches with high aspect ratio. Generally, ion beam parameters such as beam current, dwell time, and pitch (beam spot overlap percentage) are adjusted to minimize the redeposition effects. Apart from these parameters, there are some processes such as serial and parallel milling that must be taken into account to reduce the effect of redeposition. To understand serial and parallel milling or deposition by ion beam, it is necessary to understand how the ion beam undertakes milling or deposition. The beam spot size, which is generally of the range of a few tens of nanometers, moves from spot to spot (in the patterning page, it will be from pixel to pixel) to mill or to assist deposition in a certain defined area. In serial mode, the beam moves from one assigned pattern (structure) to another only after completely milling/depositing the preassigned pattern. On the other hand, in parallel mode, the beam simultaneously mills/deposits the assigned patterns.

6.4.1.3 Amorphization

When ions enter the target material, they can displace the target atoms, replace the target atoms in their sites, or reside within (implantation) the target material. These processes can lead to amorphization of a crystalline material [10] and can dominate the milling process at low beam energy and dose. It is found that the amorphization process can lead to the swelling of the target surface, to as high as tens of nanometers. The swelling occurs because of the density change, which is caused by the amorphization of the target material at its surface level.

Different types of structures that have been made using the FIB milling process are shown in Figure 6.9.

6.4.2 Ion-Beam-Induced Deposition of Various Materials

Ion-beam-induced chemical vapor deposition (IB-CVD) is the cracking of adsorbed precursor gas molecules by an ion beam on a substrate. Naphthalene [$C_{10}H_8$], methyl

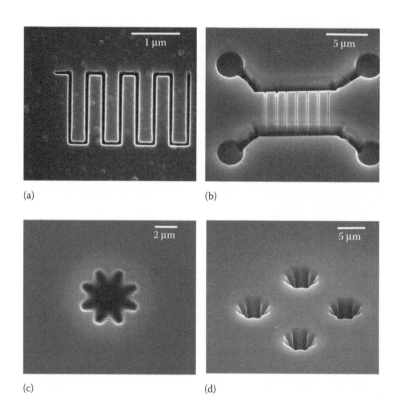

FIGURE 6.9
SEM image of (a) a template for fabricating interdigited electrodes, (b) nanochannels connecting two bigger channels, (c and d) templates for making micron/nanosize dies.

cyclopentadienyl trimethyl platinum $[(CH_3)_3Pt(CpCH_3)]$, and tungsten hexacarbonyl $[W(CO)_6]$ are the most commonly utilized precursor gases for the deposition of carbon, platinum, and tungsten nanostructures, respectively. There are other precursor gases also, which are used for depositing metallic films and semiconducting materials. Precursor gas molecules are allowed to fall on a substrate through gas injection needles. The flow rates of the precursor gases are controlled by GIS. During the gas flow for the deposition, GIS needles having diameters of about 500 μm are placed in the vicinity of about 250 μm from the substrate. A few monolayers of the gas species get adsorbed on the target. Simultaneously, ions are irradiated on the adsorbed region in a predefined patterned form. The ions crack the adsorbed molecules into volatile and nonvolatile parts (Figure 6.10). The nonvolatile part remains on the substrate, and the volatile part is evacuated by the vacuum pump. In the case of organometallic gas, the nonvolatile part is the metallic part of the compound.

During the process, there is competition between ion-induced milling and FIB-CVD. If the beam flux is not enough to crack all the adsorbed molecules, the deposition rate will be low. On the other hand, milling will occur if the number of irradiated ions is higher than the number required for cracking the adsorbed molecules. In such a case, there will be more milling rather than deposition. To avoid such a situation, proper beam current along with beam energy and beam dwell time must be used [14]. The FIB-fabricated pillars have many protrusions that can be smoothened by milling [15]. Figure 6.11a shows the

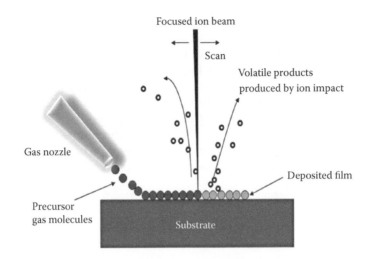

FIGURE 6.10
Schematic representation of the FIB-CVD process.

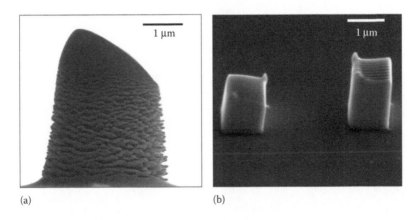

(a)   (b)

FIGURE 6.11
Fabricated Pt (a) micropillar and (b) microblocks using the FIB-CVD process.

fabrication of a Pt micropillar, and Figure 6.11b shows Pt microblocks grown by the FIB-CVD process.

6.4.2.1 Fabrication of 3D Micro/Nanostructure

Various novel strategies can be employed for the fabrication of 3D overhanging structures. In order to make various sides of the prefabricated FIB structures, innovative substrate holders can be used, and a greater number of smaller sequential steps can be incorporated in the fabrication of micro/nanostructures. Fabrication of micro/nanocantilevers, springs, pillars, and overhanging bridges have many potential applications. For example, cantilevers, springs, and pillars can be used for the measurement of elastic modulus constants of micro/nanostructures. Such a pillar can be used as an attogram (10^{-18} g) sensor and, overhanging structure as a pressure sensor (nano-Pirani gauge).

6.4.2.1.1 *Step-by-Step Fabrication of Overhanging Structure*

Fabrication of 3D micro/nanostructures can be thought of as being made layer by layer from small fragments. The smaller the layer fragments, the smoother will be the steps in the structure. The first schematic methodology of the bottom-to-top approach to fabricate overhanging structures has been shown in Figure 6.12a. To fabricate a lens-like structure, disk patterns have been drawn in the patterning page (Figure 6.12a) to a height of a few nanometers with increasing diameter until the required diameter is attained. The patterning has been done in serial mode so that the beam jumps to the next disk with different diameter only after it finishes the first deposition. After attaining the required diameter, disk patterns with successively decreasing diameters have been drawn on top of

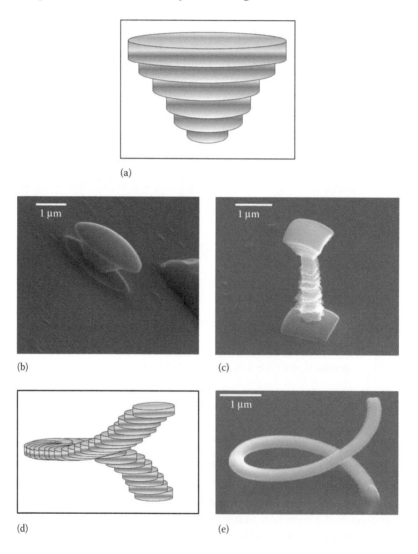

FIGURE 6.12
The SEM images show varieties of 3D structures fabricated using FIB-CVD deposition process. (a) A template of increasing diameter for fabricating, (b) a microlens, (c) a dumbbell-like structure, (d) schematic diagram template for fabricating a spring in the nanodimensions, and (e) SEM image of a fabricated nanospring using the template (d).

the fabricated structure. The result of such a fabrication process is shown in Figure 6.12b. Using the same technique, a dumbbell-like structure has been fabricated and is shown in Figure 6.12c.

Shifting this layer-by-layer growth sequentially can give rise to a twisted structure. Figure 6.12d shows the schematic representation of this shifted layer-by-layer growth methodology, and the actual spring fabricated by this method is shown in Figure 6.12e. These fabrication steps (Figure 6.12a and d) can be performed either by writing codes or by making pattern files on the patterning page of the user interface (UI) software.

6.4.2.1.2 Inclusion of Auxiliary Tilted Stub on the Sample Stage for Sideways Milling

The stage in most of the FIB systems can be tilted up to ~60°. Because of this limited tilting option, it becomes tricky to fabricate complex 3D overhanging structures, since after fabricating a wall or pillar, sideways access to these structures by the ion beam is not possible. So, if we place an auxiliary stub tilted at an angle of 45° to the sample and rotate the prefabricated structures judiciously, the sides of the walls can be made accessible to the ion beam. The methodology of these steps has been explained in Figure 6.13.

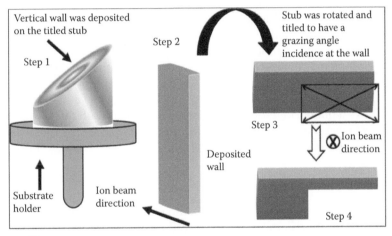

(a) ⊗ : Implies the direction into the plane of paper

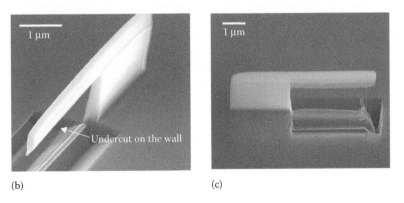

(b)   (c)

FIGURE 6.13
(a) Schematic diagram of a tilted stub for fabricating an overhanging type of structure (the details are given in the text), (b and c) SEM images of an overhanging structure using a tilted stub.

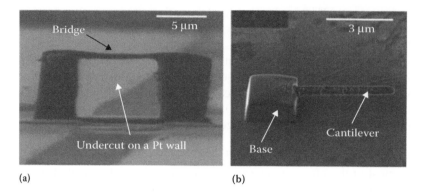

FIGURE 6.14
(a) An overhanging bridge-type structure and (b) a free hanging nanocantilever was fabricated using a tilted stub.

As shown in Figure 6.13a, a tilted aluminum stub has been used (step 1) as a part of the overhanging 3D structure fabrication method. The stage on which the stub has been fixed can be tilted to make the sample surface normal to the ion beam. Following this, a vertical wall has been deposited (Figure 6.13a, step 2). Now, if we again rotate and tilt the stage judiciously, the sample surface will be at the grazing angle (close to 90°), making the sidewalls of the previously deposited vertical wall normal to the incident ion beam (step 3), which makes milling from the sides of the vertical wall (step 4) possible. Figure 6.13b and c show the SEM images of the cantilever made by the above-mentioned technique. Figure 6.14a shows a nanobridge-type of structure fabricated with the help of a tilted stub where an undercut was made in a vertical wall. Similarly, by accessing the sides of a micro/ nanostructure, further deposition of a pillar can be performed as shown in Figure 6.14b.

6.4.3 FIB-Induced Etching Process

By using a reactive gas in the milling process, the material removal process can be accelerated. A chemical reaction is initiated between the substrate surface and gas molecules, adsorbed on the substrate, which helps in the milling process. The main advantage of using a reactive gas for the milling process is the higher material removal rate and lower redeposition.

Usually, halogen or halogen-containing compounds are used as reactive gases in the etching process. I_2, Br_2, and XeF_2 are some compounds commonly used in the FIB-etching process. As the etching process depends both on the reactive gas and on the target material, a suitable reactive gas must be used for a particular substrate. SEM images of an etching effect are shown in Figure 6.15.

6.5 Summary

FIB has found numerous applications in micro- and nanomanufacturing because of its unique feature of top-down and bottom-up approaches. The process is useful in fabricating atomic force microscope (AFM) tips, nano-Pirani gauges, and templates for making

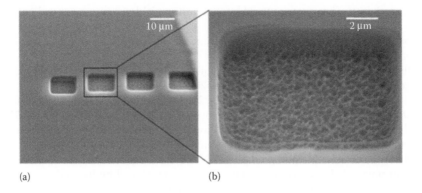

FIGURE 6.15
(a) Insulator-enhanced etching (IEE) on SiO_2 substrate and (b) magnified image of the second pattern.

dies, namely, dies for micro- and nanogears. FIB is a very useful tool for the fabrication of TEM samples and failure analysis in microcircuits. Its limitations are the inadequacy of sufficient organometallic precursor gases for various materials and Ga contamination of FIB-CVD-grown structures.

Acknowledgments

The authors are thankful to Dr. Sarvesh K. Tripathi and Sudhanshu Srivastava for their contributions. The financial support received from the Department of Science and Technology (DST) and NSTI is gratefully acknowledged.

References

1. Tseng, A.A. 2004. Recent developments in micromilling using focused ion beam technology. *J. Micromech. Microeng.* 14: R15–R34.
2. Langford, R.M., Nellen, P.M., Gierak, J., and Fu, Y. 2007. Focused ion beam micro- and nanoengineering. *MRS Bull.* 32: 417–423.
3. Fujita, J., Ishida, M., Sakamoto, T., Ochiai, Y., Kaito, T., and Matsui, S. 2001. Observation and characteristics of mechanical vibration in three dimensional nanostructures and pillars grown by focused ion beam chemical vapor deposition. *J. Vac. Sci. Technol. B* 19: 2834.
4. Orloff, J., Utlaut, M., and Swanson, L. 2003. *High Resolution Focused Ion Beams.* New York: Kluwer Academic.
5. Giannuzzi, L.A. and Stevie, F.A. 2005. *Introduction to Focused Ion Beams: Instrumentation, Theory, Techniques and Practice.* New York: Springer.
6. Ziegler, J.F., Biersack, J.P., and Littmark, U. 1985. *The Stopping and Range of Ions in Solids.* New York: Pergamon.
7. Ohring, M. 2002. *The Materials Science of Thin Films.* 2nd ed. San Diego, USA: Academic Press.
8. Feldman, L.C. and Mayer, J.W. 1986. *Fundamentals of Surface and Thin Film Analysis.* New York: Elsevier Science Publishing Co.

9. Sigmund, P. 1969. Theory of sputtering. I. Sputtering yield of amorphous and polycrystalline targets. *Physical Review* 184: 383–416.
10. Yao, N. 2007. *Focused Ion Beam Systems: Basics and Applications.* Cambridge, UK: Cambridge University Press.
11. Tripathi, S.K., Shukla, N., Rajput, N.S., and Kulkarni, V.N. 2009. The out of beam sight effects in focused ion beam processing. *Nanotechnology* 20: 275301.
12. Tripathi, S.K., Shukla, N., and Kulkarni, V.N. 2009. Exploring a new strategy for nanofabrication: Deposition by scattered Ga ions using focused ion beam. *Nanotechnology* 20: 075304.
13. Tripathi, S.K. 2009. Investigation of focused ion beam induced processes and their utilization for nanofabrication, PhD Thesis, Indian Institute of Technology Kanpur, India.
14. Tripathi, S.K., Shukla, N., and Kulkarni, V.N. 2008. Correlation between ion beam parameters and physical characteristics of nanostructures fabricated by focused ion beam. *Nuclear Instruments and Methods in Physics Research* B 266: 1468.
15. Tripathi, S.K. and Kulkarni, V.N. 2009. Evolution of surface morphology of nano and microstructures during focused ion beam induced growth. *Nuclear Instruments and Methods in Physics Research* B 267: 1381.

Section III

Nanofinishing

7

Magnetorheological and Allied Finishing Processes

Ajay Sidpara and V.K. Jain

Indian Institute of Technology Kanpur

CONTENTS

7.1 Introduction

Surfaces play a significant role in the service life of engineering components. The present technological world demands surface roughness of many products on the order of a nanometer (10^{-9} m). It is also desirable to automate the process, whereby the desired surface finish on the components can be obtained easily in the shortest possible time. It is well established that the fatigue life of a component is strongly influenced by surface integrity, including its surface finish. Therefore, surface conditions become important factors influencing fatigue strength because irregular and rough surfaces generally exhibit inferior fatigue properties [1]. As surface roughness increases, many problems such as flow resistance, wear, and optical loss increase, resulting in a decreased efficiency. Hence, low surface roughness value (or better surface finish) of engineering components is necessary to improve the following product features, among many others:

- Wear resistance
- Mechanical properties of the material such as enhanced fatigue life and toughness
- Electrical properties
- Corrosion and oxidation resistance
- Aesthetic appearance

Many components/products require surface roughness value in the nanometer range. In such cases, some surface polishing is required during or after fabrication. Recently, the need for fine finishing of materials such as semiconductors and ceramics has become important in the fields of electronics and precision machinery. A variety of polishing methods have been developed to fulfill these requirements.

Abrasive-based traditional finishing processes such as grinding, honing, etc. are not capable of producing ultrafine surfaces because the abrasives are in bonded form and a high specific cutting energy is required. Traditional finishing processes alone are therefore incapable of producing the required surface finish. In some cases, these processes can be used, but they require expensive equipment and long processing times, making them economically incompetent. Hence, advanced finishing processes are of prime importance for high-precision components.

Some advanced finishing processes such as lapping, abrasive flow finishing (AFF), and chemomechanical polishing (CMP), and some magnetic-field-assisted finishing processes such as magnetic float polishing (MFP), magnetic abrasive finishing (MAF), magnetorheological finishing (MRF), magnetorheological abrasive flow finishing (MRAFF), and rotational-magnetorheological abrasive flow finishing (R-MRAFF) have evolved in the past few decades [2].

The above-mentioned finishing processes have their distinct merits and demerits. Lapping and CMP use loose abrasives in a fluid carrier, but they can finish only flat surfaces. AFF [3] is capable of finishing any complicated geometry by extruding an abrasive-laden polymeric medium through the passage formed by the workpiece and fixture assembly. However, the rheological properties of the polymeric medium, which are responsible for transforming abrading forces to the abrasives, cannot be controlled by external means. MRF [4] is used for external finishing of optical lenses to nanometer levels. It manipulates the magnetorheological (MR) fluid properties during the finishing process by external means (i.e., magnetic field) to control machined surface characteristics. However, it cannot finish cylindrical surfaces (external or internal). Therefore, MRAFF [5] and R-MRAFF [6] have been developed for finishing internal and external cylindrical surfaces. This chapter mainly focuses on the magnetorheological (MR) fluid based finishing processes such as MRF, MRAFF, and R-MRAFF.

7.2 Magnetorheological (MR) Fluid

The magnetorheological (MR) fluid, invented by Rabinow [7] in the late 1940s, belongs to a class of smart controllable materials whose rheological behavior can be manipulated externally by the application of a magnetic field. MR fluids are known as *controllable smart materials*, since their flow properties, such as viscosity and stiffness (yield stress), can be easily manipulated using externally applied magnetic field. Under the influence of the magnetic field, an MR fluid can be transformed from a fluidlike state to a semi-solid-like state within milliseconds by the formation of chain clusters of magnetic particles along the lines of the magnetic field. This phenomenon is called the *MR effect*.

The main component of all MR-fluid-based finishing processes is an MR fluid, which consists of magnetic particles, abrasive particles, carrier fluid, and additives. Initially, MR fluids were developed/used for clutches, shock absorbers, and vibration isolators without addition of any abrasive particle. However, in the 1980s, a few researchers [8–10] demonstrated MR fluids as polishing media for glass and other materials. Nowadays, MR-fluid-based finishing processes have become an inevitable part of precision manufacturing.

7.2.1 MR Fluid Compositions

The final finished surface of a workpiece is significantly affected by the composition of the MR fluid. Stable MR fluid properties such as apparent viscosity, temperature, and homogeneity are critical for controlling material removal [4]. A typical MR fluid consists of micron-sized magnetic carbonyl iron particles (CIPs), nonmagnetic polishing abrasives suspended in a carrier liquid, and some stabilizers. CIPs are responsible for creating high stiffness of the MR fluid in the presence of a magnetic field. The polishing abrasive particles are necessary for high material removal rate (MRR). The carrier liquid is responsible for establishing the correct polishing chemistry; the stabilizers reduce sedimentation of solid particles and increase fluid stability. Table 7.1 shows the constituents of MR fluids that are used for finishing of various materials.

Some essential features of an MR polishing fluid are as follows [11]:

- Optimum concentration of magnetic particles and abrasive particles
- High yield stress under magnetic field
- Low off-state viscosity
- Less agglomeration and good redispersibility
- Resistance to corrosion
- Stability against static sedimentation (the resistance of a fluid to form a hard sediment that is difficult to redisperse)
- High polishing efficiency (high finishing rate without surface damage)

TABLE 7.1

Constituents of MR Fluids

Magnetic Particle	Abrasive Particle	Carrier Fluid	Stabilizers
Carbonyl iron particle (CIP)	Cerium oxide	Water	Glycerol
Electrolyte iron powder	Diamond powder	Oil	Grease
Iron–cobalt alloy powder	Aluminum oxide		Oleic acid
	Silicon carbide		Xanthan gum
	Boron carbide		

7.2.2 Magnetic Particles

Magnetic field and magnetic particles are the most significant parameters for MR effect in MR-fluid-based finishing processes. High magnetic saturation and low magnetic coercivity are necessary to obtain good rheological properties in MR fluids [12]. The important properties of the magnetic particles are their size and hardness. If the magnetic particles are too large, they will rapidly settle in the suspension. If they are too small, the MR effect begins to diminish. The main functions of magnetic particles are to form a dense chain structure under a magnetic field and to hold abrasive particles during finishing. Without a magnetic field, magnetic particles are dispersed randomly in the MR fluid as shown in Figure 7.1a. Under the influence of a magnetic field, magnetic particles are aligned along the magnetic flux lines (from N-pole to S-pole) as shown in Figure 7.1b.

Many types of magnetic particles are available for MR fluid preparation, namely, CIPs, electrolyte iron powder, and iron–cobalt alloy powder (Table 7.1). CIPs are made by chemical vapor deposition (CVD) by the decomposition of iron pentacarbonyl, while electrolytic iron powder is prepared using the electrolytic or spray atomization process. However, CIPs are preferable to other magnetic particles because they are magnetically soft and chemically pure, and have a high saturation magnetization (2.1 T at room temperature). The low level of chemical impurity (<1%) means less domain-pinning defects, and the spherical shape helps minimizes magnetic shape anisotropy [13].

7.2.3 Abrasive Particles

Abrasives are nonmagnetic particles in MR fluid responsible for material removal. The quality of the finished surface is affected significantly by the granularity and hardness of the abrasive particles [12]. Abrasives such as Al_2O_3, SiC, cerium oxide, and diamond powder are commonly used. Polishing abrasives such as alumina and diamond powder have been used for finishing optical materials other than glasses [14,15]. Diamond powder has high finishing efficiency, but it is too expensive, while the hardness of Al_2O_3 is low. However, a commonly used polishing abrasive and carrier liquid combination for MR-fluid-based finishing of optical glasses including many crystals is that of cerium oxide and water.

7.2.4 Carrier Liquid

The function of a carrier liquid is primarily to provide a medium in which the solid particles (CIPs and abrasive particles) are suspended. A carrier liquid should be noncorrosive

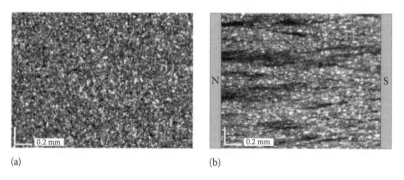

(a) (b)

FIGURE 7.1
Distribution of magnetic particles (a) without a magnetic field and (b) with a magnetic field.

in nature, and it should have a high boiling temperature and low freezing point [13]. The off-state viscosity of the MR fluid should be as low as possible so that it can be circulated easily in the device. The carrier liquid should not show significant variation in viscosity at a given temperature.

Oil or water is generally used as a carrier liquid. Nonaqueous carrier fluids have also been used for the finishing of water-soluble crystals [16]. Oil-based carrier liquids have high viscosity, so MR fluids show good performance in stability and sedimentation. From the rheological property point of view, water-based MR fluids have high yield stress. Moreover, water hydroxylates the surface of glass to improve finishing efficiency. In addition, water imparts a cooling effect during finishing, and solid particles disperse uniformly in the presence of water. Water has been proved to be the most suitable carrier liquid for finishing silicon or silicon-based materials, whereas oil-based MR fluids significantly reduce the finishing efficiency [17].

7.2.5 Stabilizers/Additives

MR fluids also contain a stabilizer such as glycerol, grease, oleic acid, or xanthan gum. The stabilizer is used to disperse the magnetic particles and abrasive particles uniformly in suspension. Stabilizers create a coating on the particles so that the MR fluid can easily redisperse. Solid particles in the MR fluid settle down at the bottom if it is kept in storage. A stabilizer helps prevent the dense magnetic particles from settling too fast and reduces the formation of hard sediment, which is difficult to remix. Moreover, stabilizing agents are also used to impart durability and corrosion resistance. Some stabilizers are necessary in water-based MR fluids to retard oxidation of the magnetic particles (CIPs) and prevent agglomeration. The stability of a water-based MR fluid can be improved by addition of glycerol because it works as a corrosion inhibitor and also provides some steric stabilization [11]. Use of acid-free cerium oxide compounds has been shown to be as important as that of the stabilizers in improving the stability of the MR fluid [18]. Hence, some additive agents are necessary to formulate a stable MR fluid.

Three MR-fluid-based finishing processes are discussed in the following sections, namely, MRF, MRAFF, and R-MRAFF.

7.3 Magnetorheological Finishing (MRF)

The MRF process is based on MR fluids that finish high-precision optical components effectively, replacing the manual technology that previously took hours or weeks. An MR fluid temporarily hardens when exposed to a magnetic field and conforms to the surface of the workpiece to be polished, making it ideal to polish many types of components. It offers various ways to overcome many of the limitations of conventional polishing.

MRF was developed for polishing optics by a team of scientists led by William Kordonski at the Luikov Institute of Heat and Mass Transfer in Minsk, Belarus in 1988. However, the concept of using MRF as an automated process to polish high-precision optics was first introduced by the Center for Optics Manufacturing (COM) at the University of Rochester in 1993 in collaboration with William Kordonski. Later in 1996, the MRF technology was commercialized by QED Technologies.

MRF [4] is a deterministic finishing process based on an MR fluid consisting of non-magnetic polishing abrasives and magnetic CIPs in water or some other carrier. The MR fluid forms a polishing tool that is perfectly conformal and therefore can polish a variety of shapes, including flats, spheres, aspheres, and prisms, with a variety of aperture shapes. With the appropriate combination of MR fluid and other finishing parameters, MRF has been successfully used to polish a variety of materials to subnanometer surface roughness values.

In the MRF process, as schematically shown in Figure 7.2a, a convex, flat, or concave workpiece is positioned below a rotating carrier wheel. An MR fluid ribbon is deposited on the carrier wheel. Two MR fluid pumps (peristaltic or centrifugal) are used for delivery and suction of the MR fluid from the carrier wheel. By applying a magnetic field using an electromagnet or a permanent magnet, the stiffened region forms a transient work zone or finishing spot. Surface finishing, removal of subsurface damage, and figure corrections are accomplished by providing feed in the X–Y direction to the workpiece. A fluid storage system is used to control the MR fluid temperature to prevent sedimentation of solids through mixing and to monitor/add water lost by evaporation. Peristaltic pumps deliver the fluid through a delivery nozzle onto the surface of the carrier wheel just ahead of the

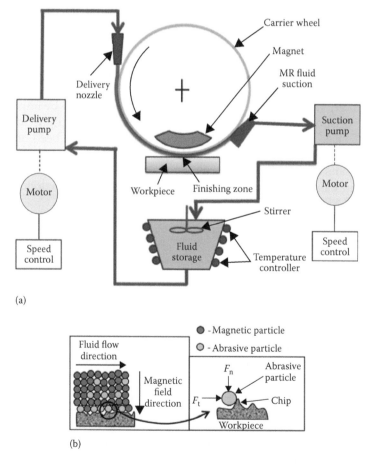

FIGURE 7.2
(a) Schematic diagram of the MR-fluid-based finishing process and (b) magnified view of the finishing zone.

polishing zone, as shown in Figure 7.2a. The fluid is recovered by a collector using a suction pump and returned to the fluid storage system. Material removal takes place through the interaction of the MR fluid and the workpiece when the MR fluid ribbon is dragged into the converging gap (finishing zone) by rotation of the carrier wheel. The zone of contact is restricted to a spot that conforms to the local topography of the part. Figure 7.2b shows a magnified view of the finishing zone where an abrasive particle interacts with the workpiece surface under the influence of the magnetic field. Material is removed from the workpiece by normal and tangential forces applied by magnetic particles through abrasive particles and by rotation of the carrier wheel.

Figure 7.3a shows the random distribution of CIPs and abrasive particles in the absence of the magnetic field. When the magnetic field is applied, the CIPs get magnetized and move toward the rotating wheel, where magnetic field strength is higher, and most of the abrasive particles move away from the carrier wheel, that is, toward the workpiece surface because of the magnetic levitation force as shown in Figure 7.3b. Figure 7.3c shows that the abrasive particles are in contact with the workpiece surface but intact with the fluid (ribbon) and that the CIPs are closer to the rotating wheel. The normal magnetic force (or penetrating force) is transferred to the work surface through the abrasive particles, and results in abrasive penetration into the work surface. Material removal is caused by the relative motion between the abrasive particles and the workpiece surface.

Figure 7.4a and b shows a schematic of the normal force acting on the workpiece by an abrasive in conventional polishing and in the MRF process, respectively. In the conventional polishing process, the normal load on the workpiece is transmitted from the

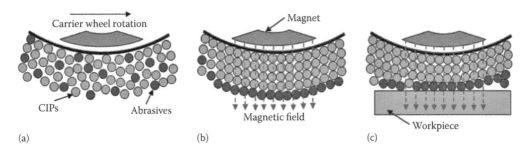

(a)   (b)   (c)

FIGURE 7.3
Schematic diagram of the material removal mechanism in MRF.

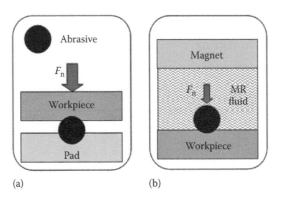

(a)   (b)

FIGURE 7.4
Normal load acting in (a) conventional polishing and (b) the MRF process.

polishing pad (typically made from polyurethane or pitch) through the abrasive particle (Figure 7.4 a). In this configuration, the normal load on the abrasive during polishing has been approximated to be on the order of millinewtons [19].

Unlike conventional polishing in which the particle's normal load can result in scratching, MRF creates a relatively low normal load on the abrasive particles. Figure 7.4b shows a schematic representation of the normal loads applied to the workpiece by an abrasive particle in the MRF process. The normal load on an abrasive particle is generated by the flow of the MR fluid over the workpiece [20]. This normal force can be estimated by the sum of the magnetic levitation force and the squeezing force normal to the workpiece surface. This force has been approximated to be several orders of magnitude lower than that of conventional polishing [21]; hence, the MRF process results in a smooth, damage-free surface finish.

7.3.1 Process Parameters in MRF

There are several process parameters that affect the MRF process in terms of surface finish and MRR. Figure 7.5 shows a fishbone diagram of MRF process parameters.

7.3.1.1 Effect of Magnetic Particle Concentration

At low volume concentration, magnetic particles align as thin and weak chains that are not capable of supporting abrasive particles effectively during finishing. However, as the concentration of the magnetic particles increases, the chains grow in length as well as in width, and eventually aggregate in the form of a thick and strong MR fluid ribbon. Abrasives that are firmly gripped by magnetic particle chains remove material efficiently. However, concentration of magnetic particles beyond a certain limit results in lower finishing efficiency. It is also observed that total solid-particle (magnetic particles and abrasive particles) concentration should not exceed 45%–50% by volume in the MR fluid. There are a few problems associated with a high concentration of solid particles such as the following:

- Low fluidity of the MR fluid (high off-state viscosity)
- Difficulty in circulation of the MR fluid through the pipes and pumps
- Difficulty in remixing the MR fluid (low redispersibility)

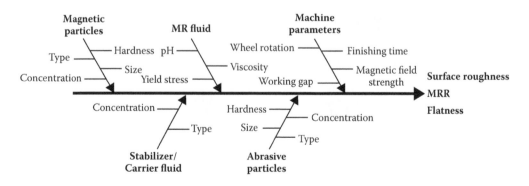

FIGURE 7.5
Fishbone diagram of MRF process parameters.

The magnetic interaction force (F) is the attracting force between two magnetic particles as given by Equation 7.1 [22]. This force increases with the increasing size of magnetic particles. It means the larger the magnetic particle, the higher the attracting force between them, as shown in Figure 7.6; the thickness of the arrow indicates the magnitude of the force, and circle diameter indicates particle size. This results in greater stiffness of the MR fluid and subsequently higher MRR.

$$F = \frac{\mu_0 \pi}{144} \left(\frac{D^2 CM}{S} \right)^2$$

(7.1)

where μ_0 is the permeability of free space, D is the diameter of a CIP, S is the distance between the centers of two CIPs, M is the intensity of magnetization, and C is a constant.

7.3.1.2 Effect of Abrasive Particle Concentration

Variation in concentration of abrasive particles significantly changes the finishing mechanism in an MR-fluid-based finishing process. For clear visibility and understanding of the chain structure of CIPs mixed with abrasive particles, bigger abrasive particles and magnetic particles are chosen, as shown in Figure 7.7. Electrolytic iron powder is selected as the magnetic particle in place of CIPs, and aluminum oxide (Al_2O_3) is selected as the abrasive because of its white appearance. Figure 7.7a and b shows the CIP chain structure at low and high concentration of the abrasive particles, respectively. Figure 7.7 shows that the abrasive particles are distributed nonuniformly. At low concentration of abrasive particles,

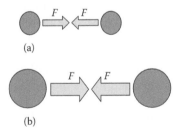

(a)

(b)

FIGURE 7.6
Effect of the size of a magnetic particle on the magnetic interaction force (F).

(a)   (b)

FIGURE 7.7
CIP chain structure at (a) low and (b) high abrasive concentrations.

a few abrasives are trapped in the CIPs chains structure, so fewer discontinuous chains of CIPs are observed under the magnetic field (Figure 7.7a). Hence CIP chains are capable of gripping abrasive particles more firmly during the finishing action. At higher abrasive particle concentration, the amount of abrasives entangled in CIP chains is increased (Figure 7.7b), and this reduces the strength of the fluid under shear flow.

As given by Equation 7.1, the magnetic interaction force (F) reduces with increasing distance between magnetic particles (S). This means that if any abrasive particle is trapped between two CIPs (Figure 7.7), the distance between the CIPs increases depending on the size of the abrasive particle and, as a result, F decreases. The reduction in F is higher if the abrasive particle size is larger (Figure 7.7b) as compared to a small abrasive particle (Figure 7.7a). Owing to low F, the bonding strength between CIPs reduces, and subsequently they are not able to grip the abrasives properly. F also decreases if any larger abrasive is trapped within the CIP chain as shown in Figure 7.8. Furthermore, a higher abrasive concentration also results in more discontinuous chains, which further reduces the magnetic interaction force (F). Therefore, beyond an optimum concentration of abrasive particles, MRR and finishing rate decrease and the difference in surface finish (ΔRa = initial Ra value − final Ra value) obtained within the prescribed time becomes lower.

7.3.2 Force Analysis in MRF Process

Knowledge of the forces acting during any finishing process is important to understand the mechanism of material removal. These forces have direct impact on the quality of the finished surface and its accuracy. Study of these forces is also necessary to predict the MRR as well as the final achievable surface roughness value. In this section, the forces acting on the workpiece in the MRF process are discussed.

In the MRF process, mainly two types of forces act on an abrasive particle, namely, the normal force (F_n) responsible for abrasive particle penetration inside the workpiece and the tangential force (F_t) responsible for removal of material in the form of micro/nanochips as shown in Figure 7.9. The resultant force (F_r) removes material from the workpiece. The normal force is a combination of the magnetic levitation force applied by the magnetic particles and the squeezing force applied by the hydrodynamic drag force due to flow of the MR fluid in the converging gap.

The magnetic field gradient causes CIPs to be attracted toward the magnet (toward the carrier wheel surface) while the abrasives move toward the workpiece surface owing to the magnetic levitation force as shown in Figure 7.10. Magnetic levitation is defined

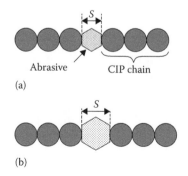

(a)

(b)

FIGURE 7.8
Effect of abrasive size on the magnetic interaction force (F).

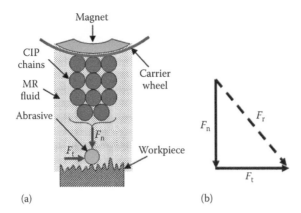

(a)   (b)

FIGURE 7.9
(a) Forces acting on an abrasive particle and (b) force diagram in the MRF process.

as the force exerted on nonmagnetic bodies by a magnetic fluid. The magnetic levitation force helps in indenting the abrasive particle into the workpiece surface, which results in material removal. A few abrasive particles are trapped in the CIP chains and stay there, as shown in Figure 7.10. The magnitude of the magnetic levitation force on an abrasive particle is determined by the magnetic field strength and the magnetic properties of the MR fluid. The levitation force (F_m) [23] is given by

$$F_m = -V \mu_0 M \nabla H$$
(7.2)

where V is the volume of the nonmagnetic body, M is the intensity of magnetization of the magnetic fluid, μ_0 is the permeability of free space, and ∇H is the gradient of a magnetic field.

Holding of abrasive particles in the CIP chain structure depends on the rheological properties (yield stress and viscosity) of the MR fluid under a magnetic field [24]. Stiffness of the MR fluid under a magnetic field is measured by yield stress. When the yield stress

FIGURE 7.10
Side view of CIP chains with abrasive particles (white in color).

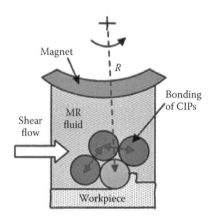

FIGURE 7.11
Schematic view of an abrasive particle held under shear flow.

of the MR fluid is high, CIPs chains can hold abrasive particles for a long time under shear force, which is necessary for efficient material removal.

A tangential force is exerted on the workpiece by the shear flow of the MR fluid, pushing the abrasive forward. Before the MR fluid yields under shear, the CIPs chains structure is intact (bonds between CIPs are strong enough to withstand shear flow). This means the abrasives, which are gripped by the bonded structure of the CIPs as shown in Figure 7.11, would apply force on the workpiece without loosening their own orientation.

7.3.3 Applications of MRF Process

Calcium fluoride and fused silica lenses are specifically used for lithographic steppers and scanners. In these applications, shape and accuracy requirements are stringent. Therefore, MRF is an appropriate process for finishing of these lenses.

Free-form optical surfaces are difficult to finish by conventional polishing techniques owing to constantly changing curvature and lack of symmetry. In this case, MRF is a suitable process because the MR fluid adapts to the local surface curvature in such a way that it does not have problems with tool/part shape mismatch [25].

Silicon-on-insulator (SIO) wafer technology needs stringent tolerances. Global flatness and thickness uniformity are the most important parameters for SIO wafers. The MRF process is able to improve both the parameters after finishing without affecting other features [25].

Single-point diamond turning (SPDT) is commonly used for fabrication of optics for infrared (IR) applications. However, SPDT leaves turning marks or grooves on the optics that reduce optical functionality. MRF can be very effective at both improving surface figure and removing the diamond turning marks in materials such as silicon and calcium fluoride [25]. The MRF process can significantly reduce the surface roughness (initially from 960 to 12 nm) of single crystal silicon as shown in Figure 7.12.

Polymer optics are generally manufactured by injection molding, compression molding, or diamond turning. Polymer optics are soft and have a high linear expansion coefficient and poor thermal conductivity. Therefore, cold working or similar processes are not preferable to improve surface finish or figure. Furthermore, conventional grinding and polishing processes usually result in scratching, embedding of abrasive particles,

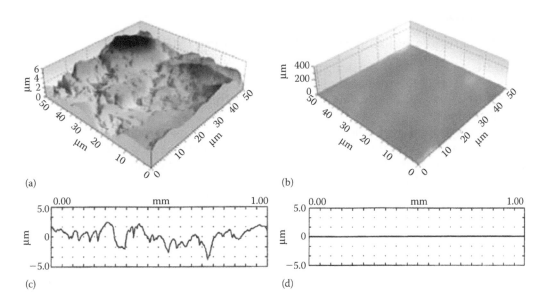

FIGURE 7.12
(a and b) Atomic force microscope images and (c and d) surface roughness plot of (a and c) initial surface (960 nm) and (b and d) final surface (8 nm) of single crystal silicon by the MRF process.

FIGURE 7.13
Selected applications of MRF process finishing of (a) 450-mm BK7 mirror, (b) 400-mm-diameter concave silicon, and (c) 300-mm-diameter lightweight ULE glass. (From Jones, A., et al., *Deterministic Fabrication of Large and Astronomical Optics using Magnetorheological Finishing (MRF®)*. QED Technologies, New York, USA, 2006.)

and degradation of surface figure [26]. In this case, MRF is a suitable finishing process to improve surface finish and figure correction of polymer optics.

MRF can also improve surface integrity by removing microcracks and subsurface damage, removing residual stresses, and improving glass resistance to laser damage [27]. Some applications of the MRF process are shown in Figure 7.13 [28].

From the above discussion and the literature, it is found that MRF is an important and efficient process for finishing concave, convex, and flat components. However, it is not suitable for finishing of internal or external surfaces of cylindrical components. Hence, there is need for a suitable process similar to MRF for cylindrical components. Other processes such as MRAFF and R-MRAFF have been developed and are discussed in the following sections.

7.4 Magnetorheological Abrasive Flow Finishing (MRAFF)

MRAFF is a new precision finishing process that has been developed by taking advantage of both AFM and MRF for nanofinishing of parts even with complicated geometries [5]. MRAFF is a fine-finishing process that uses abrasives mixed with the MR fluid in a controlled manner.

MR fluids exhibit real-time controllable change in flow properties of the fluid, enabling in-process control of finishing forces through the magnetic field. Any complex geometrical surface, internal or external, inaccessible to existing finishing processes can be finished by the MRAFF process by suitably designing the workpiece–fixture assembly. In the MRAFF process, the MR fluid is a homogeneous mixture of CIPs and abrasive particles in a base medium of liquid paraffin and grease. When an external magnetic field is applied, the CIPs in the fluid form chainlike structures along the lines of the magnetic field between two magnets, as shown in Figure 7.14. Nonmagnetic abrasive particles are embedded into the CIP chains, and these chains give bonding strength to the abrasive particles. When the MR fluid is extruded back and forth through the passage formed by the workpiece or fixture, abrasive particles embedded into/between the chains of CIPs and in contact with the workpiece surface perform finishing by shearing off peaks of surface undulations.

Among the three main variables, magnetic flux density, extrusion pressure, and number of finishing cycles, magnetic flux density is most significant in the MRAFF process that influences improvement of surface finish [5]. Furthermore, MR polishing fluid composition also plays an important role in process performance [29,30] for finishing of hard materials such as stainless steel.

7.4.1 Rotational-Magnetorheological Abrasive Flow Finishing (R-MRAFF)

The concept of R-MRAFF has evolved from the MRAFF process. In R-MRAFF, rotational motion is also provided to the magnets, which rotate the MR fluid inside the cylindrical

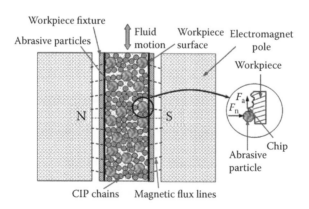

FIGURE 7.14

Mechanism of finishing in the MRAFF process. (From Jha, S., et al., *International Journal of Advanced Manufacturing Technology* 33, 725–729, 2006.)

workpiece. This rotation of magnets provides a circular motion to the MR fluid in addition to the axial motion provided as in the MRAFF process. A unique attribute of the R-MRAFF process is the determinism attained through the use of a precisely controlled magnetic field and guided flow of the MR fluid to minimize known surface irregularities and defects. The computer-controlled R-MRAFF process will demonstrate the ability to meet high standards of surface accuracy by overcoming many of the fundamental limitations inherent in traditional finishing techniques.

R-MRAFF [31] employs a rotating magnetic field that is applied vertically with respect to the cylindrical workpieces in order to best utilize the fluid characteristics. The working principle of R-MRAFF process is shown in Figure 7.15. The polishing mechanism involves abrasive particles aggregating at a certain section of the polished wall. When the magnets rotate with respect to the axis of the cylindrical workpieces along with the extruding motion of the MR fluid through the hydraulic unit, abrasives shear off roughness peaks uniformly from the internal surface of the workpiece (similar to a honing operation but in a precisely controlled manner).

Figure 7.16a and b shows movement of the abrasive inside the workpiece in the MRAFF and R-MRAFF processes, respectively. In MRAFF, the abrasive moves parallel to the cylindrical workpiece because the MR fluid has only reciprocating motion, while in case of R-MRAFF, the abrasive moves in a helical path that follows the additional circular motion of the MR fluid provided by rotation of the magnets. So, the abrasive has greater interaction length/area with the workpiece in the case of R-MRAFF compared to MRAFF. In the R-MRAFF process, the path of the abrasive followed at the surface of the cylindrical fixture due to the resultant motion is helical (Figure 7.16b). Therefore, it generates a crosshatch pattern (similar to a honing operation) on the finished surface. Figure 7.17a and b shows the unfinished and finished internal surfaces of a stainless steel workpiece, respectively. It is clear from the figure by comparing the reflection of the letters "IITK" on the surfaces that R-MRAFF can finish even hard materials efficiently. Surface roughness of the stainless steel workpiece is reduced from 330 (Figure 7.17c) to 16 nm (Figure 7.17d) [32].

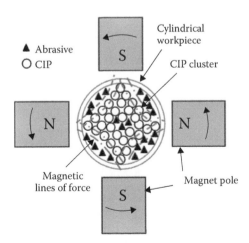

FIGURE 7.15
Principle of the R-MRAFF process. (From Das, M., et al., *ASME 2009 International Manufacturing Science and Engineering Conference (MSEC 2009)* October 4–7, 2009, West Lafayette, Indiana, USA, 251–260, 2008.)

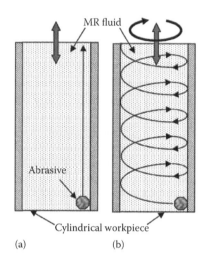

FIGURE 7.16
Abrasive movement in (a) MRAFF and (b) R-MRAFF processes.

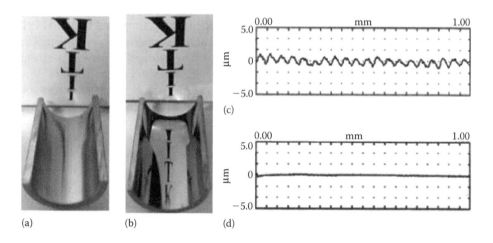

FIGURE 7.17
Photograph (a and b) and surface roughness plot (c and d) of the internal surface of the stainless steel workpiece (a and c) before finishing (Ra = 330 nm) and (b and d) after finishing (Ra = 16 nm) with R-MRAFF. (From Das, M., et al., *Machining Science and Technology* 14(3), 365–389, 2010.)

7.4.2 Process Parameters in MRAFF and R-MRAFF

The MRF, MRAFF, and R-MRAFF processes use an MR fluid as a finishing medium. Therefore, there are many common process parameters that affect the quality of the surface produced on the workpiece. Furthermore, the difference between MRAFF and R-MRAFF is only due to the rotation of the magnet. That is, the R-MRAFF process has one additional parameter (i.e., rotation of the MR fluid due to rotation of the magnet) over the MRAFF process, as given in the form of a fishbone diagram (Figure 7.18).

Because of a greater interaction length/area between the abrasive and the workpiece in R-MRAFF, more material removal is observed in the case of stainless steel and brass

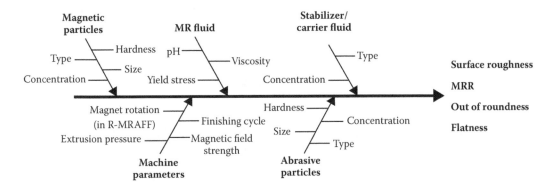

FIGURE 7.18
Fishbone diagram of MRAFF and R-MRAFF parameters.

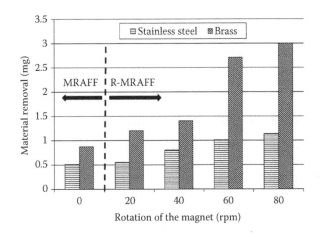

FIGURE 7.19
Material removal in MRAFF and R-MRAFF processes. (From Das, M., et al., *ASME 2009 International Manufacturing Science and Engineering Conference (MSEC 2009)* October 4–7, 2009, West Lafayette, Indiana, USA, 251–260, 2008.)

workpieces as shown in Figure 7.19 [31]. This is because as the rotation of the magnet increases, the distance traveled by an abrasive on the workpiece surface also increases.

7.4.3 Forces in MRAFF and R-MRAFF Processes

In the MRAFF process, normal (F_n) and axial (F_a) forces act on the workpiece through an abrasive as shown in Figure 7.20a. The axial force acts parallel to the axis of the cylindrical workpiece owing to reciprocation of the MR fluid (caused by extrusion pressure) over the workpiece. The normal force includes two forces, namely, the radial force (F_{rad}) and the magnetic force (F_{mg}). The radial force on the abrasive grain is generated by extrusion of the MR fluid through the cylindrical workpiece. An abrasive also exerts a magnetic force through surrounding CIPs. F_a and F_{rad} are proportional to the extrusion pressure and medium viscosity, and can be calculated by simulating the MR fluid inside the workpiece fixture.

The MR field strength and properties of a magnetic particle determine the magnitude of the magnetic force acting on it. The magnetic force (F_{mg}) on a small magnetic particle of mass m in a magnetic field is given by [33]

$$F_{mg} = \frac{m\chi}{\mu_0} B \nabla B$$

(7.3)

where m is the mass of the magnetic particle, B is the magnetic flux density, μ_0 is the permeability of free space, ∇B is the gradient of magnetic flux density, and χ is the magnetic susceptibility of the magnetic particle.

In the case of R-MRAFF, an additional rotational motion is provided to the magnet compared to the MRAFF process. Because of this additional rotation of the magnet, the MR fluid rotates inside the cylindrical workpiece along with a reciprocating movement. So, two more forces act on the workpiece, namely, a tangential force (F_t), tangent to the workpiece surface, and a centrifugal force (F_c), normal to the workpiece surface. Both these additional forces are proportional to the rotational speed of the magnets.

The resultant force carries out a cutting action by shearing the peaks in the form of microchips in MRAFF (F_r in Figure 7.20b) and R-MRAFF (F_{rr} in Figure 7.20c). Figure 7.20a and b also shows that the direction of the resultant force is different in the MRAFF and R-MRAFF processes, and this changes the cutting direction of an abrasive particle.

7.4.4 Applications of MRAFF and R-MRAFF Processes

MRAFF can be utilized for finishing internal surfaces of cylindrical components up to nanometer levels.

In many applications such as the piston–cylinder assembly and ball bearing, two mating surfaces of the assembly require continuous lubrication to reduce friction and wear. In such applications, mating surfaces should be smooth as well as properly lubricated. The honing process is widely used for making a crosshatch pattern in the cylinder of the engine for oil retention as shown in Figure 7.21. In this case, the R-MRAFF process can be used to finish internal surfaces at the nanometer level and can also generate crosshatch patterns similar to the honing process at nanoscale for oil retention as shown in Figure 7.22c. It is also apparent that the surface roughness value of initial grinded workpiece (250 nm in

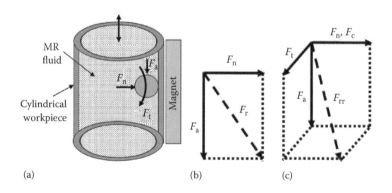

(a)   (b)   (c)

FIGURE 7.20
(a) Forces acting on an abrasive particle in MRAFF and R-MRAFF, (b) force diagram in an MRAFF process, and (c) R-MRAFF.

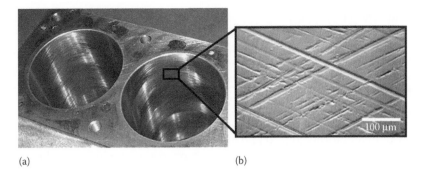

(a)   (b)

FIGURE 7.21
(a) Internal surface of a cylinder and (b) crosshatch pattern generated by the honing process.

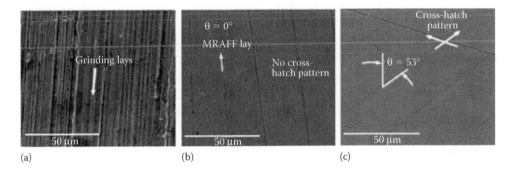

(a)   (b)   (c)

FIGURE 7.22
SEM images of a brass workpiece (a) initial surface (Ra = 0.25 μm), (b) without magnet rotation (Ra = 0.17 μm), and (c) with magnet rotation at 80 rpm (Ra = 0.08 μm). (From Das, M., et al., *ASME 2009 International Manufacturing Science and Engineering Conference (MSEC 2009)* October 4–7, 2009, West Lafayette, Indiana, USA, 251–260, 2008.)

Figure 7.22a) is reduced to 170 nm (Figure 7.22b) by MRAFF process. However, the surface roughness value is further reduced to 80 nm (Figure 7.22c) with cross-hatch patterns by R-MRAFF process owing to rotation of the magnet.

The R-MRAFF process can also be used when a small portion of any cylindrical workpiece is required to be finished at a nanometer surface roughness value without affecting/damaging the rest of the part as shown in Figure 7.23. In such cases, the magnets are rotated by placing them close to the surface to be finished and the MR fluid is extruded through the cylindrical workpiece. The MR fluid will be more aggressive under the influence of the

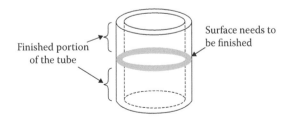

FIGURE 7.23
Finishing of a small portion by the R-MRAFF process.

magnetic field. Therefore, finishing takes place at the desired location without affecting the other areas of the workpiece.

7.5 Concluding Remarks

An overview of magnetorheological and allied finishing processes is given in this chapter. These processes are used effectively for finishing complex surfaces where traditional finishing processes are uneconomical and incompetent. These processes are deterministic, and the forces can be controlled externally by changing magnetic field strength. Using a proper combination of MR fluid constituents, the desired surface finish on different types of materials can be achieved. MR-fluid-based finishing processes have large potential for finishing free form or sculptured surfaces.

References

1. Bayoumi, M.R. and Abdellatif, A.K. 1995. Effect of surface finish on fatigue strength. *Engineering Fracture Mechanics* 51(5): 861–870.
2. Jain, V.K. 2008. Abrasive-based nano-finishing techniques: An overview. *Machining Science and Technology* 12(3): 257–294.
3. Rhoades, L.J. 1988. Abrasive flow machining. *Manufacturing Engineering* 101: 75–78.
4. Jacobs, S.D., Golini, D., Hsu, Y., Puchebner, B.E., Strafford, D., Kordonski, W.I., Prokhorov, I.V., Fess, E., Pietrowski, D., and Kordonski, V.W. 1995. Magnetorheological finishing: A deterministic process for optics manufacturing. *SPIE 2576: International Conference on Optical Fabrication and Testing*, ed. T. Kasai. Bellingham: SPIE. pp. 372–282.
5. Jha, S. and Jain, V.K. 2004. Design and development of the magnetorheological abrasive flow finishing process. *International Journal of Machine Tools and Manufacture* 44: 1019–1029.
6. Das, M., Jain, V.K., and Ghoshdastidar, P.S. 2009. Parametric study of process parameters and characterization of surface texture using rotational-magnetorheological abrasive flow finishing (R-MRAFF). *Proceedings of the ASME 2009 International Manufacturing Science and Engineering Conference, MSEC 2009 October 4–7, 2009.* West Lafayette, IN.
7. Rabinow, J. 1948. The magnetic fluid clutch. *AIEE Transactions* 67: 1308–1315.
8. Tani, Y., Kawata, K., and Nakayama, K. 1984. Development of high-efficient fine finishing process using magnetic fluid. *CIRP Annals—Manufacturing Technology* 33(1): 217–220.
9. Kurobe, T. and Imanaka, O. 1984. Magnetic field assisted fine finishing. *Precision Engineering* 6(3): 119–124.
10. Suzuki, H., Kodera, S., Hara, S., Matsunaga, H., and Kurobe, T. 1989. Magnetic field-assisted polishing—Application to a curved surface. *Precision Engineering* 11(4): 197–202.
11. Kordonski, W.I. and Golini, D. 1999. Fundamentals of magnetorheological fluid utilization in high precision finishing. *Journal of Intelligent Material Systems and Structures* 10: 683–689.
12. Hongjun, W., Ailing, T., Qian, T., Zhili, C., and Bingcai, L. 2007. Research on rheological properties of magnetorheological fluid. *Proceedings of the SPIE—The International Society for Optical Engineering* 6722: 672230.
13. Phulé, P.P. 2001. Magnetorheological (MR) fluids: Principles and applications. *Smart Materials Bulletin* 2001(2): 7–10.
14. Jacobs, S.D. 1996. Nanodiamonds enhance removal in magnetorheological finishing. *Finer Points* 7: 47–54.

15. Jacobs, S.D., Yang, F., Fess, E.M., Feingold, J.B., Gillman, B.E., Kordonski, W.I., Edwards, H., and Golini, D. 1997. Magnetorheological finishing of IR materials. *SPIE 3134: Optical Manufacturing and Testing II*, ed. H.P. Stahl. Bellingham, WA: SPIE. pp. 258–269.

16. Arrasmith, S.A., Jacobs, S.D., Romanofsky, H.J., Gregg, L.L., Shorey, A.B., Kozhinova, I.A., Golini, D., Kordonski, W.I., Hogan, S., and Dumas, P. 1999. Studies of material removal in magnetorheological finishing (MRF) from polishing spots. *Symposium D: Finishing of Advanced Ceramics and Glasses, Ceramic Transactions*, Vol. 102, eds R. Sabia, V.A. Greenhut, and C. Pantano. Westerville, OH: American Ceramic Society.

17. Sidpara, A. and Jain, V.K. 2010. Effect of fluid composition on nano-finishing of single crystal silicon by magnetic field assisted finishing process. *International Journal of Advanced Manufacturing Technology*. DOI: 10.1007/s00170-010-3032-5.

18. Kozhinova, I.A., Arrasmith, S.R., Gregg, L.L., and Jacobs, S.D. 1999. Origin of corrosion in magnetorheological fluids used for optical finishing. *Symposium HH: Soft Condensed Matter— Fundamentals and Applications, MRF Spring Meeting*. San Francisco, CA.

19. Bulsara, V.H., Ahn, Y., Chandrasekar, S., and Farris, T.N. 1998. Mechanics of polishing. *ASME Journal of Applied Mechanics* 45: 410–416.

20. Golini, D., Jacobs, S.D., Kordonski, W., and Dumas, P. 1997. Precision optics fabrication using magnetorheological finishing. *Advanced Materials for Optics and Precision Structures*, eds M. Ealey, R.A. Paquin, and T.B. Parsonage, *Critical Reviews of Optical Science and Technology*, Vol. CR67. Bellingham, WA: SPIE. pp. 251–274.

21. DeGroote, J.E., Marino, A.E., Wilson, J.P., Bishop, A.L., Lambropoulos, J.C., and Jacobs, S.D. 2007. Removal rate model for magnetorheological finishing of glass. *Applied Optics* 46(32): 7927–7941.

22. Huang, J., Zhang, J.Q., and Liu, J.N. 2005. Effect of magnetic field on properties of MR fluid. *International Journal of Modern Physics B* 19: 597–601.

23. Rosensweig, R.E. 1985. *Ferrohydrodynamics*. New York: Dover.

24. Sidpara, A., Das, M., and Jain, V.K. 2009. Rheological characterization of magnetorheological finishing fluid. *Journal of Materials and Manufacturing Processes* 24(12): 1467–1478.

25. Tricard, M., Dumas, P.R., and Golini, D. 2004. New industrial applications of magnetor-heological finishing (MRF). *Frontiers in Optics*, OSA Technical Digest (CD) (Optical Society of America), OMD1.

26. DeGroote, J.E., Romanofsky, H.J., Kozhinova, I.A., Schoen, J.M., and Jacobs, S.D. 2004. Polishing of PMMA and other optical polymers with magnetorheological finishing. Optical Manufacturing and Testing V (Proceedings of SPIE 5180: *The International Society for Optical Engineering*), ed. H.P. Stahl. San Diego, CA, USA. pp. 123–134.

27. Kordonski, W.I. and Golini, D. 1999. Progress update in magnetorheological finishing. *International Journal of Modern Physics B* 13: 2205–2212.

28. Jones, A., Shorey, A., Hall, C., O'Donohue, S., and Tricard, M. 2006. *Deterministic Fabrication of Large and Astronomical Optics Using Magnetorheological Finishing (MRF®)*. New York, USA: QED Technologies.

29. Jha, S., Jain, V.K., and Komanduri, R. 2006. Effect of extrusion pressure and number of finishing cycles on surface roughness in magnetorheological abrasive flow finishing (MRAFF) process. *International Journal of Advanced Manufacturing Technology* 33: 725–729.

30. Das, M., Jain, V.K., and Ghoshdastidar, P.S. 2008. Analysis of magnetorheological abrasive flow finishing (MRAFF) process. *International Journal of Advanced Manufacturing Technology* 38: 613–621.

31. Das, M., Jain, V.K., and Ghoshdastidar, P.S. 2008. Parametric study of process parameters and characterization of surface texture using rotational-magnetorheological abrasive flow finishing (R-MRAFF) process. *ASME 2009 International Manufacturing Science and Engineering Conference (MSEC2009) October 4–7, 2009*. West Lafayette, IN. pp. 251–260.

32. Das, M., Jain, V.K., and Ghoshdastidar, P.S. 2010. Nano-finishing of stainless-steel tubes using rotational magnetorheological abrasive flow finishing process. *Machining Science and Technology* 14(3): 365–389.

33. Stradling, A.W. 1993. The physics of open-gradient dry magnetic separation. *International Journal of Mineral Processing* 39: 19–29.

8

Magnetic Abrasive Finishing (MAF)

D.K. Singh and S.C. Jayswal

Madan Mohan Malaviya Engineering College

V.K. Jain

Indian Institute of Technology Kanpur

CONTENTS

8.1 Introduction

With the advancements in manufacturing technology, fine surface finish is in high demand in a wide spectrum of industrial applications. The parts used in industries related to the manufacturing of semiconductors, atomic energy, medical instruments, and similar others require very low values of surface roughness (Yan and Hu, 2005). In recent years, the establishment of micromanufacturing methods for realizing further precision, smoothness, and miniaturization has become increasingly essential in the fabrication/manufacture of various microdevices as functional structures. This demand is particularly strong in the areas of electronics, mechatronics, biology, and medicine. Micromanufacturing technologies play an increasingly decisive role in the miniaturization of products ranging from biomedical devices to chemical microreactors and microsensors. Microslots, complex surfaces, and microholes with micro- to nanosurface finish are produced in large numbers, sometimes in a single workpiece, especially in the electronics industry.

Various micro- or macrocomponents of advanced engineering materials, such as silicon nitride, silicon carbide, magnesium alloys, ceramics, glasses, superalloys (titanium and nickel based), and aluminum oxide with low surface roughness value and high form accuracy, are in demand, especially in ultraprecision manufacturing industries such as the microelectronics industry. These advanced engineering materials are being used because of their high strength-to-weight ratio, hardness, toughness, heat resistance, wear resistance, and corrosion resistance. These materials play an important role in those areas where specific characteristics are needed. Applications of these materials are rapidly increasing in electronics, computers, aerospace, automotives, and semiconductor industries (Jain, 2009a,b; McGeough, 1988).

Finishing of advanced engineering materials at micro- to nanolevels is the most critical, usually uncontrollable, and expensive phase of the overall production processes. Abrasive finishing is invariably used in the production of components of the highest quality in terms of form accuracy and surface integrity. Surface roughness has a vital influence on functional properties, namely, wear resistance and power loss due to friction, in most engineering devices (William and Rajurkar, 1992). To achieve nanolevel surface finish in these materials by conventional finishing techniques is very difficult and uneconomical. The techniques that have been recently developed are advanced abrasive fine finishing techniques (Jain, 2009a,b). The basic reason for using micron-sized abrasive particles in these finishing processes is twofold: first, because of low force and small cutting edge, the penetration is very small in the submicrometer range, and second, a large number of random cutting edges with indefinite orientation and geometry remove material simultaneously. Since material removal in fine abrasive finishing processes is extremely small, that is, on the order of submicrons (i.e., nanometer), Taniguchi in 1974 coined the term *nanotechnology* for target precision processing (Komanduri et al., 1997). Because of the micrometer/nanometer-sized chips being produced during these processes, they yield better surface finish and closer dimensional tolerances and permit machining of harder and difficult-to-machine materials (Inasaki et al., 1993).

Microfeatures and microparts with rough surfaces, micro/nanocracks, recast metal, micro/nanoburrs, metallurgical phase changes, subsurface damage, and residual stresses are some attributes that result in rejection or failure of a device during its use. Microgrinding, electric discharge micromachining, laser beam micromachining, and other similar processes can produce such unwanted side effects. However, the magnetic abrasive finishing (MAF) process can remove these defects and produce micro- to nanolevel finished surfaces. Grinding, lapping, and honing may also remove the recast layer, but they may impart surface damage, which is unacceptable in critical applications, especially in microelectronic devices.

In the nanofinishing of advanced engineering materials by conventional methods, the common manufacturing practice is to use processes such as grinding followed by lapping or honing depending on the product shape. In practice, it takes considerable time to finish advanced engineering materials. Long processing time and use of expensive abrasive particles, such as diamond, result in high production cost. Furthermore, use of a diamond abrasive at high loads can result in surface and subsurface damage to the workpiece. In addition, these conventional finishing techniques are being pushed to their limits of performance and productivity in general, and in advanced engineering materials and components of complicated shapes in particular. The introduction of many difficult-to-machine materials, such as hardened steel, ceramics, glasses, and titanium- and nickel-based superalloys has posed a challenge for manufacturing engineers to shape, size, and finish these materials accurately. It is necessary to process these materials under very "gentle" conditions. A relatively new finishing method, MAF, is one such advanced machining process in which the cutting force is primarily controlled by the magnetic field. This process can be used to efficiently produce good surface quality on the order of a few nanometers surface finish (~8 nm) on flat surfaces as well as on internal and external cylindrical surfaces. The method can machine not only ferromagnetic materials such as steel but also nonferromagnetic materials such as stainless steel and brass. In addition, the MAF process provides many attractive advantages, namely, self-sharpening of abrasives, self-adaptability of the medium, and controllability. Further, the finishing tool requires neither compensation nor dressing (Yan and Hu, 2005). The MAF principle has been successfully used for deburring as well. Hybridization of the MAF process is also quite common, for example, ultrasonic-assisted magnetic abrasive finishing (UAMAF) and electrolytic magnetic abrasive finishing (EMAF).

The MAF process is a nanofinishing technique that can be employed to produce optical, mechanical, and electronic components with micrometer or submicrometer form accuracy and surface roughness within the nanometer range with practically no surface defects. Finishing of bearing components (balls, rollers), precision automotive components, shafts, and artificial hip joints made of oxide ceramics and cobalt alloys are some components for which this process can be applied. It can produce surface finish on the order of a few nanometers (Jain, 2002). This method was originally introduced in the Soviet Union, with further basic fundamental research in Japan, the United States, and India.

8.2 Working Principle of Magnetic Abrasive Finishing

In MAF, magnetic forces play a dominant role in the formation of a flexible magnetic abrasive brush (FMAB), which enables abrasives to polish the workpiece surface. Figure 8.1a shows the working principle of the MAF of plane surfaces. Abrasive particles, magnetic iron powder, and mineral oil are homogeneously mixed. This mixture is known as a *medium*, and it fills the gap between the bottom face of the magnet and the top face of the workpiece (Figure 8.1a). As the electromagnet gets energized, the ferromagnetic particles (magnetic iron powder) get aligned along the magnetic lines of force and form a chainlike structure known as a *flexible magnetic abrasive brush*. The abrasive particles (nonmagnetic in nature) get entangled within and between these chains. As the bond of the magnetic medium is soft and flexible, it can easily adapt to any minor nonuniformity in the shape of the surface being finished. The magnitude of force exerted by an abrasive particle in the magnetic field has been found to be in the micronewton range (Jayswal, 2005). However, in place of a mixture of magnetic and nonmagnetic abrasive particles and mineral oil,

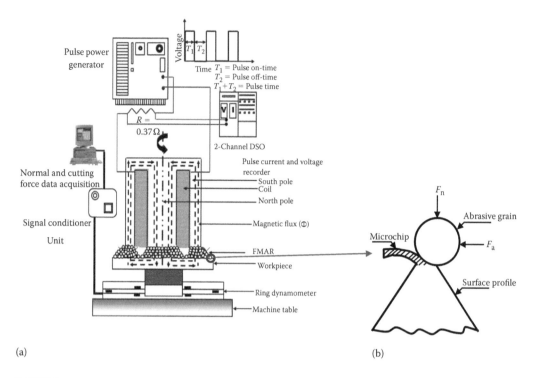

FIGURE 8.1
Schematic diagram of (a) DC-MAF setup and (b) an abrasive particle producing a microchip.

the sintered ferromagnetic abrasive particles are more effective, as discussed elsewhere. As shown in Figure 8.1b, two forces act on an abrasive particle. Normal force (F_n) is responsible for penetration of the abrasive particle inside the workpiece surface, and axial force (F_a) removes the material in the form of microchips. If the surface of the workpiece is softer than that of the ferromagnetic particles, these particles will remove the material as do the abrasive particles.

The magnetic-field-assisted nanofinishing processes are based on the electromagnetic behavior of the magnetic abrasive particles (MAPs). The electromagnetic behavior of the MAPs can be either static or dynamic depending on the excitation of the electromagnet. Excitation of the electromagnet can be done by using either a DC power supply or a pulsed DC power supply. When DC power is used, the process is called *direct current magnetic abrasive finishing (DC-MAF)*, and the magnetic brush so formed is called a *static-flexible magnetic abrasive brush (S-FMAB)*. The process in which the medium is excited by pulsed DC is known as *pulse current magnetic abrasive finishing (PC-MAF)* and the brush so formed is termed a *pulsating flexible magnetic abrasive brush (P-FMAB)* (Singh, 2005).

8.2.1 Magnetic Abrasive Particles (MAPs)

The abrasive particles used in MAF are of different sizes and materials, such as silicon carbide (SiC) or alumina (Al_2O_3), which come in contact with and finish the part's surface. The abrasive and ferromagnetic particles are mixed with a small amount of metalworking fluid, such as mineral oil, SAE30 motor oil, or kerosene. The fluid helps in loosely bonding constituent particles, adds lubricity, and cools the part. These particles can be used in

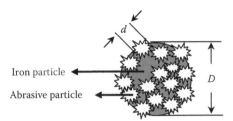

FIGURE 8.2
A bounded magnetic abrasive particle (D, diameter of the iron particle; d, diameter of the abrasive particle).

the form of either bounded magnetic abrasive particles (BMAPs, in which abrasives are held in a ferromagnetic matrix formed by sintering or other techniques) or unbounded magnetic abrasive particles (UMAPs, a homogeneous mechanical mixture of ferromagnetic particles, abrasive particles, and additives). Figure 8.2 shows a schematic diagram of a micron-sized BMAP (where D is the diameter of a single magnetic particle and d is the diameter of an abrasive particle held by an iron particle). In the case of BMAPs, the abrasive particles become an integral part of the magnetic material, and hence the movement of the abrasive particle becomes constrained. In the latter case (UMAPs), under the influence of a magnetic field, the abrasive particles are loose and may move around freely, however constrained by the ferromagnetic particles holding them. It has been reported by Fox et al. (1994) that the material removal rate (MRR) with UMAPs is higher than that with BMAPs. The higher removal rate is attributed to the availability of free abrasives that can scratch deeper than the bonded abrasives. The material removal mechanism also appears to be different in the case of BMAPs and UMAPs. When using this technique for finishing, one should use UMAPs if the initial finish is poor. This will enable a higher removal rate, which can be followed by finishing using BMAPs. If the initial surface is semifinished (e.g., a ground surface), then also BMAPs would be preferable for nanofinishing, as shown in Figure 8.3.

The working gap is filled with a homogeneous mixture of the medium. The effect of the mesh size of the particles on the decrease in surface roughness (Ra) is less compared to that of the current and working gap in DC-MAF (Singh et al., 2004). Hence, silicon carbide abrasive (mesh size of 1000), oil, and ferromagnetic particles are homogeneously mixed in a predetermined ratio by weight.

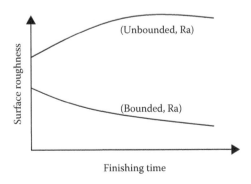

FIGURE 8.3
Variation of the surface roughness (Ra) with finishing time for BMAPs and UMAPs.

8.2.2 Finishing of Cylindrical Surfaces

Figure 8.1a shows the schematic diagram of an MAF setup having an electromagnet, power supply, and a dynamometer interfaced with PC through a data acquisition (DAQ) card (Singh, 2005). The salient features of this process are the use of a controllable magnetic field (or magnetic force) and the flexibility of the brush to adapt to the contour of the workpiece surface to be finished. The flexible nature of the brush permits finishing components that are otherwise difficult to finish, such as inside of flexible pipes and complex bent tubes (Yamaguchi and Shinmura, 2004).

Figure 8.4 is the schematic illustration of the principle of the finishing of an external cylindrical surface. Here, the workpiece rotates and vibrates along its axis. The MAPs are filled on both sides in the gap between the workpiece and the magnetic poles. Yamaguchi and Shinmura (2004) have achieved a surface roughness of 0.02 mm from an initial surface roughness of 2.4 mm of the inner surface of ceramic tubes. Experimental results, achieved using magnetic-field-assisted finishing processes, are reported by Komanduri (1996), Komanduri et al. (1994, 1997), Tani and Kawata (1984), Umehara et al. (1995), Khairy (2001), and Singh et al. (2005). Stowers et al. (1988) have also compared surface roughness and removal rate achievable by various advanced processes such as float polishing, CNC polishing, elastic emission machining, magnetic field assisted fine finishing processes, etc.

8.2.3 Finishing of Flat Surfaces

A magnetic abrasive nanofinishing setup designed and fabricated in-house for plane surfaces, as shown in Figure 8.5a, was used (Singh, 2005) to carry out an experimental study. The soft FMAB formed during the finishing process is shown in Figure 8.5b. The schematic diagram for a similar type of setup is shown in Figure 8.1. A round, flat-faced electromagnet with a ferromagnetic center pole with an electrical coil surrounded by an outer shell was designed. When activated by a power supply, the magnetic field in the center pole radially emanates and returns to the outer shell with minimum leakage of the field. The end result is a strong, concentrated magnetic field at the center pole and an evenly distributed weak magnetic field at the outer shell. The magnetic flux (Φ) is generated by the coil, and a part of it passes through the ferromagnetic workpiece, completing the magnetic circuit. The magnetic field generates an effective magnetic force to develop abrasion pressure for nanofinishing. The magnetic flux density and current supplied to the electromagnet are almost related to each other in the machining gap within the range of parameters shown in Figure 8.6.

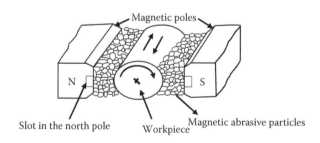

FIGURE 8.4
Schematic diagram of the principle of external finishing of a cylindrical workpiece by MAF.

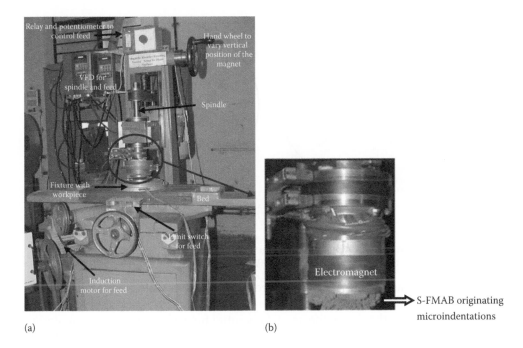

(a) (b)

FIGURE 8.5
Photograph of (a) the experimental setup and (b) a flexible magnetic abrasive brush (FMAB) in the machining gap.

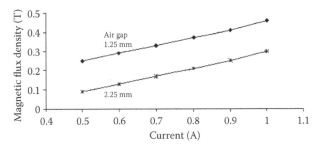

FIGURE 8.6
Relationship between magnetic flux density and current in the electromagnet coil at different air gaps between the midportion of the central pole and the ferromagnetic workpiece (alloy steel). (From Singh, D.K., Experimental investigations into magnetic abrasive finishing of plane surfaces, PhD Thesis, Indian Institute of Technology Kanpur, India, 2005.)

It is the magnetic force (F_m) that plays a dominant role in the formation and strength of the FMAB, and it initiates micro/nanoindentation of abrasive particles into the workpiece material.

The iron particles in UMAPs are attracted toward each other along the magnetic lines of force because of dipole–dipole interaction and form an FMAB (Figure 8.5b) in the machining gap. If a sculptured surface has gentle curvatures (concave or convex), the surface can be finished with a CNC (computer numerical control) MAF machine because of the adaptability of the FMAB. However, the finishing rates may be slightly different in each zone, depending on the working gap in that zone. The magnetic force acting on the ferromagnetic abrasive particle is transferred to the workpiece (or through the abrasive particles or

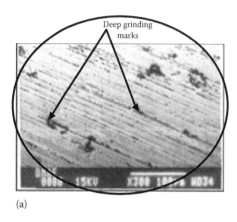
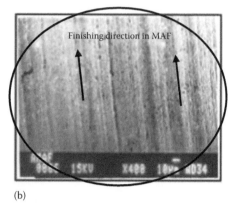

(a) (b)

FIGURE 8.7
Scanning electron microscopic views of a workpiece (a) before MAF and (b) after MAF, at voltage = 11.5 V, machining gap = 1.25 mm, rpm = 180, mesh number = 800, machining time = 20 min.

iron particles in the case of UMAPs). This force can be resolved into two components: one in the normal direction and the other in the radial direction. The first is responsible for penetration, while the other helps the axial/tangential force in the removal of material in the form of micro/nanochips. When the strength of the FMAB is such that it is just equal to the total resistance offered by the workpiece for deformation, it is called the *equilibrium condition*. When the normal force and cutting force (F_c) are greater than the resistance offered for deformation by the work material, the abrasive particles remove material from the workpiece. Figure 8.7 shows scanning electron microscopic (SEM) views of a ground surface having closely spaced feed marks left by the grinding operation. Figure 8.7b shows the parallel stripes generated by the rotation of the FMAB. It shows that the width of the scratches on the workpiece lies in the range 3–14 μm and that they are in the direction of rotation of the FMAB (Singh et al., 2004, 2005a).

Figure 8.8a illustrates the atomic force micrograph of the workpiece after grinding, showing that heights of peaks are approximately 500 nm. In Figure 8.8b, the peaks have been sheared off after MAF, resulting in heights in the range 100–200 nm. Smaller peaks indicate the area in which abrasive particles of the FMAB were held strongly and hence could reduce the peak heights more compared to the areas where the brush was comparatively weak. Better flatness and smoothness at micro/nanolevels has been achieved over the whole surface by giving translatory motion to the workpiece, as shown in Figure 8.8c (Singh et al., 2005b).

8.3 Allied and Hybrid MAF Processes

8.3.1 Magnetic Abrasive Deburring (MADe)

The quality of precision parts can be evaluated by the surface and edge qualities (Park et al., 2005). The quality of an edge is evaluated after the deburring process is over by assessing how well the deburring or microdeburring of the edge been carried out and what the final edge radius is (Gillespie LaRoux, 2008). Madarkar and Jain (2007) have investigated the effectiveness and validity of the magnetic abrasive deburring method for the microdeburring of drilled holes on brass material, as shown in Figure 8.9.

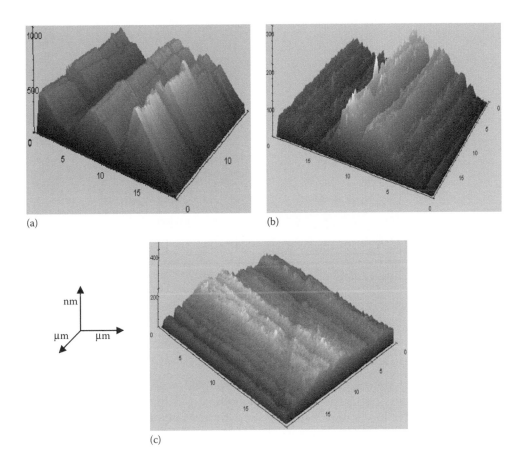

(a)   (b)

(c)

FIGURE 8.8
Atomic force micrographs of a workpiece (a) after grinding; (b) after MAF at current = 0.88 A, machining gap = 1.25 mm, rpm = 90, lubricant = 2%, time = 30 min; (c) after MAF at current = 0.75 A, machining gap = 1.75 mm, grain mesh number = 800. (From Singh, D.K., et al., *Wear*, 259 (7–12), 1254–1261, 2005b.)

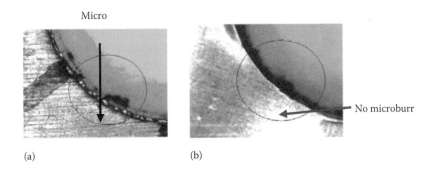

(a)   (b)

FIGURE 8.9
(a) A hole edge having microburrs before deburring and (b) after magnetic abrasive deburring (From Madarkar, R. and Jain, V.K., Investigation into magnetic abrasive micro deburring. *Proceedings of the 5th International Conference on Precision*, Dec 13–14, 307–312, 2007.)

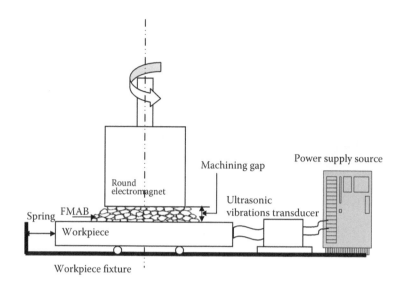

FIGURE 8.10
Schematic diagram of an ultrasonic-assisted plane MAF setup.

8.3.2 Ultrasonic-Assisted Magnetic Abrasive Finishing (UAMAF)

UAMAF is a hybrid process in which MAF is coupled with ultrasonic vibration. It (Figure 8.10) combines the use of ultrasonic vibration and the MAF process to finish surfaces to nanometer levels in a relatively short time. Malik and Pandey (2011) found experimentally that the UAMAF process gives a better performance compared to the simple MAF process for the same input conditions. The surface roughness value obtained by UAMAF was as low as 22 nm on a hardened AISI 52100 steel workpiece using UMAP (SiC abrasives). Shaohui and Shinmurra (2004) developed a vibration-assisted MAF process in which they reported the effects of three modes of vibration (X-mode, Z-mode, and compound mode) on the polishing characteristics.

8.3.3 Electrolytic Magnetic Abrasive Finishing (EMAF)

The EMAF system (Yan et al., 2003) is a new hybrid process in which electrochemical machining (ECM) and MAF are combined together. The setup consists of an electrolyte supply subsystem, a DC power source, and electrodes. A pair of electromagnetic/permanent magnetic poles and ferromagnetic abrasive particles are used as in MAF. During processing, the electrode gap between the electrode and the workpiece is filled with the electrolyte, and the workpiece is connected to the positive terminal (anode), while the tool electrode is connected to the negative terminal (cathode) of the DC power source. A passive film is produced on the work surface during electrolytic processing because of the electrochemical reaction. The softness of the passive film makes removal by MAF easier as compared to the original metal surface. EMAF has a better finishing performance than simple MAF, especially at a high electrolytic current.

The parameters, namely, electrode gap, magnetic flux density, and electrolytic current, must be appropriately selected for fast production of a passive film. The rate of workpiece rotation must also be matched to the rate of passive film formation to remove it rapidly and efficiently to obtain a superior surface in EMAF. Under the conditions of high electrolytic current and low rate of workpiece rotation, some steel grits and abrasives come out of the working gap of MAF, enter the interelectrode gap during electrolytic machining, and retard the electrolytic dissolution process.

Increasing both the electrolytic current and the rate of workpiece revolution increases finishing efficiency, and the surface roughness improves rapidly. However, a high electrolytic current severely disturbs the distribution of the existing magnetic field and facilitates the extraction of the steel grit from the working gap. Therefore, the use of high electrolytic current is limited.

8.3.4 Ultrasonic- and Magnetic-Field-Assisted Electrochemical Micromachining

Pa (2009) proposed an ultrasonic- and magnetic-field-assisted electrochemical micro-machining system (USM + MAF + ECM). It is a superfinishing hybrid process that uses ultrasonic and magnetic fields to assist in the discharging of dregs from the interelectrode gap during electrochemical micromachining. The selected experimental parameters are frequency and power level of the ultrasonic field, magnetic field strength, distance between two magnets, rotational speed of the magnetic system, current density, on/off period of the pulsed current, and electrode geometry.

8.4 Pulsating Current Magnetic Abrasive Finishing (PC-MAF)

Two basic problems encountered during finishing with the DC power supply to the electromagnet are that the heating of the coil may damage the magnet and that the abrasives held in the FMAB may not get stirred/refreshed/mixed. If by some means these abrasive particles that are intact in the S-FMAB can be refreshed or mixed within the brush without recharging, the finishing rate (μm/s) would be enhanced. Hence, the use of pulse power supply to energize the electromagnet was considered to study the effect of controlled formation and partial destruction of the magnetic abrasive brush in the finishing zone (Jain et al., 2008).

The direct current magnetic abrasive microfinishing process hardly provides any stirring effect on the S-FMAB. As a result, the surface obtained has a comparatively non-uniform finish (Figure 8.8b). Continuous use of the S-FMAB and the absence of stirring lead to dullness of the cutting edges of abrasive particles, resulting in a low finishing rate. To overcome this weakness of the process, the FMAB has been made pulsating using a pulsed DC power supply, and the process has been termed as *pulsed current magnetic abrasive finishing*. Kurobe and Imanaka (1984) conducted experiments on a glass workpiece by periodically turning on/off the DC power supply to the electromagnet and compared the material removal achieved with that obtained during DC-MAF. The on and off periods were taken as equal in magnitude. Material removal for a longer pulse on-time was found to be higher compared to that during DC-MAF. But the on/off control of DC power supply in this manner was inaccurate and difficult. Yamaguchi et al. (2003) developed a precision internal finishing process using an alternating magnetic field and studied its effects on the finishing characteristics. They concluded that the extent of plastic deformation controls surface integrity, including increase in hardness and compressive residual stresses.

The schematic diagram of a PC-MAF setup is shown in Figure 8.1a. The setup used in these experiments is the same as that for DC-MAF except that the DC power supply is replaced by a pulsed power supply to the electromagnet coils. A nominal resistance

(0.37 Ω) is connected in series to the coil (R = 12.0 Ω) of the magnet to measure the pulse current across it using a two-channel digital storage oscilloscope (DSO). The ferromagnetic abrasive particles of the FMAB line up along the magnetic lines of force. When pulsed DC is supplied to the electromagnet, the shape and size of the FMAB change, to some extent, with the current pulses. Hence, the FMAB under these conditions is called *pulsating flexible magnetic abrasive brush (P-FMAB).*

The average pulse voltage across the electromagnet is calculated as

$$\text{Avg. voltage} = \text{Duty cycle} \times \text{Pulse generator voltage}$$

where the duty cycle (G) is defined as

$$\Gamma = \frac{t_{\text{on}}}{t_{\text{on}} + t_{\text{off}}}$$

here, t_{on} is the on-time, and t_{off} is the off-time of a pulse.

To understand the advantage of PC-MAF over DC-MAF, let us understand the effect of the duty cycle on the percentage change in surface roughness:

$$\%\Delta\ \text{Ra} = \left(\frac{\text{Initial Ra} - \text{Final Ra}}{\text{Initial Ra}}\right) \times 100$$

8.4.1 Effect of Duty Cycle on Percentage Decrease in Surface Roughness

The relationship between percentage decrease in Ra and the duty cycle (on-time = 2.0 ms) has been established as shown in Figure 8.11. It shows that as the duty cycle decreases for a constant on-time, the percentage decrease in surface roughness (% ΔRa) increases significantly. It can be seen from Figure 8.12a and b that as the pulse off-time starts,

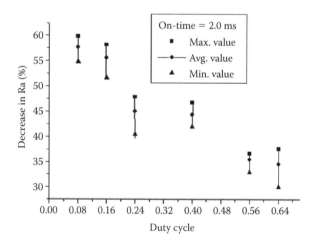

FIGURE 8.11
Relationship between percentage decrease in Ra and duty cycle at on-time = 2.0 ms, working gap = 1.50 mm, rpm = 200, grain mesh number = 1000, finishing time = 15 min.

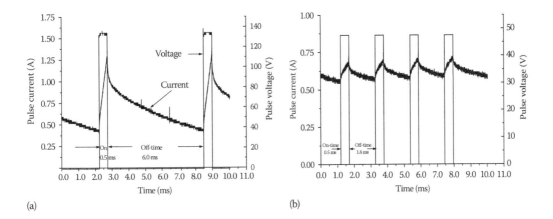

FIGURE 8.12
Variation of current and voltage (a) at duty cycle = 0.08, on-time = 0.5 ms, off-time = 6.0 ms and (b) at duty cycle = 0.24, on-time = 0.5 ms, off-time = 1.6 ms.

the current in the electromagnet starts falling, and before it attains zero value a fresh current pulse starts and again attains a rising trend. The FMAB becomes weaker as the current starts falling. During on-time when the magnetic force is high owing to peak current (Figure 8.12), the depth of indentation (t_h) is high. What happens to this depth of indentation (t_h) when the magnetic force becomes low (or a weak FMAB) during pulse off-time? Because of the weakening of the FMAB during off-time and depending on the depth of indentation attained during on-time of a pulse, some abrasive particles may not be able to withstand the resistance offered by the workpiece for material removal. This situation would lead to either rotation of the abrasive particles or jumbling of the ferromagnetic and abrasive particles in the case of UMAPs. What happens during a cycle is governed by the ratio of the peak current during on-time to the minimum current during off-time (Figure 8.12), which affects the ratio of the normal force and the cutting force (Jain et al., 2008).

Optical microscopic views of the magnetic abrasive finished workpiece by the P-FMAB are shown in Figure 8.13a and b. Qualitatively, it is obvious from the photographs that with a lower duty cycle, the surface finish obtained is better, and thus the deep grinding marks can be removed more effectively by rotating the P-FMAB.

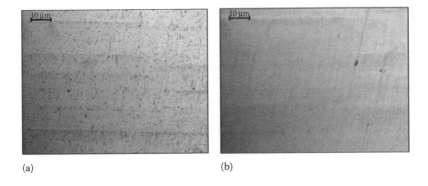

FIGURE 8.13
Optical microscopic views obtained at on-time = 2.0 ms, working gap = 1.50 mm, rpm = 200, abrasive mesh size = 1000, finishing time = 15.0 min. A finished workpiece by a pulsating FMAB at duty cycle (a) 0.40 and (b) 0.08.

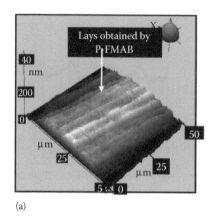 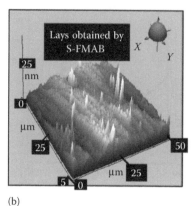

(a)   (b)

FIGURE 8.14
AFM images of a surface obtained by (a) pulsating FMAB at duty cycle = 0.08, on-time = 2 ms, gap = 1.5 mm, rpm = 200, finishing time = 15.0 min, abrasive mesh no. = 1000, feed = 0; (b) static FMAB at duty cycle = 1.0 (smooth DC), current = 0.75 A, gap = 1.25 mm, rpm = 180, finishing time = 45.0 min, abrasive mesh no. = 1000, feed = 8.0 mm/min.

The percentage decrease in Ra alone does not present a complete picture of the capabilities of abrasive finishing. Therefore, to study the surface texture at micro/nanolevels, atomic force microscopic (AFM) images (Figure 8.14a and b) of the workpiece after finishing were taken to understand and compare the generated surface texture. The atomic force micrograph of the ground surface (Figure 8.8a) shows periodic peaks and valleys, whose characteristics are a measure of the surface roughness value. The grinding lays shown in Figure 8.8a have been replaced by very fine lays obtained by P-FMAB at a 0.08 duty cycle (Figure 8.14a). The AFM image of the surface texture obtained by S-FMAB (Figure 8.14b) is quite different from the surface texture of the surface obtained by P-FMAB (Jain et al., 2008). At isolated locations, some peaks are quite high compared to others because the distribution of abrasive particles in the FMAB does not seem to be homogeneous and remixing of abrasive and ferromagnetic particles does not occur as in the case of a P-FMAB. Further, the finishing time to obtain this surface by S-FMAB is almost three times more than that by P-FMAB for the same level of surface roughness.

8.4.2 Forces and Brush Formation during PC-MAF

In the case of UMAPs, a magnetic force keeps abrasive particles intact in the working zone and develops a finishing pressure. To maintain a constant duty cycle, an increase in the on-time also requires an increased off-time. In a particular case of high off-time, the value of the current changes from a maximum of 0.70 A to a minimum of 0.38 A. This current ratio (maximum current in a pulse/minimum current in the same pulse) seems to be responsible for the intermixing of the UMAPs. For a constant duty cycle and a low on-time, the value of the pulse frequency is high. The magnetic force fluctuates more at a lower duty cycle than at a higher duty cycle. Figure 8.15a shows the variation of the forces at a low duty cycle, and Figure 8.15b shows the same with smooth DC (duty cycle = 1.0). High current results in high normal magnetic force, and hence high depth of indentation. However, soon enough, a decrease in the current [during off-time (Figure 8.12a)] leads to a lower strength of the FMAB, and it partially breaks. Again, during on-time when a new FMAB is formed, some fresh abrasive particles come in contact with the workpiece and

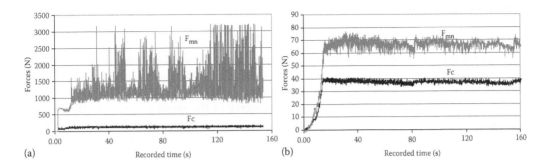

(a) Recorded time (s) (b) Recorded time (s)

FIGURE 8.15
Forces at working gap = 1.50 mm, on-time = 2.0 ms, rpm = 200, finishing time = 15.0 min (a) duty cycle = 0.08; (b) duty cycle = 1.00, current (smooth DC) = 0.75 A, working gap = 1.25 mm, rpm = 180.

the FMAB again gives a better performance. Furthermore, some active abrasive particles that have microindented substantially may rotate and reduce the depth of indentation to remove material or separate from the brush without removing any material. This is how the intermixing of UMAPs as well as BMAPs during pulse power supply takes place. The higher the current ratio, the better is the mixing, and hence the higher the percentage decrease in Ra. This is why the highest improvement in surface roughness is attained at 8% duty cycle (Figure 8.11).

As discussed above, magnetization and demagnetization of the FMAB take place during P-FMAB. The frequency of forming and breaking of the brush becomes almost zero as the duty cycle approaches close to 1, that is, the smooth DC. The fluctuation in normal magnetic force keeps reducing as the duty cycle increases until it becomes a smooth DC. The value of the peak current also goes up in the case of a lower duty cycle (Figure 8.12a). That is why the magnitude of a magnetic force with a lower duty cycle is higher (Figure 8.15).

A model to explain the formation of the FMAB during on-time and its partial breaking during off-time is shown in Figure 8.16. During on-time, the electromagnet gets energized and forms a strong FMAB, whereas during off-time, its strength falls but not to zero. This is why a part of the brush remains intact. The periodic formation and partial breaking of the FMAB promote stirring within itself, resulting in the interaction of fresh cutting edges with the workpiece. Hence, the surface roughness is significantly reduced or the percentage decrease in Ra is substantially increased at a lower duty cycle.

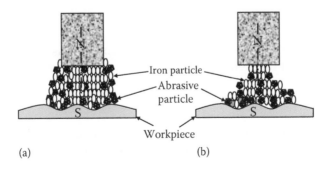

(a) (b)

FIGURE 8.16
Schematic illustration of formation and partial breaking of an FMAB. N, north pole; S, south pole. (a) Formation of FMAB during on-time and (b) breaking/falling of FMAB during off-time.

8.5 Finite Element Analysis of MAF Process

It is important to understand the mechanism of material removal and the modeling of surface finish during MAF at micro/nanolevels. The effects of various process parameters on MRR and surface quality have been evaluated (Jayswal et al., 2005b). Analysis of the MAF process is difficult because of the irregular geometry of the working gap and the heterogeneous properties of the medium. Numerical analysis techniques are usually employed to solve problems that are difficult to solve with analytical methods. Among various numerical techniques available, the finite element method (FEM) and finite difference (FD) method are normally employed to solve such engineering problems. FEM provides more accurate solutions to problems with complex geometries and nonhomogeneous material properties. Therefore, FEM is used to understand the MAF process in depth.

In FEM, the governing equations of the process are solved in terms of primary variables, namely, magnetic potential, intensity of the magnetic field, and magnetic forces at the interface of the FMAB and the workpiece surface. Figure 8.17 shows a schematic diagram of the front and top views of the MAF process. An enlarged view of the abrasive–workpiece interface shows the magnetic force (F_m) acting on the workpiece through the MAP. The normal component (F_{mn}) is responsible for the penetration of the abrasive particle into the workpiece, and the tangential component (F_{mt}) and the mechanical force due to its (FMAB) rotation are responsible for removal of material in the form of microchips.

The MAF process is based on the electromagnetic behavior of MAPs. MAPs acquire magnetic polarization and join each other along the lines of the magnetic force under the influence of the magnetic field, forming an FMAB (Figure 8.17a). Knowledge of the distribution of magnetic forces on the workpiece surface is essential to determine the surface quality to be produced. To evaluate the distribution of magnetic forces, the governing equation of the process is expressed in terms of the magnetic potential.

The electromagnetic field can be described in terms of the electric field vector (**E**) and the magnetic field vector (**H**). These vectors are related to the electric (**D**) and magnetic (**B**) flux densities, as well as the field source and electric current density **J**. Their interrelationships are expressed in the classical Maxwell's equations (Jefimenko, 1966):

$$\nabla \times \mathbf{E} = -\frac{\partial \mathbf{B}}{\partial t}$$

$$\nabla \times \mathbf{H} = \mathbf{J} + \frac{\partial \mathbf{D}}{\partial t}$$

$$\nabla \bullet \mathbf{B} = 0$$

There is no current source in the working gap. This process is assumed to be steady, and therefore the intensity of magnetic field **H** can be expressed as a gradient of the magnetic scalar potential ϕ:

$$\mathbf{H} = -\nabla \phi$$

(8.1)

On the basis of certain assumptions (Jayswal et al., 2005a) and considering the axisymmetric form of the problem, the governing equation of the process becomes

$$\frac{1}{r}\frac{\partial}{\partial r}\left[r\mu_r \frac{\partial \phi}{\partial r}\right] + \frac{\partial}{\partial z}\left[\mu_r \frac{\partial \phi}{\partial z}\right] = 0$$

(8.2)

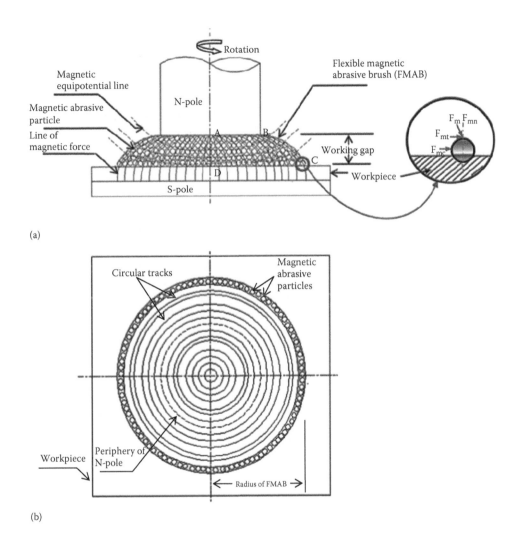

(a)

(b)

FIGURE 8.17
Magnetic abrasive finishing setup: (a) front and (b) top views.

where ϕ is a magnetic scalar potential and μ_r is the relative permeability of MAPs. MAPs are composed of ferromagnetic and abrasive materials. The permeability of most ferromagnetic materials is not constant and varies with the field. This implies that Equation 8.2 is nonlinear.

8.5.1 Application of Finite Element Method

In brief, the various steps followed during FEM are described as follows:

1. Conversion of a set of governing equations into an equivalent system of integral equations
2. Descretization of the solution domain into several small elements of chosen shape and size

3. Approximation of the unknown field variables in terms of assumed shape functions at the specified points, known as *nodes*

4. Evaluation of the integral governing the equations for each element

5. Assembly of all elemental integrals to obtain final matrix equations

6. Solution of the final matrix equations for the entire domain to get values of the primary variable at the nodes

For the conversion of the governing partial differential equations into equivalent integral equations, Galerkin's method (Reddy, 1993), which involves two basic steps, is used.

1. Assume an approximation function for the dependable field variables involving known shape functions and unknown nodal values of field variables. Substitution of this approximation into the differential equations and boundary conditions results in some error at all locations, which is known as the *residue*.

2. Now, minimize this residue over the entire domain so as to obtain an approximate solution. The solution domain (Figure 8.18) is axisymmetric in the present case, and only the right half (ABCD in Figure 8.17a) is used as a solution domain for the analysis. The solution domain is divided into two parts, namely, the main zone (ABED) and the fringing zone (BCE). From the computational experiments, it is concluded that the length of the fringing zone (distance EC) is approximately two times the working gap (EC $\approx$ 2AD). The total boundary consists of four parts: C_1, C_2, C_3, and C_4. The essential boundary conditions to be specified on the boundaries C_1 (ϕ_w) and C_3 (ϕ_n) are

$$\phi_w = \frac{B_n A_n l_{tw}}{\mu_0 \mu_{rw} A_w} \qquad \text{on } C_1$$

$$\phi_n = \frac{B_n l_g}{\mu_0} \qquad \text{on } C_3$$

where B_n is the magnetic flux density at the N-pole, l_g is the working gap, μ_0 is the permeability of free space, l_{tw} is the thickness of the workpiece, A_w is the cross-sectional area of the workpiece, A_n is the cross-sectional area of the N-pole, and μ_{rw} is the relative permeability of the workpiece.

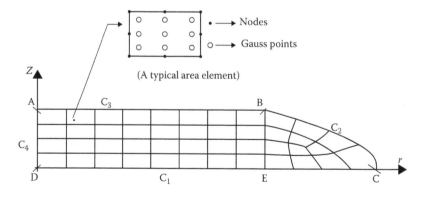

FIGURE 8.18
Finite element mesh of the domain of interest.

The natural boundary conditions contain derivatives of an unknown function of the primary variable. These are as follows:

$$\frac{\partial \phi}{\partial n} = 0 \quad \text{on } C_2 \text{ and } C_4$$

Galerkin's method is used to convert the governing differential equation (Equation 8.2) into algebraic equations by using appropriate weight functions. The residue (I) can be evaluated as follows:

$$I = -\int_A \mu_r \left(\frac{\partial \tilde{\phi}}{\partial r} \frac{\partial w}{\partial r} + \frac{\partial \tilde{\phi}}{\partial z} \frac{\partial w}{\partial z} \right) 2\pi r \, dr \, dz \tag{8.3}$$

where w is the weight function, which belongs to a set of arbitrary but known functions. The function ϕ is an approximation to the primary variable that satisfies the essential boundary conditions exactly. In vector form, Equation 8.3 can be written as

$$I = -\int_A \mu_r \{\nabla w\}^T \{\nabla \tilde{\phi}\} \, 2\pi r \, dr \, dz \tag{8.4}$$

In the present analysis, an eight-noded quadrilateral element is used. To obtain the finite element equation from Equation 8.4, the domain is discretized into ne number of elements (Figure 8.18). Further, over a typical element, the magnetic potential is approximated ($\tilde{\phi}$) using the unknown nodal values (ϕ_i^e) and known shape functions N_i^e. Thus,

$$\tilde{\phi} = \sum_{i=1}^{8} N_i^e \phi_i^e = \{N\}^{eT} \{\phi\}^e = \{\phi\}^{eT} \{N\}^e \tag{8.5}$$

where $\{N\}^e$ is the vector of the shape function and $\{\phi\}^e$ is the vector of the elemental degree of freedom.

In Galerkin's formulation, the weight functions w are expressed using the same shape functions as that for the primary variable ϕ. $\{\nabla \tilde{\phi}\}$ and $\{\nabla w\}^T$ can be written as

$$\{\nabla \tilde{\phi}\} = [B]^e \{\phi\}^e \tag{8.6}$$

$$\{\nabla w\}^T = \{w\}^{eT} [B]^{eT} \tag{8.7}$$

where matrix $[B]^e$ contains derivatives of the shape functions.

After substituting Equations 8.6 and 8.7 into Equation 8.4, we get

$$I = \sum_{e=1}^{ne} \int_{A_e} \mu_r \{w\}^{eT} [B]^{eT} [B]^e \{\phi\}^e \, 2\pi r \, dr \, dz \tag{8.8}$$

where ne is the number of elements and A_e is the domain of integration for a typical element.

The residue is made small by setting the residual expression (*I*) to 0. It can be expressed in terms of the global coefficient matrix [K] and global magnetic potential vector {ϕ}:

$$[K]\{\phi\} = 0$$

(8.9)

The value of [K] is obtained after assembling the elemental coefficient matrices over all the elements. The typical elemental coefficient matrix [K]^e is given by

$$[K]^e = \int_{A_e} \mu_r [B]^{eT} [B]^e \, 2\pi r \, dr \, dz$$

(8.10)

All the elemental matrices are evaluated using the Gauss–Legendre quadrature with three Gauss points (Figure 8.18) in each direction. The value of μ_r of the MAPs depends on the field intensity **H**, which, in turn, depends on the magnetic scalar potential. Therefore, the values of μ_r at the Gauss points are found iteratively. In every iteration, these equations are solved by the Gauss elimination method (Balagurswamy, 2000) after imposing the essential boundary conditions.

8.5.1.1 Evaluation of Secondary Variables

The solution of the problem discussed above gives nodal values of the magnetic potential. From these nodal values, the secondary variables, namely, intensity of the magnetic field and the magnetic forces, are computed as follows.

The *r* and *z* components of the intensity of the magnetic field (**H**) are expressed as

$$H_r = -\frac{\partial \phi}{\partial r}$$

(8.11)

$$H_z = -\frac{\partial \phi}{\partial z}$$

(8.12)

Derivatives of H_r and H_z are evaluated using the FD method.

The volume of an MAP is very small (micron-sized particles), hence the magnetization **M** and the magnetic field **H** can be considered as uniform throughout its volume. Then the force **F** can expressed as (Stradling, 1993)

$$\mathbf{F} = \frac{\mu_0}{2} v \, \nabla (\mathbf{M} \bullet \mathbf{H})$$

(8.13)

For magnetic particles,

$$\mathbf{M} = \chi_r \mathbf{H}$$

(8.14)

where χ_r is the susceptibility of the MAPs. Substituting the value of **M** from Equation 8.14 in Equation 8.13, and resolving **F** into the radial and axial directions, Equation 8.13 becomes

$$F_r = \frac{\mu_0}{2} v \frac{\partial}{\partial r} (\chi_r \mathbf{H} \bullet \mathbf{H})$$

(8.15)

$$F_z = \frac{\mu_0}{2} v \frac{\partial}{\partial z} (\chi_r \mathbf{H} \bullet \mathbf{H})$$

(8.16)

where *v* is the volume of the MAP.

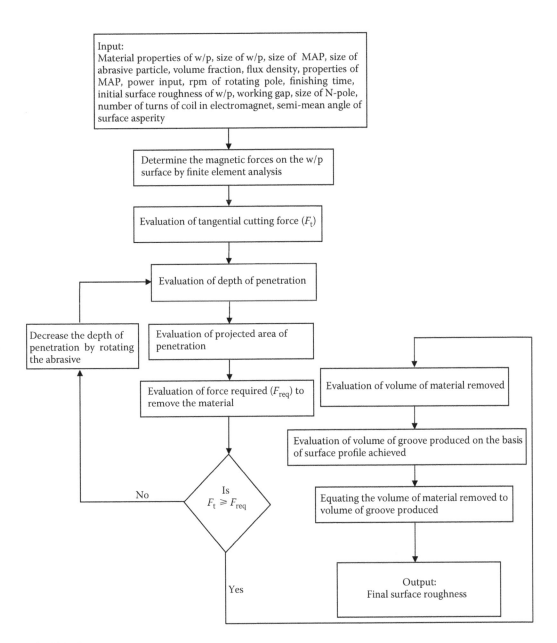

FIGURE 8.19
Flowchart for determination of surface roughness during MAF simulation.

Combining the various steps of the calculation of forces, a full system can be developed to calculate the surface roughness value of the component finished by the MAF process. The sequence of these steps and other details are given in the flowchart in Figure 8.19.

8.5.2 Modeling for Evaluation of Material Removal and Surface Roughness during MAF

In MAF, an FMAB behaves like a multipoint cutting tool for the finishing operation. In MAF, the FMAB rotates and acts as a flexible grinding wheel. It is usually assumed that there is no slip between the rotating N-pole and the FMAB. During analysis, the assumed arrangement of MAPs is shown in the last track of Figure 8.17b.

Material removal takes place as a result of the mechanical interaction of the cutting edges of the abrasive particles with the workpiece surface, leading to microcutting (Figure 8.1b). The volume of the chips produced is equal to the volume of grooves formed on the workpiece surface during the MAF process. Magnetic and mechanical energies are utilized in the MAF process. The electric motor rotates the FMAB in the finishing zone. It generates the tangential force on the cutting edges of MAPs. There is also a mechanical force in the radial direction acting on a cutting particle. The net radial force (mechanical and magnetic) has to provide radial acceleration. The magnetic force acting on the MAPs can be resolved in the normal and radial directions. The normal magnetic force creates a compressive reaction on the surface of the workpiece, which is responsible for the penetration of cutting edges into the workpiece. The mechanical tangential force on the cutting edges of MAPs removes material from the workpiece.

The tangential force (F_t) on a cutting edge (Jayswal et al., 2007) can be evaluated by

$$F_t = \frac{P_{act}}{\left(\sum_{tr=1}^{n_t} \frac{4\pi^2 R_{(tr)}^2 N_{rs} n_a}{D_{map}} \right)}$$

(8.17)

where P_{act} is the finishing power available at the N-pole, $R_{(tr)}$ is the radius of the tr^{th} track, N_{rs} is the rotational speed of the magnetic pole, D_{map} is the diameter of the MAP, n_t is the total number of tracks, and n_a is the number of active cutting edges on the MAP.

The force required (F_{req}) to remove material from the workpiece depends on the shear strength (τ_s) of the workpiece material and the projected area of penetration (A_p in Figure 8.20). Thus,

$$F_{req} = \tau_s A_p$$

(8.18)

If the cutting force is smaller than the resistance offered for deformation by the workpiece (which depends on the depth of microindentation), the abrasive particle may rotate (Jayswal et al., 2005a) without removing any material, leading to a reduced depth of indentation (Figure 8.20). This reduction in the depth of indentation will be to the extent that this particle just starts removing the material.

Thus, during the MAF process, one of the following three situations may occur:

1. $F_{req} = F_t$ (8.19a)

    This is the equilibrium condition. It indicates the start of the finishing operation.

2. $F_{req} < F_t$ (8.19b)

    Under this condition only, material is removed.

3. $F_{req} > F_t$ (8.19c)

This shows the no-cutting condition due to larger depth of penetration. Under this condition, the depth of penetration of a cutting edge is adjusted by the rotation of the

MAP (Figure 8.20) until the force required for cutting becomes equal to or less than the force available. Thus,

$$\overline{F}_{req} \leq F_t \tag{8.19d}$$

where $\overline{F}_{req}$ is the modified required force for cutting at the reduced depth of penetration hs' as shown in Figure 8.20, and it is given by

$$\overline{F}_{req} = \tau_s A'_p \tag{8.20}$$

where A'_p is the modified projected area of penetration of an abrasive particle after reduction of depth of penetration to hs'.

The tool (FMAB) finishes the desired area of the workpiece by providing feed to the workpiece in the x- and y directions. Owing to the rotation of the MAPs, circular groove tracks (Figure 8.17a) are formed on the workpiece surface. Surface roughness is determined on the basis of the surface profile generated by equating the volume of material removed to the volume of grooves produced. The volume of material removed depends on the projected area of penetration and the length of contact of the MAPs with the surface profile. The volume of grooves produced is evaluated from the geometry of the surface profile after finishing. The workpiece surface is divided into small square cells. Each cell is specified by the Cartesian coordinates of its center. The surface roughness in a cell (i,j) after the nth revolution is expressed as (Jayswal et al., 2005a)

$$R_{max(i,j)}^{(n)} = R_{max}^0 - R_{max}^0 - \left[R_{max(i,j)}^{(n1)}{}^2 + \left(\frac{\Delta V_{(i,j)}^{(n)} R_{max}^0 D_{map}}{\pi r_{(tr)} l_c R_{(tr)} n_a} \right) \right]^{\frac{1}{2}} \tag{8.21}$$

where $R_{max(i,j)}^{(n)}$ is the maximum surface roughness after the nth revolution, $\Delta V_{(i,j)}^{(n)}$ is the volume of material removed in the nth revolution, D_{map} is the diameter of the MAP, l_c is

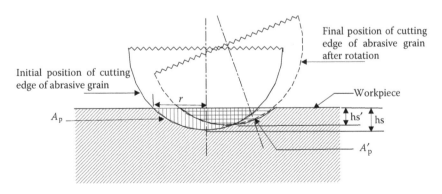

FIGURE 8.20
Rotation of a magnetic abrasive particle: hs, depth of penetration before rotation; hs', depth of penetration after rotation; A_p, projected area of penetration before rotation; A'_p, projected area of penetration after rotation of the abrasive particle.

the length of the cell, n_a is the number of active cutting edges on the MAP, $r_{(tr)}$ is the radius of the projected area of indentation, and $R_{(tr)}$ is the radius of the track. After finishing for a definite time, the reduction in surface roughness value in one cycle $\left(\frac{(\text{Initial Ra})_i - (\text{Final Ra})_{i+1}}{\text{Finishing cycle time}}\right)$ reaches a stage where the rate of decrease of surface roughness value attains a value approaching zero (here i indicates the ith cycle). This is called the *critical surface roughness value*.

The validity of the surface roughness model is demonstrated by comparing the simulated results with the experimental results of Shinmurra et al. (1985) using the same machining parameters and material properties. It is found that the simulated values in general are higher than the experimental results, but the trend is the same as that of the experimental results.

8.5.3 Parametric Analysis

A series of numerical experiments were performed (Jayswal et al., 2007) to investigate the effects of process parameters such as flux density, working gap, and size of MAPs on the finishing performance of the MAF process. Figure 8.21 shows the effects of these parameters on the reduction in surface roughness value (R_{max}) during plane MAF. It is evident that the surface roughness value (R_{max}) increases with increase in flux density (Figure 8.21a)

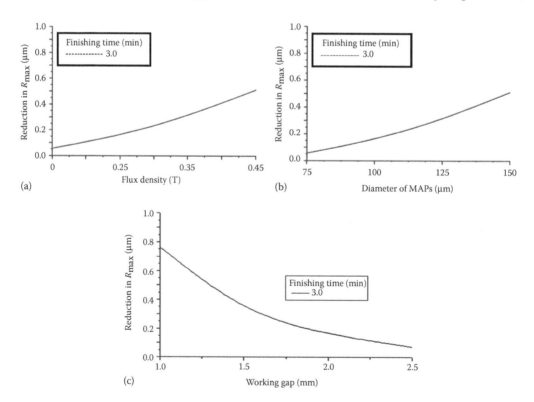

FIGURE 8.21

Effect of process parameters on the reduction in surface roughness value for volume fraction = 0.7, initial R_{max} = 1.5 μm, rotation speed = 200 rpm, working gap = 2 mm, diameter of MAP = 100 μm. (a) Effect of flux density on reduction in R_{max}; (b) effect of diameter of MAP on reduction in R_{max}; (c) effect of working gap at current input = 0.8 A.

and size of MAPs (Figure 8.21b) but decreases with increase in working gap (Figure 8.21c). Such trends are observed experimentally as well (Singh, 2005).

Most researchers have been using an electromagnet with a slot in it to improve the performance of the process, but hardly any information is available in the literature about how it affects the process performance. A magnet having a circular slot was considered (Jayswal et al., 2004) to evaluate its effects on the magnetic forces and surface quality produced during the MAF process.

The normal magnetic force is calculated at the interface of the MAPs and workpiece from Equation 8.16 using the machining conditions. The variation in magnetic force with the distance from the center of the tool is shown in Figure 8.22 for the cases with slot and without slot. It can be seen (Figure 8.22) that the magnitude of the force gets enhanced around the slot and the direction of the force changes to negative in the area under the slot. This means that no machining would occur in the region where the magnitude of the force is negative. However, in the area toward the center of the tool with a slot, the magnitude of the force acting on the particles is higher than the case where there is no slot on the tool (or N-pole). Further, the force (F_z) is still higher at the edges of the N-pole as compared to that anywhere else. Since, the tool covers the entire area of the workpiece to be finished (by giving feed in the x- and y directions) the use of a slotted pole, to some extent, becomes an advantage. Otherwise, the area of the workpiece under the slot

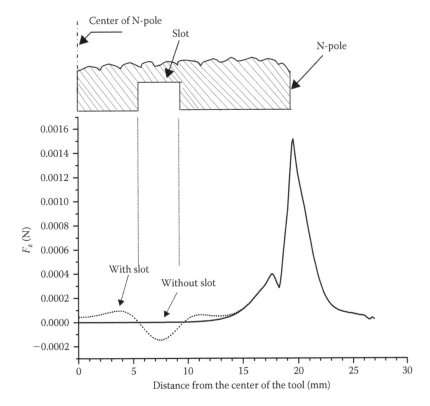

FIGURE 8.22
Computed variation of a normal magnetic force with distance from the center of the tool at workpiece gap = 4 mm, rotational speed of N-pole = 200 rpm, current input = 1.12 A, applied voltage = 15 V, diameter of MAP = 100 μm.

experiences a negative force; hence, no machining would take place in the region under the slot.

Surface roughness is computed for the cases of slotted and smooth surfaces of the pole. It has been found that a magnet with a slot gives better surface finish and higher finishing rate than one without a slot. To investigate the advantages of having more than one slot, numerical experiments have been performed (Jayswal, 2005) by taking two slots on the surface of the pole. It is found that the change in surface roughness is not significant with an increase in the number of slots. The optimum number of slots suggested is 1.

Nomenclature

A_n	Cross-sectional area of N-pole (m^2)
A_p	Projected area of the penetration (m^2)
A_w	Cross-sectional area of workpiece (m^2)
B	Magnetic flux intensity (T)
D	Electric flux density (C/m^2)
D_{map}	Diameter of MAP (m)
E	Electric field vector
E_m	Magnetic potential energy (J)
F_r	Radial force (N)
F_t	Tangential force (N)
F_z	Normal force (N)
H	Magnetic field intensity (A/m)
H_m	Hardness of workpiece material (N/m^2)
I	Input current to electromagnet (A)
J	Current density (A/m^2)
k^e	Elemental coefficient matrix
l_c	Length of cell (m)
l_g	Working gap (m)
l_{tw}	Thickness of workpiece (m)
M	Magnetization of magnetic abrasive particle (A/m)
n_a	Number of active cutting edges on a magnetic abrasive particle
N_{rs}	Rotational speed of the tool (rpm)
P_{act}	Power available at N-pole (kW)
r and z	Coordinates
R^0_{max}	Initial maximum surface roughness of workpiece (μm)
R_{max}	Maximum surface roughness of workpiece (μm)
$r_{(tr)}$	Radius of projected area of indentation (m)
$R_{(tr)}$	Radius of tr^{th} track (m)
v	Volume of magnetic abrasive particle (m^3)
$\Delta V_{(i,j)}^{(n)}$	Volume of material removed in nth revolution in a cell (i, j) (m^3)
χ_r	Susceptibility of magnetic abrasive particle
μ_0	Permeability of free space (H/m)
μ_r	Relative permeability of magnetic abrasive particle
ϕ	Magnetic potential (At)
τ_s	Shear strength of workpiece material (N/m^2)

Subscripts

tr	Track
(i,j)	Cell

Superscripts

e	Element
T	Transpose

References

Balagurswamy, E. 2000. *Numerical Methods*. New Delhi: Tata McGraw-Hill.

Fox, M., Agrawal, K., Shinmura, T., Komanduri, R. 1994. Magnetic abrasive finishing of rollers. *Annals of CIRP* 43(1): 181–184.

Gillespie LaRoux, K. 2008. Using magnetic abrasive finishing for deburring producing parts. *Allure Deburring* 60(4): 50–56.

Inasaki, I., Tonsoff, H.K., and Howes, T.D. 1993. Abrasive machining in future. *Annals of CIRP* 42: 723–732.

Jain, V.K. 2002. *Advanced Machining Processes*. New Delhi: Allied Publishers.

Jain, V.K. 2009a. Magnetic field assisted abrasive based micro-/nano-finishing. *Journal of Materials Processing Technology* 209: 6022–6038.

Jain, V.K. 2009b. *Introduction to Micromachining*. New Delhi: Narosa.

Jain, V.K., Singh, D.K., and Raghuram, V. 2008. Analysis of performance of pulsating flexible magnetic abrasive brush (P-FMAB). *Machining Science and Technology* 12(1): 53–76.

Jayswal, S.C. 2005. Modeling and simulation of magnetic abrasive finishing process, PhD Thesis, Indian Institute of Technology Kanpur, India.

Jayswal, S.C., Jain, V.K., and Dixit, P.M. 2004. Analysis of magnetic abrasive finishing with slotted magnetic pole. *The 8th International Conference on Numerical Methods in Industrial Forming Process, Columbus, June 13–14*. Ohio, USA.

Jayswal, S.C., Jain, V.K., and Dixit, P.M. 2005a. Modeling and simulation of magnetic abrasive finishing process. *International Journal of Advanced Manufacturing Technology* 26: 477–490.

Jayswal, S.C., Jain, V.K., Dixit, P.M. 2005b. Abrasive finishing process—A parametric analysis. *Advanced Manufacturing System* 4(2): 131–150.

Jayswal, S.C., Jain, V.K., and Dixit, P.M. 2007. Simulation of surface roughness in MAF using non-uniform surface Roughness. *Materials and Manufacturing Processes* 22: 256–270.

Jefimenko, O.D. 1966. *Electricity and Magnetism*. New York: Meredith Publishing Company.

Khairy, B.A. 2001. Aspects of surface and edge finish by magnetoabrasive particles. *Journal of Materials Processing Technology* 116(1): 77–83.

Komanduri, R. 1996. On material removal mechanisms in finishing of advanced ceramics and glasses. *Annals of CIRP* 45(1): 509–514.

Komanduri, R., Lucca, D.A., and Tani, Y. 1997. Technological advances in fine abrasive processes. *Annals of CIRP* 46(2): 545–580.

Komanduri, R., Umehara, N., and Raghunandan, M. 1994. On the possibility of chemo-mechanical polishing of silicon nitride by magnetic float polishing. *Transactions of the ASME, Journal of Tribology* 118: 721–727.

Kurobe, T. and Imanaka, O. 1984. Magnetic field assisted fine finishing. *Precision Engineering* 6: 119–124.

Madarkar, R. and Jain, V.K. 2007. Investigation into magnetic abrasive micro deburring. *Proceeding of the 5th International Conference on Precision, Dec 13–14*. Jadhavpur, West Bengal. pp. 307–312.

Malik, R.S. and Pandey, P.M. 2011. Ultrasonic assisted magnetic abrasive finishing of hardened AISI 52100 steel using unbonded SiC abrasives. *International Journal of Refractory Metals and Hard Materials* 29(1): 68–77.

McGeough, J.A. 1988. *Advanced Methods of Machining*. London: Chapman and Hall.

Pa, P.S. 2009. Super finishing with ultrasonic and magnetic assistance in electrochemical micro-machining. *Electrochimica Acta* 54: 6022–6027.

Park, J.I., Ko, S.L., Hanh, Y.H., and Baron, Y.M. 2005. Effective deburring of micro burr using magnetic abrasive finishing method. *Key Engineering Materials* 291–292: 259–264.

Reddy, J.N. 1993. *An Introduction to the Finite Element Method*. 2nd ed. New Delhi: McGraw-Hill.

Shaohui, Y. and Shinmurra, T. 2004. A comparative study: Polishing characteristics and its mechanisms of three vibration modes in vibration-assisted magnetic abrasive polishing. *International Journal of Machine Tools and Manufacture* 44(4): 383–390.

Shinmurra, T., Takazawa, K., and Hatano, E. 1985. Study on magnetic abrasive process—Application to plane finishing. *Bulletin of the Japan Society of Precision Engineering* 19(40): 289–291.

Singh, D.K. 2005. Experimental investigations into magnetic abrasive finishing of plane surfaces, PhD Thesis, Indian Institute of Technology Kanpur, India.

Singh, D.K., Jain, V.K., and Raghuram, V. 2004. Parametric study of magnetic abrasive finishing process. *Journal of Materials Processing Technology* 149(1–3): 22–29.

Singh, D.K., Jain, V.K., and Raghuram, V. 2005a. On the performance analysis of flexible magnetic abrasive brush. *Machining Science and Technology* 9(4): 601–619.

Singh, D.K., Jain, V.K., and Raghuram, V. 2005b. Analysis of surface texture generated by a flexible magnetic abrasive brush. *Wear* 259(7–12): 1254–1261.

Stowers, I.F., Komanduri, R., and Baird, E.D. 1988. Review of precision surface generating processes and their potential applications to the fabrication of large optical components. *SPIE* 0966: 62–73.

Stradling, A.W. 1993. The physics of open-gradient dry magnetic separation. *International Journal of Mineral Processing* 39: 1–18.

Tani, Y. and Kawata, K. 1984. Development of high-efficient fine finishing process using magnetic fluid. *Annals of CIRP* 33(1): 217–220.

Umehara, N., Kobayashi, T., and Kato, K. 1995. Internal polishing of tube with magnetic fluid grinding. Part 1, fundamental polishing properties with taper-type tools. *Journal of Magnetism and Magnetic Materials* 149: 185–187.

William, R.E. and Rajurkar, K.P. 1992. Stochastic modelling and analysis of abrasive flow machining. *Transactions of the ASME—Journal of Engineering for Industry* 114: 74–80.

Yamaguchi, H. and Shinmura, T. 2004. Internal finishing process for alumina ceramic components by a magnetic field assisted finishing process. *Precision Engineering* 28: 135–142.

Yamaguchi, H., Shinmura, T., and Takenaga, M. 2003. Development of a new precision internal machining process using an alternating magnetic field. *Precision Engineering* 27: 51–58.

Yan, B.-H., Chang, G.-W., and Cheng, T.J. 2003. Electrolytic magnetic abrasive finishing. *International Journal of Machine Tools and Manufacture* 43: 1355–1356.

Yan, W. and Hu, D. 2005. Study on the inner surface finishing of tubing by magnetic abrasive finishing. *International Journal of Machine Tools and Manufacture* 45: 43–49.

9

Abrasive Flow Finishing (AFF) for Micromanufacturing

M. Ravisankar

Indian Institute of Technology Guwahati

J. Ramkumar and V.K. Jain

Indian Institute of Technology Kanpur

CONTENTS

9.1 Introduction

Abrasive flow finishing (AFF) was developed around the 1960s, and this process is used to deburr, polish, and radius difficult-to-machine material components having difficult-to-reach surfaces such as intricate geometries and edges. Four different types of AFF machines have been designed based on their applications [one-way, two-way, orbital, and micro-abrasive flow finishing (μ-AFF)]. Because two-way AFF process is simpler to understand compared to others, hence, in this chapter, the analysis of the process and the mechanism of material removal have been discussed with reference to two-way AFF process only. The key components of an AFF process are the machine, tooling, and viscoelastic medium (medium). The medium possesses easy flowability, good self-deformability, and fine abrading capability. The layer thickness of the material removed is of the order of 1–10 μm. The best surface finish that has been achieved by the Extrude Hone system is 50 nm, and the tolerances achieved are ±0.5 μm. AFF reduces surface roughness by 75%–90% on cast and machined surfaces. This process can deburr holes as small as 0.2 mm and radius edges from 0.025 to 1.5 mm. It can process dozens of holes or multiple passage parts simultaneously with uniform results. Also, air cooling holes on turbine disks and hundreds of holes in a combustion liner can be deburred and finished in a single operation. An important feature that differentiates the AFF process from other finishing processes is that it is possible to control and select the intensity and location of abrasion through fixture design, medium selection, and process parameters. Although there are many other advantages, it has a few drawbacks also, such as low finishing rate (FR) and inability to correct the form geometry. Thus, researchers have proposed various AFF setup configurations and hybrid AFF processes to minimize the effect of limitations of the process.

9.2 Abrasive Flow Finishing (AFF) Process

9.2.1 Working Principle

In the AFF process, a polymer-based viscoelastic abrasive medium (medium) is extruded back and forth between two vertically opposed cylinders. This medium tries to finish the workpiece surface selectively while extruding through the passage formed by the workpiece and tooling (Figure 9.1) [1]. Because of its flow through a restricted, narrow passage in the workpiece region, the medium polymer chains hold abrasive particles flexibly and orient themselves in the direction of the applied extrusion pressure (along the length of the passage). Thus, the medium acts as a multipoint flexible cutting tool and starts abrading the workpiece surface. Extrusion pressure, number of cycles, initial surface roughness of the workpiece, and medium viscosity are the important variable parameters that have an impact on the final surface roughness achievable. When appropriate extrusion pressure is applied on the viscoelastic medium, it moves along the direction of the applied pressure with axial velocity (V_a) and axial force (F_a) because of the viscous component. Because of the elastic component of the medium, it exerts a radial force (F_r). This helps the abrasive particles to indent into the workpiece surface (Figure 9.1).

An AFF system consists of three basic elements, that is, machine, tooling, and medium. The machine consists of a hydraulic system and medium cylinders that are held in line by a

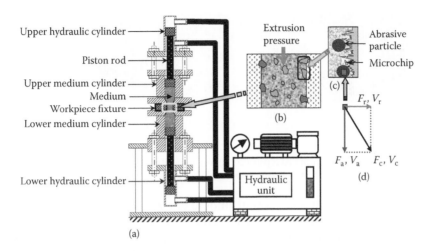

FIGURE 9.1
(a) Schematic diagram of an AFF setup; (b) magnified view showing the interaction between abrasive particles (APs) and the workpiece; (c) further magnified view to show APs and a microchip; (d) forces and velocities in AFF process.

supporting frame. The machine controls the extent of abrasion in the finishing region. The hydraulic system is a positive displacement system that forces the medium through the tooling at a selected pressure and flow rate. Normally, the hydraulic system operates in the pressure range of about 1–20 MPa with a maximum flow rate of about 18.90 L/min. Tooling is one of the major elements of the AFF system. The major functions of tooling are to direct the flow of the medium and precisely control the area of abrasion [2,3]. In the AFF process, finishing action on the workpiece surface takes place by forcing the abrasive medium across the workpiece surface. Material removal is caused by extruding the medium back and forth from one medium cylinder to the other across the passage formed by the workpiece and tooling.

The medium is the most important element of an AFF process as it is the primary flexible cutting tool of the process, also called a *self-deformable stone*. It should possess three basic properties, that is, good flowability, self-deformability, and abrading ability, to finish the given surface to a nanoscale. The medium should also possess enough degree of cohesion and tenacity to drag the abrasive particles along with it through various passages or regions. It is important to know the composition of the medium and its rheological properties to understand the role of the medium in the mechanism of material removal [4–6].

9.2.2 Medium Composition and Rheology

A medium is comprised of a base carrier (viscoelastic polymer), rheological additives, and abrasive particles. The base carrier is a high-molecular-weight material, with dominant elastic properties and poor viscous properties. To improve the viscous properties of the base carrier, rheological additives such as plasticizers and softeners are mixed into it. The plasticizers as well as softeners are low-molecular-weight materials and can easily diffuse into the high-molecular-weight base polymer carrier when mixed. This diffusion of the low-molecular-weight material forces the polymer chains of the base polymer carrier apart and increases the viscous properties of the medium. Abrasive particles are held by the matrix material in this mixture to keep them uniformly suspended in a less dense base carrier (low-viscosity medium). A high-viscosity (stiff) medium (Figure 9.2) is rich in

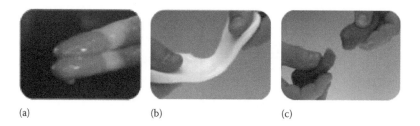

(a)   (b)   (c)

FIGURE 9.2
Types of abrasive medium used in the AFF process: (a) low viscosity medium, (b) medium viscosity medium, and (c) high viscosity medium. (From Kennametal Extrude Hone Corp.)

base polymer carrier content with a small amount of plasticizer. A high-viscosity medium possesses relatively high elastic properties, and thus this medium yields high material removal per cycle. High elastic properties help in applying a large radial force, which governs the depth of penetration of an abrasive particle in the workpiece material. However, a low-viscosity medium (rich in plasticizer content compared to the base polymer carrier) behaves like a fluid that can easily pass through holes of very small diameter (minimum of 0.2 mm).

The viscoelastic medium when left at rest slowly flows like a fluid because of the natural gravitational force. When dropped on the ground, it bounces back like a ball. It behaves like a solid elastic ball, but when stretched rapidly, it breaks in a manner similar to a solid plastic piece. These unique properties demonstrate the importance of measuring the rheological properties of the medium to understand its behavior during the finishing process. The rheological properties of the medium determine the pattern and aggressiveness of the abrasive action. A variation in the composition of the medium can change the flow and deformation properties (rheological properties) of the medium.

As the AFF medium is viscoelastic in nature, the viscous component assists flow in the direction of the extrusion pressure [axial force (F_a) and axial velocity (V_a)]. Because of its rubberlike elastic nature, it acts as the backbone of the medium, holds the abrasive particles against the surface being finished, and exerts a radial indentation force (F_r) (Figure 9.3) [7]. It is important to know the amount of viscous and elastic components in different media.

The creep recovery test gives the quantitative values of viscous and elastic components in different media. In a creep recovery test (Figure 9.4), during the creep phase, the complete viscoelastic (viscous + elastic) behavior (AB) is known, while in the recovery phase,

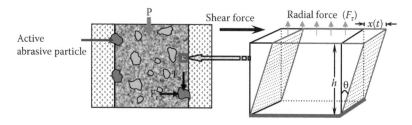

FIGURE 9.3
Left: Forces F_a and F_r acting on a single active abrasive particle in an AFF process. Right: Viscoelastic medium material behavior of exerting radial force normal to the direction of the applied shear force. (From Ravi Sankar, M., et al., *Wear*, 267 (1–4), 43–51, 2009.)

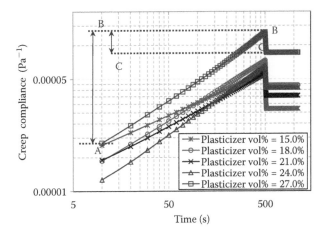

FIGURE 9.4
Variation of creep compliance with time ($\tau_0 = 100$ Pa) (logarithmic scale on both x- and y-axes).

the elastic component (BC) is recovered. So, subtraction of the elastic recovery component from the complete viscoelastic part gives the medium's viscous component.

The base polymer carrier possesses predominantly elastic properties and less of viscous properties. The major function of a plasticizer in the medium is to increase the viscous component. So, as the plasticizer content increases, the viscous component in the medium also increases. Thus, a medium with 15% and 27% of plasticizer by volume possesses low and high viscous components (Figure 9.4), respectively [8,9].

9.2.3 Material Removal Mechanism and AFF Forces

In the AFF process, material removal is only by mechanical abrasion. Finishing forces at the workpiece surface are low and, as a result, the workpiece surface remains stress free or is subjected to minimal compressive residual stress. During the AFF process, abrasive particles penetrate into the workpiece surface because of the radial force (F_r). The axial force (F_a) tries to push the indented abrasive particle in the axial direction to remove the material in the form of a microchip. The mechanism of material removal follows one of the three conditions shown in Figure 9.5.

$$F_a > F_{req} \tag{9.1}$$

$$F_a < F_{req} \tag{9.2}$$

$$F_a \geq F_{req} \tag{9.3}$$

In Figure 9.5, h is the depth of indentation, h' is the high depth of indentation due to increased extrusion pressure $F_a < F'_{req}$, h'' is the reduced depth of indentation due to abrasive particle rotation, F_a is the axial force acting on the single abrasive particle, F_r is the radial force acting on the single abrasive particle, and F_{req} is the force required on the single abrasive particle to remove the material in the form of a microchip [10]. F_{req} can be evaluated as below:

$$F_{req} = \tau_s A_p \tag{9.4}$$

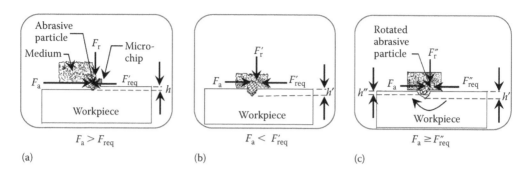

FIGURE 9.5

Material removal mechanism in the AFF process: (a) $F_a > F_{req}$, (b) $F_a < F'_{req}$, and (c) $F_a \geq F''_{req}$. (From Ravi Sankar, M., Nano finishing of metal matrix composites using rotational abrasive flow finishing (R-AFF) process, PhD Thesis, IIT Kanpur, India, 2011.)

where τ_s is the shear strength of the workpiece material and A_p is the projected area of penetration.

If $F_a > F_{req}$ (Figure 9.5a), material is removed in the form of microchips; however, under certain circumstances, plowing may take place. If $F_a < F'_{req}$ there is no material removal and only indentation occurs on the workpiece surface (Figure 9.5b). Under this condition, the abrasive particle may rotate and reduce the depth of penetration till it satisfies Equation 9.1 and then remove the material. If $F_a \geq F''_{req}$ material removal is just about to start (Figure 9.5c) [1].

Change in surface roughness (ΔRa = initial Ra − final Ra) varies nonlinearly with increase in extrusion pressure during finishing of metal matrix composite (MMC) (Figure 9.6a). At low pressure (4 MPa), ΔRa is low. As extrusion pressure increases from 4 to 6 MPa, ΔRa increases. This is because with increase in extrusion pressure the number of active abrasive particles increases (active abrasive particles are those that interact with the workpiece and remove material). An increase in the extrusion pressure also leads to increased F_a and F_r on the work surface through abrasive particles, resulting in a higher ΔRa. For extrusion pressures up to 6 MPa, $F_a \geq F_{req}$. Because increase in F_r increases the depth of penetration, ΔRa increases up to a certain point. At or near 6 MPa pressure, an optimum is observed.

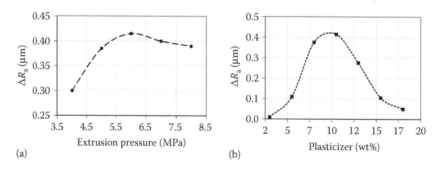

FIGURE 9.6

Variation of change in roughness on an Al alloy/SiC (10%) MMC workpiece with (a) extrusion pressure and (b) weight percentage of the plasticizer in the AFF medium. (From Ravi Sankar, M., et al., *Wear*, 266 (7–8), 688–698, 2009.)

An increase in extrusion pressure beyond this optimum value marginally decreases ΔRa (Figure 9.6a), or ΔRa becomes constant.

In AFF, force ratio is estimated as the ratio of radial force to axial force (F_r/F_a). At high extrusion pressures, the force ratio is found to be higher and F_r increases to F'_r, hence the depth of indentation becomes more ($h' > h$ in Figure 9.5b). At this depth of indentation, the axial force acting on the abrasive particle becomes proportionately lesser than the force required to remove the material ($F_a < F'_{req}$) in the form of microchip. As a result, the abrasive particle rotates inside the medium till its depth of penetration h' is reduced to h'' (F'_r reduces to F''_r) such that the condition $F_a \geq F_{req}$ is satisfied. Then the material is removed in the form of microchip. This may lead to the marginal deterioration in ΔRa value as compared to the ΔRa value at the preceding extrusion pressure (Figures 9.5a and 9.6a). If $F_a \leq F''_{req}$, the indentation phenomenon is more dominant and creates a deep cavity without removing material in the form of chip and it leads to the deterioration of surface finish. Another argument in the support of decrease in ΔRa beyond 6 MPa extrusion pressure is as follows. Since the medium is compressible, at high extrusion pressures (≥ 6 MPa) and hence higher radial force, the number of abrasive particles decreases due to penetration of abrasive particles inside the medium. Thus, ΔRa starts decreasing after 6 MPa.

At low plasticizer content, ΔRa is low as the medium is stiff and behaves like a soft grinding wheel. As the plasticizer content increases, ΔRa gradually increases owing to the higher viscous component of the medium that allows the medium to hold the abrasive particles more strongly and withstand higher axial force without shearing the medium itself. It also imparts better flowability, self-deformability, and flexible abrading ability. At about 10% plasticizer content, an optimum is observed. With a further increase in the plasticizer content, the gap between polymer chains increases and the abrasive-holding ability decreases. As a result, instead of shearing the surface peaks, the medium simply flows over the surface peaks or goes through the mechanism shown in Figure 9.5c. Thus, ΔRa gradually starts decreasing as shown in Figure 9.6b [1]. The number of finishing cycles is another important parameter controlling ΔRa. As the number of finishing cycles increases, the total number of times that the abrasive particles participate in finishing also increases. So, the surface roughness value gradually decreases as the number of cycles increases (Figure 9.7). This decrease continues up to the critical surface finish beyond which it becomes stable with minor oscillations depending on the homogeneity of the mixture and its constituents in different cycles and locations.

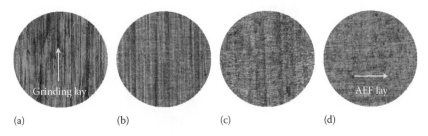

(a)   (b)   (c)   (d)

FIGURE 9.7
Variation of surface finish on Al alloy/SiC MMC (10%) with number of cycles (abrasive mesh size = 220, extrusion pressure = 6 MPa, and plasticizer wt% = 10%): (a) Ra = 0.70 μm, $N = 0$; (b) Ra = 0.58 μm, $N = 50$; (c) Ra = 0.41 μm, $N = 100$; and (d) Ra = 0.29 μm, $N = 200$.

9.3 Types of AFF Processes and Their Capabilities

On the basis of the complexity of finishing components, AFF (some researchers call it *abrasive flow machining*, AFM) machines are classified into four types, one-way AFF, two-way AFF, orbital AFF, and μ-AFF processes. A brief discussion of these processes is given below.

9.3.1 One-Way AFF Process

The one-way AFF process extrudes the medium unidirectionally and the medium comes out of the workpiece that is being finished (Figure 9.8a). This process is good for components with complex internal shapes in which forward flow of the medium is possible and reverse flow is difficult [11]. This process is capable of finishing 3D complex surfaces that are difficult to analyze.

The one-way AFF process apparatus is provided with a hydraulically actuated reciprocating piston and a medium chamber that is adapted to receive and extrude the medium unidirectionally across the internal surfaces of the workpiece, as shown in Figure 9.8a. A fixture directs the flow of the medium from the medium chamber into the internal passages of the workpiece, while the medium collector (Figure 9.8a) collects the medium as it extrudes out from the internal passage. The extrusion medium chamber is provided with an access port to periodically receive the medium from the collector. The hydraulically actuated piston intermittently withdraws from its extruding position to open the medium chamber access port to permit the extruded medium to enter the medium chamber. When the extrusion medium chamber has been charged with the working medium, the operation is resumed (Figure 9.8a) [11]. This process can be used for finishing various types of parts such as automobile engine blocks and multiple holes in cylinders (Figure 9.8b).

9.3.2 Two-Way AFF Process

The two-way AFF process (Figure 9.1) extrudes the medium in the to and fro directions. It is normally used to finish cylindrical (internal and external) surfaces, flat surfaces, and complex dies that possess through holes (Figure 9.9). It is comparatively easy to study the mechanism of material removal in the two-way AFF process and to model the same [12,13].

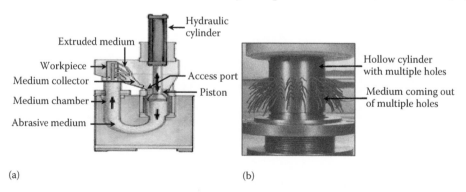

(a)   (b)

FIGURE 9.8
(a) One-way AFF process and (b) multiple holes in a cylinder finished by the one-way AFF process. (From Kennametal Extrude Hone Corp.)

FIGURE 9.9
Various types of extrusion dies that can be finished by the two-way AFF process. (From Kennametal Extrude Hone Corp.)

The two-way AFF machine has two hydraulic cylinders and two medium cylinders. The medium is extruded, hydraulically or mechanically, through the restricted passageway formed by the workpiece surface to be finished and the tooling (Figure 9.1). Typically, the medium is extruded back and forth between the chambers for the prefixed number of cycles. Counterbores, recessed areas, and extrusion dies can be easily finished (Figure 9.9).

9.3.3 Orbital AFF Process

The orbital AFF process is normally used to finish blind and shallow-cavity workpieces such as coins and dies. It is a combination of form grinding and the two-way AFF process. Here, the medium reciprocates in two ways and orbital motion is given to the form tool (Figure 9.10a) [14].

In the orbital AFF process, relative motion between the workpiece and the tool is provided to accomplish positive displacement of the viscous medium (Figure 9.10). This movement can be gyratory, rotating, orbital, reciprocatory, or any combination of these. The simplest

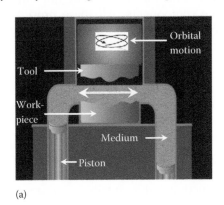

(a)

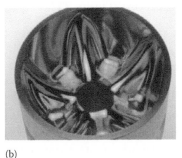

(b)

FIGURE 9.10
(a) Sectional schematic diagram of the orbital AFM process and (b) blind bottle-making dies finished by the orbital AFF process. (From Kennametal Extrude Hone Corp.)

way is to utilize only orbital relative motion, if the workpiece is a die-casting mold. The tool, having a profile smaller than the cavity, can be inserted within the cavity to provide a medium chamber formed between the entire surface of the cavity and the displacer.

The medium is filled in the medium chamber and sealed by a sealing ring. The tool is suitably inserted within the cavity. With the displacer and sealing ring biased against the medium, a relative orbital motion is provided to the workpiece with respect to the displacer. This relative orbital motion causes a relative translational motion between the medium and the contacting surface of the workpiece and displacer; thereby, the abrasives mixed in the medium abrade the workpiece surface. The relative orbital motion is continued until the workpiece is finished to the extent desired. Figure 9.10b shows a die used to make the base of a bottle. This die has been finished by the orbital AFF process.

9.3.4 Micro-AFF (μ-AFF)

The μ-AFF process is similar to the two-way AFF process. The difference between the processes is that μ-AFF uses a medium having very low viscosity and very fine abrasive particles compared to the macro-AFF process. This helps the medium to easily flow through micro-sized holes (50–750 μm). This process is capable of performing all the three types of operations at the same time, that is, deburring, radiusing, and finishing. A high-viscosity medium is more suited for simultaneous processing of multiple parts, while a low-viscosity medium is ideal for better flow rate in a single-step process. The better flow rate of the low-viscosity medium helps in radiusing along with finishing. μ-AFF consists of sophisticated precision sensing devices in a closed loop system that monitor and accurately maintain various aspects of the process ranging from medium viscosity and medium pressure to shear force history and temperature [15,16].

The μ-AFF process finishes automobile spray nozzle holes, and it gives a much smoother and higher quality surface finish compared to the surface finish of holes made by electric discharge machining (EDM; Figure 9.11a and b). This process helps in radiusing the entrance, which contributes to flow capacity and fuel spray distribution (Figure 9.11c). This smooth radiusing also improves the fatigue strength of the injection nozzle. Essentially, the process produces a preaged spray nozzle with improved fuel atomization for increased engine performance, reduced power requirement, and reduced emissions.

In many applications, microdeburring, edge radiusing, and surface finishing of micro-holes decide the overall performance of the components. All these processes (micro-deburring, edge radiusing, and nanofinishing of microhole surface) can be done in one go by the μ-AFF process (Figure 9.12).

9.4 Advancements of AFF Process

Although there are many merits to the AFF process, it has a few limitations also—mainly a low FR and inability to correct form geometries. Because of the low FR, the finishing time is high. Hence, apart from low productivity, the rheological properties of the abrasive medium also degrade with its use. Thus, the medium loses its effective finishing capabilities. Many researchers have been working on improving FR, surface integrity, and compressive residual stresses produced on the workpiece surface. Some efforts made in this direction are discussed in the following sections.

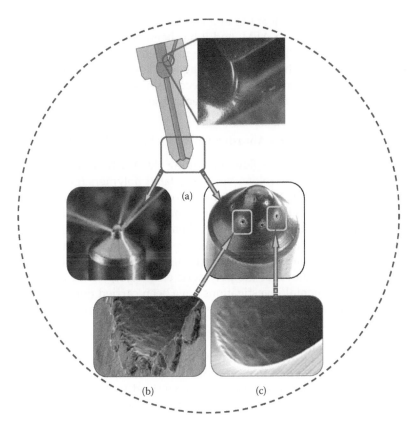

FIGURE 9.11
(a) Overview of a fuel spray injector nozzle surface; (b) microelectro-discharge drilled holes in the fuel spray injector nozzle; and (c) a finished orifice hole using the μ-AFF process. (From Kennametal Extrude Hone Corp.)

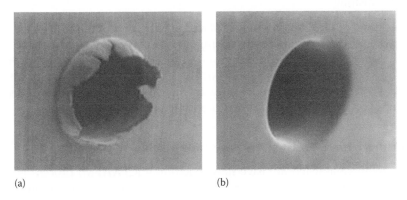

FIGURE 9.12
(a) Microhole with burrs after drilling; and (b) microdeburring and edge radiusing by the μ-AFM process. (From Kennametal Extrude Hone Corp.)

9.4.1 Magnetic AFM

Singh and Shan [17,18] used a medium made of a silicon polymer base carrier, hydrocarbon gel, and magnetic abrasive particles (Brown Super Emery, containing 40% ferromagnetic constituents, 45% Al_2O_3, and 15% Si_2O_3) in a magnetic AFM setup. The magnetic field is applied around the workpiece in AFM, and it has been observed that the magnetic field affects material removal and change in surface roughness. This would work better for a low-viscosity medium in which migration of magnetic abrasive particles toward the magnets is possible. However, abrasive motion becomes restricted as the viscosity of the medium increases, and hence the process efficiency goes down.

9.4.2 Electrochemically Assisted Abrasive Flow Machining (ECAFM)

Electrochemically assisted abrasive flow machining (ECAFM), developed by Dabrowski et al. [19,20], uses polymeric electrolytes such as gelated polymers (polypropylene glycol) and water-gels (sodium iodide salt and polyethylene glycol with potassium cyanide) as base carrier. The ion conductivity of electrolytes is many times lower than the conductivity of electrolytes employed in ordinary electrochemical machining (ECM). Addition of inorganic fillers (abrasives) to the electrolyte decreases the conductivity even more. The polymeric electrolyte medium is forced through the small interelectrode gap. This in turn results in greater flow resistance of the polymeric electrolyte, which takes the form of a semiliquid paste. Experimental investigations have been carried out on flat surfaces, and ECAFM showed technological and economical advantages compared to the AFM process.

9.4.3 Ultrasonic Flow Polishing

Ultrasonic flow polishing is a combination of AFM and ultrasonic machining. The medium pumped down the center of the ultrasonically energized tool flows radially relative to the axis of the tool. The combination of medium radial flow and tool vibration results in abrasion of the workpiece surface. Ultrasonic flow polishing is claimed to be capable of improving surface finish by a factor of 10 [21].

9.4.4 Spiral Polishing

In spiral polishing, a spiral fluted screw is placed at the center of the hole in the workpiece to be finished, and the screw is rotated using an external energy source. The rotational motion of the screw lifts the medium from lower medium cylinder to the upper medium cylinder and tries to finish the hole while passing through it [22,23]. This process showed good results for high-viscosity media only. In the case of a low-viscosity medium, the medium cannot be scooped up and down as a lump.

9.4.5 Centrifugal-Force-Assisted Abrasive Flow Machining Process

The centrifugal-force-assisted abrasive flow machining (CFAAFM) process is one of the advanced finishing processes. In this process, a centrifugal-force-generating (CFG) tiny rod (triangular, spline, square, or rectangular; Figure 9.13) is placed at the center of the medium slug in the workpiece finishing region. In this region, the rotating rod strikes the abrasives that come in contact with it. The angle at which the abrasive particles move depends on the rotational speed and shape of the rod. Placing the CFG rod in the center and providing rotation to it increases the finishing rate by 70%–80% [24]. This process holds good for low-viscosity media only. In the case of a high-viscosity medium, the force with which an abrasive particle should be hit by the CFG rod to move it toward the workpiece surface should be very high. Otherwise, the abrasive particle will not be able to reach

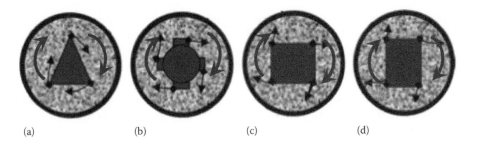

FIGURE 9.13
Cross section of different shapes of CFG rods used in the CFAAFM process: (a) triangle, (b) spline, (c) square, and (d) rectangle.

the workpiece surface and remove the material. There are viscous losses in the mixture of carrier and rheological additives. Hence, the efficiency of this process becomes low as the viscosity of the medium increases.

9.4.6 Drill-Bit-Guided AFF (DBG-AFF) Process

Drill-bit-guided abrasive flow finishing (DBG-AFF) is one of the advanced versions of the AFF process in which a freely rotatable drill bit is placed with the help of special fixture plates in the workpiece finishing zone (Figure 9.14). A combination of medium reciprocation, flow through the drill bit flute and scooping flow across the flute imparts dynamic motion to the abrasive particles. This makes the abrasives move in an inclined path rather than along the shortest straight line path. Turbulence at the center also causes frequent

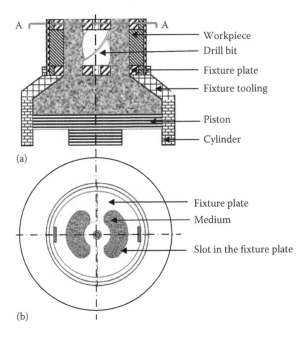

FIGURE 9.14
(a) Sectional front view of tooling in the finishing region in the DBG-AFF process and (b) top view of medium splitting through a twin slotted fixture plate.

reshuffling of active abrasive particles. Inclined motion and frequent reshuffling cumulatively enhance the material removal rate, FR, as well as change in surface roughness [25].

9.4.7 Rotational AFF

In the rotational abrasive flow finishing (R-AFF) process, the workpiece fixture is rotated externally by a variable frequency drive. Owing to workpiece rotation and medium reciprocation (Figure 9.15a), the active abrasive particle moves in a helical path. The number of surface peaks sheared by the active abrasive particle per unit time increases considerably in the R-AFF process compared to the AFF process. Thus, both FR and material removal rate increase in the R-AFF process. In addition, rotation of the workpiece imparts force (tangential force F_t) and velocity (tangential velocity V_t) along with the forces (axial force F_a and radial force F_r) as well as velocity (axial velocity V_a) acting during the AFF process (Figure 9.15b and c) [7–9].

The resultant finishing velocity (V_c) in R-AFF is the vector sum of V_t and V_a. Theoretically, the direction of V_c is at the semi-crosshatch angle (α) (Equation 9.5) to the tangential motion:

$$\alpha = \tan^{-1}\left(\frac{V_a}{V_t}\right)$$

(9.5)

Assistance from additional force and velocity due to rotation of the workpiece results in the shearing of surface peaks becoming easier (the depth of indentation depends on F_r, while F_a and $F_t \left(F_c = \sqrt{F_a^2 + F_t^2}\right)$ are responsible for removing material in the form of microchips). Thus, the finishing force (F_f) is a vector sum of all the three forces involved in the finishing action (Equation 9.6). Hence, as the rotational speed increases, the change in surface roughness increases (Figure 9.16a). This process finishes to nanolevels and also creates nano/micro-crosshatch patterns on the finished surface (Figure 9.16b).

$$F_f = \sqrt{F_r^2 + F_a^2 + F_t^2}$$

(9.6)

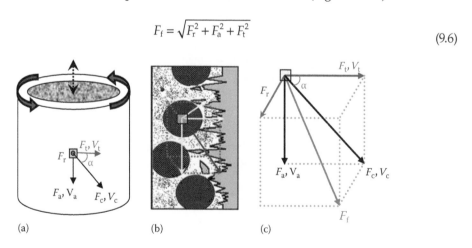

(a)   (b)   (c)

FIGURE 9.15
Forces and velocity components in the R-AFF process in the finishing region: (a) various forces and velocities acting on an abrasive particle during AFF; (b) enlarged view of the shearing of roughness peaks on the workpiece surface by abrasive particles; and (c) free body diagram of forces and velocities in the R-AFF process.

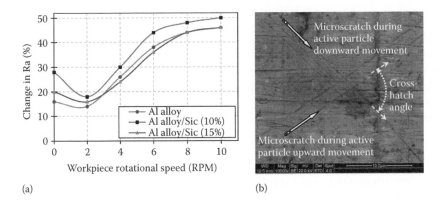

(a)   (b)

FIGURE 9.16
(a) Effect of workpiece rotational speed on percentage change in Ra and (b) microscopic view of a nano-crosshatch pattern (magnification 10,000×).

9.5 Applications of AFF Process

9.5.1 Automotives

The demand for the AFF process is increasing among car and two-wheeler manufacturers, as it is capable of finishing the surfaces to a high level for improved air flow as well as for better performance. The AFF process is used to enhance the performance of high-speed automotive engines.

9.5.2 Dies and Molds

In the AFF process, the abrading medium can flow through complex geometries and finish them easily. Dies are ideal workpieces for the AFF process, as they restrict the flow of the medium, typically eliminating fixturing requirements. The uniformity of stock removal by AFF permits accurate "sizing" of undersized precision die passages. The original EDM finish of 2 μm Ra is improved to 0.2 μm Ra with a stock removal of (EDM recast layer) 0.025 mm per surface (Figure 9.17).

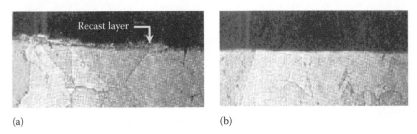

(a)   (b)

FIGURE 9.17
Finishing of the recast layer that forms during electrical discharge machining of a die: (a) before AFF and (b) after AFF. (From Kennametal Extrude Hone Corp.)

9.5.3 Aerospace

Housings and impellers of turbochargers and superchargers can be fine-finished with μ-AFF. Such processing increases the performance of the devices by decreasing friction. Superior surface finish on turbochargers gives greater strength and improved reliability. Cooling holes in turbine blades are finished by the μ-AFF process (Figure 9.18a). The turbulent cooling holes (Figure 9.18b) developed at IIT Kanpur for better cooling in turbine blades can be easily finished using the μ-AFF process.

9.5.4 Finishing and Deburring of Medical Technology Products

Orthopedic prostheses (hip joint implants) are commonly made of stainless steel or titanium. Presently, the life span of a hip joint implant is between 15 and 20 years. Because of the rough surface, it attracts the proteins that are responsible for blood clotting. So, these hip joint implants are finished to nanoscale by the AFF process to give them a much longer life (Figure 9.19a). This process would not affect the metallurgical properties of the parent material. To be on the safe side, these AFF-finished implants are coated with a totally biocompatible diamond nanofilm.

The AFF process is also used to finish and deburr many more human body implants such as heart valve implants and knee joint implants. Similarly, a knee joint implant finished

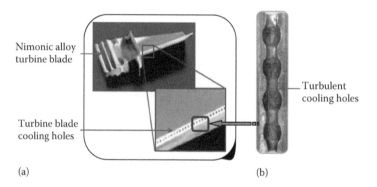

(a) (b)

FIGURE 9.18
(a) Finishing of cooling holes in a turbine blade (From Kennametal Extrude Hone Corp.) and (b) turbulent cooling holes produced in Nimonic superalloy. (Product of IIT Kanpur.)

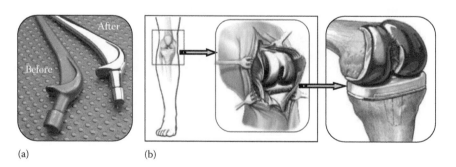

(a) (b)

FIGURE 9.19
(a) Surface finish improvement before and after AFF on a hip joint and (b) knee joint implant placed in the human knee, which can be finished using the AFF process. (From Kennametal Extrude Hone Corp.)

using the AFF process gives longer and better life and provides free movement of the leg (Figure 9.19b).

References

1. Ravi Sankar, M., Ramkumar, J., and Jain, V.K. 2009. Experimental investigation and mechanism of material removal in nano finishing of MMCs using abrasive flow finishing (AFF) process. *Wear* 266(7–8): 688–698.
2. Rhoades, L.J. 1991. Abrasive flow machining: A case study. *Journal of Material Processing Technology* 28: 107–116.
3. Loveless, T.R, Willams, R.E., and Rajurkar, K.P. 1994. A study of the effects of abrasive flow finishing on various machined surfaces. *Journal Material Processing Technology* 47: 133–151.
4. Jain, V.K. 2002. *Advanced Machining Processes*. New Delhi: Allied Publishers Pvt. Limited.
5. Jain, V.K. 2010. *Introduction to Micromachining*. New Delhi: Narosa Publishing House Pvt. Limited.
6. Ravi Sankar, M., Jain, V.K., Ramkumar, J., and Joshi, Y.M. 2010. Rheological characterization and performance evaluation of a new medium developed for abrasive flow finishing. *International Journal of Precision Technology* 1: 302–313.
7. Ravi Sankar, M., Jain, V.K., and Ramkumar, J. 2009. Experimental investigations into rotating workpiece abrasive flow finishing. *Wear* 267(1–4): 43–51.
8. Ravi Sankar, M., Jain, V.K., and Ramkumar, J. 2010. Rotational abrasive flow finishing (R-AFF) process and its effects on finished surface topography. *International Journal of Machine Tools and Manufacture* 50(7): 637–650.
9. Ravi Sankar, M., Jain, V.K., Ramkumar, J., and Joshi, Y.M. 2010. Rheological characterization of styrene–butadiene based medium and its finishing performance using rotational abrasive flow finishing process. *International Journal of Machine Tools and Manufacture* 51(12): 947–957.
10. Ravi Sankar, M. 2011. Nano finishing of metal matrix composites using rotational abrasive flow finishing (R-AFF) process, PhD Thesis, IIT Kanpur, India.
11. Rhoades, L.J., Kohut, T.A., Nokovich, N.P., and Yanda, D.W. 1994. Unidirectional abrasive flow machining, US patent number 5,367,833.
12. Rhoades, L.J. and Kohut, T.A. 1991. Reversible unidirectional AFM, US patent number 5,070,652.
13. Walch, W.L. 2005. Abrasive flow machining apparatus and method, US patent number 6,905,395.
14. Rhoades, L.J. 1990. Orbital and or reciprocal machining with a viscous plastic medium, International patent number WO 90/05044.
15. Greenslet, J.M. and Rhoades, L.J. 2005. Method and apparatus for measuring flow rate through and polishing a workpiece orifice, US patent number 6,953,387.
16. Walch, W.L., Greenslet, M.J. Jr, Rusnica, E.J., Abt, R.S., and Voss, L.J. 2002. High precision abrasive flow machining apparatus and method, US patent number 6,500,050.
17. Singh, S. and Shan, H.S. 2002. Development of magneto abrasive flow machining process. *International Journal of Machine Tools and Manufacture* 42: 953–959.
18. Singh, S., Shan, H.S., and Kumar, P. 2002. Wear behavior of materials in magnetically assisted abrasive flow machining. *Journal of Materials Processing Technology* 128: 155–161.
19. Dabrowski, L., Marciniak, M., and Szewczyk, T. 2006. Analysis of abrasive flow machining with an electrochemical process aid. *Proceedings of the IMechE, Part B: Journal of Engineering Manufacture* 220: 397–403.
20. Dabrowski, L., Marciniak, M., Wieczorek, W., and Zygmunt, A. 2006. Advancement of abrasive flow machining using an anodic solution. *Journal of New Materials for Electrochemical Systems* 9: 439–445.
21. Jones, A.R. and Hull, J.B. 1998. Ultrasonic flow polishing. *Ultrasonics* 36: 97–101.

22. Yan, B.-H. Tzeng, H.-J., Yuan Huang, F., Lin, Y.-C., and Chow, H.-M. 2007. Finishing effects of spiral polishing method on micro lapping surface. *International Journal of Machine Tools and Manufacture* 47: 920–926.

23. Wei-Chan, C., Biing-Hwa, Y., and Shin-Min, L. 2010. A study on the spiral polishing of the inner wall of stainless bores. *Advanced Materials Research* 126–128: 165–170.

24. Walia, R.S., Shan, H.S., and Kumar, P. 2006. Abrasive flow machining with additional centrifugal force applied to the media. *Machining Science and Technology* 10: 341–354.

25. Ravi Sankar, M., Mondal, S., Ramkumar, J., and Jain, V.K. 2009. Experimental investigations and modeling of drill bit guided abrasive flow finishing (DBG-AFF) process. *International Journal of Advanced Manufacturing Technology* 42(7–8): 678–688.

Section IV

Microjoining

10

Laser Microwelding

N.J. Vasa
Indian Institute of Technology Madras

CONTENTS

10.1 Introduction

In recent years, interest in the field of microfabrication of electronic, optical, and biomedical devices ranging from sensor to computation and control systems has grown tremendously. Biomedical welding applications range from the construction of precision instruments to the joining of human tissue. The electronics and biomedical industries in particular have recognized the opportunities for miniaturization that precision laser-based microjoining techniques have to offer. There is also an increasing interest in adopting micro/nanofabrication techniques for various devices such as microelectromechanical systems (MEMS) and other devices involving miniaturization and functional integration. These devices include the gyroscope chip, which can sense tilt, motion, shock, acceleration/deceleration, and vibration; the MEMS-based accelerometer deployed in airbags; pressure sensors and microphones; optical switches; micro total analysis systems; or lab-on-chips.

The development of integrated microsystems is accompanied by the advances in assembly and joining techniques. Regardless of the configuration, the parts must be joined together to form a device. When focusing on the assembling of small parts, presently, technologies such as adhesive bonding, brazing, electron beam welding, and laser microjoining are applied (Zhou et al., 2009). Considering a further miniaturization of components and increasing application of metallic structures, the usefulness of adhesive bonding might reach a limit. In brazing, the process requires fluxes and additional filler material, even for small components. Electron beam welding is useful for its low thermal load on the components, noncontact method, and the beam manipulation possibilities. In the field of microsystem technology, the possibility of exactly focusing beams of diameter of just a few micrometers makes the application of this method useful at microdimensions. Currently available electron beam machines with a high power ranging from 100 W to several kilowatts are not suitable for the welding of microfine components. Hence, a scanning electron microscope (SEM), which operates according to the same principle, has been modified and extended as a welding tool. However, it is necessary to maintain a clean environment and vacuum conditions in the welding chamber. Laser welding is used as an industrial manufacturing process for joining a variety of metallic and nonmetallic materials. Table 10.1 describes the important features of different microjoining processes.

Laser beam microwelding with lasers of the latest generation with high beam quality offers several advantages over the processes listed in Table 10.1 (Guo, 2009). Lasers are well suited for microwelding because they can deliver a controlled amount of energy to very small components with a high degree of precision. Laser microwelding involves joining of parts with at least one dimension <100 μm, and the focused laser spot size is approximately this size or smaller. High energy density in combination with good focus ability enables processes with high reproducibility. A wide range of joint geometries from single spot to several types of seam welds can be realized. The welds are characteristically of low distortion and show a small heat-affected zone (HAZ). Almost all materials can be welded, and welding of dissimilar materials is also possible. Owing to the noncontact nature of laser processing, a high degree of automation is possible in the typical industrial environment. The laser microjoining technique is used in various areas such as electronics, sensors, medical technology, and MEMS technology. In this chapter, laser microwelding processes, thermal model, microwelding systems, and welding defects are explained. Further, applications of different laser sources and recent advances in laser microwelding are also discussed.

10.2 Laser Welding Process

When a laser beam is used to irradiate the surface of a material, the absorbed energy causes the heating, melting, and/or evaporation of the material depending on the absorbed laser power density (Chryssolouris, 1991, Chapter 3; Miyamoto and Knorovsky, 2008; Dahotre and Harimkar, 2008, Chapter 10). The general precondition of laser welding is to create a pool of molten material (weld pool) at the overlapping workpiece surfaces. There are two general approaches for laser welding processes, namely, conduction welding and deep penetration (keyhole) welding. Figure 10.1 shows the schematic of conduction and deep penetration laser welding modes.

TABLE 10.1

Competing Methods of Microjoining

Joining Techniques	Features of Different Joining Techniques
Resistance welding	Used for welding devices with sheet thickness or wire diameter ranging approximately from 20 to 400 μm. Techniques: spot or seam welding
	Electrode force and welding current are <100–200 N and <2000–4000 A, respectively. Weld speed is around 5 mm/s and the equipment cost is low
	It is only suitable for materials that conduct electricity. The oxide layer on the metal makes the process unstable
	Electrode deformation may necessitate a post-weld finishing operation and access from both sides is required, which limits design possibilities
Ultrasonic welding	Used for sheet in the range of 0.1–1 mm
	One part of the assembly is vibrated, generating friction heat with the static part. Particularly suitable for joining polymers
Adhesive bonding	Avoids distortion and metallurgical changes because of the low levels of energy required compared with fusion welding
	Joint with reduced/no stress and bonding of dissimilar materials is possible However, surfaces need to be cleaned and the bonding possesses limited thermal stability with low peel strength
Soldering	Requires a high degree of operator skill and is relatively slow
	Wave soldering is a method of mass soldering printed circuit boards (PCBs)
Electron beam welding	Performed under high vacuum (<10^{-5} bar). No flux is required
	The process is useful where components require an internal vacuum
Laser beam welding	Laser welding is competitive primarily when high productivity, high joint quality, and low distortion are required
	Access is only required from one side
	A continuous laser weld in a three-dimensional framework is stiffer than the equivalent made using resistance sport weld, which enables cheaper or thinner materials to be used
	Laser beam can be positioned with high accuracy and precise control over the weld bead location
	The low energy input of laser conduction welding results in limited and predictable distortion. The rapid thermal cycle reduces the segregation of embrittled elements such as sulfur and phosphorus. Further, beneficial fine solidification microstructures are formed
	Laser soldering approach allows lead-free replacement for the conventional Sn–37Pb solder. Different types of metallic solders have been shown to exhibit equivalent properties to the conventional solder when fused using a laser beam
	Laser brazing can be carried out without the need for a vacuum, providing greater flexibility and higher productivity, and can be adapted to a wide range of part shapes and sizes

10.2.1 Conduction Welding

In conduction welding, the laser processing conditions are such that the surface of the weld pool remains unbroken as shown in Figure 10.1a. In this approach, the energy transfer into the depth of the material takes place by thermal conduction. The laser beam is focused to give a power density on the order of 10^3 W/mm^2, which is used to fuse material to create a joint without significant vaporization. Conduction welds can be made in a wide range of metals and alloys in the form of wires and thin sheets in various configurations using CO_2 (10.6 μm), Nd^{3+}:YAG (1.064 μm), and diode lasers (0.8–1.1 μm, 1.5 μm) with power levels of the order of tens of watts. Conduction welding can be carried out in two principal modes:

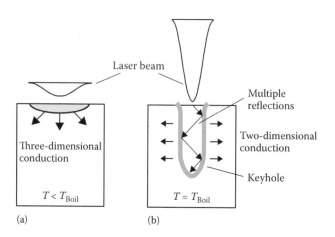

FIGURE 10.1
Laser welding processes: (a) conduction welding and (b) deep penetration (keyhole) welding. T is the temperature of the irradiated area and T_{Boil} is the boiling temperature.

direct heating and energy transmission. Conduction heating occurs in two modes: direct heating and energy transmission.

The mechanism of direct heating involves absorption of beam energy by the material surface, and heat flow is governed by classical thermal conduction from a surface heat source; the weld is made by melting a certain portion of the base material. A hemispherical weld bead and HAZ are formed in a manner similar to the conventional arc fusion welding process. Conduction-limited welds, therefore, exhibit a low depth-to-width ratio (aspect ratio), which is often required when limited penetration in the thickness direction is desired. Spot welds are made by pulsing the laser beam to melt sufficient material to form the joint. Continuous welds can be made by overlapping pulsed spot welds or by using a continuous wave (CW) beam.

The energy transmission mode of conduction welding is used with materials that transmit near-infrared (NIR) radiation, notably polymers. An absorbing ink is placed at the interface of a lap joint. The ink absorbs the laser beam energy, which is conducted into a limited thickness of surrounding material to form a molten interfacial film that solidifies as the welded joint. Thick section lap joints can thus be made without melting the outer surfaces of the joint. Butt welds can be made by directing the energy toward the joint line at an angle through material at one side of the joint or from one end if the material is highly transmissive.

The approximate laser power needed to melt the surface can be estimated by the following approach. When a material is irradiated by a stationary Gaussian laser source, the surface temperature at the center of the heat source $T(t)$ can be estimated as (Miyamoto and Knorovsky, 2008; Ion, 2005, Chapters 13 and 16; Ready, 1971)

$$T(t) = \frac{Aq}{\mu^{3/2}kr}\tan^{-1}\left(\frac{4\alpha t}{r^2}\right)^{1/2}$$

(10.1)

where A is the absorptivity of the work material, q is the laser beam power, k is the thermal conductivity of the work material, r is the laser beam radius where $q = q_{max}/e$, q_{max} is the peak power in a Gaussian distribution, α is the thermal diffusivity, $k/\rho C_p$, of the work material, ρ is the density, C_p is the specific heat, and t is the time. The minimum power required to melt (q_{melt}) the metal surface can be estimated as

$$q_{\text{melt}} = \frac{2\sqrt{\pi}krT_{\text{melt}}}{A}$$

(10.2)

where T_{melt} is the melting temperature. Conduction welding is a stable, low energy input process, which can be performed using a laser that does not need to have a high beam quality. A relatively low-power, low-cost laser can therefore be used. Material properties are relatively unimportant—hardness does not affect the process and the material need not be an electrical conductor. However, if conduction welding is used to join thicker sections, the energy input per unit length of weld is relatively high, resulting in a large weld bead and high levels of distortion.

In contrast to metals, laser beams deeply penetrate transparent materials, such as polymers and glass, with very little laser energy being reflected or absorbed. One approach involves laser transmission welding. Laser transmission welding involves joining of two parts, where heating of the plastic and the joining take place simultaneously. One of the joining parts must have high transmittance and the other must possess a high absorbance in the range of the laser wavelength. The laser beam is transmitted through the transparent joining component and the laser beam is absorbed in a thin surface layer of the second component, thereby heating and melting the layer. Owing to the heat conduction, the transparent joining part becomes plastified and joining takes place under the influence of the external force. Mostly, carbon black is added to the joining component in order to achieve sufficient absorption of the laser radiation.

10.2.2 Deep Penetration Welding

In this case, a "keyhole" is created in the weld pool as shown in Figure 10.1b. Generally, the transition from the conduction mode to the deep penetration welding is associated with an increase in laser power intensity or irradiation time such that surface vaporization at the molten weld pool begins. The resulting evaporation-induced recoil pressure forms a small depression in the weld pool, which subsequently develops into a keyhole by the upward displacement of molten material sideways along the keyhole walls. Subsequent ionization of the vapor results in the formation of the plasma plume. The laser energy entering the keyhole wall is determined by the attenuation due to absorption of laser energy in the plasma plume. As the depression deepens, the laser light is scattered repeatedly within it so that Fresnel absorption occurs at the keyhole walls too, thus increasing the coupling of laser energy into the workpiece. As the keyhole develops, the power of the source can now be absorbed at greater depths, not just at the surface. Some absorption or scattering of the power also occurs in the plasma vapor within the keyhole that can emerge as a plume that can obstruct or defocus the beam (especially with CO_2 lasers). However, the plume can also radiate energy back to the specimen. While the laser energy is applied, surface tension and gravitational forces have the effect of closing the keyhole, whereas vapor pressure and ablation help keep it open. Laser welding is usually realized at speeds much higher than conventional processes. Thus, the keyhole plays an important role in transferring and distributing the laser energy deep into the material. The surface vaporization continues at the keyhole wall during laser welding to maintain the cavity. The melt accelerated upward flows continuously out of the cavity.

10.3 Laser Welding Practice

10.3.1 Thermal Model for Laser Welding

During laser microwelding, the interaction of the heat source and the material results in rapid heating, melting, and circulation of molten metal in the weld pool aided by surface tension gradient and buoyancy forces. The resulting flow of liquid metal and heat transfer determine the change in temperature with time, that is, the thermal cycles, and the resulting structure and properties of the welds. The basic thermal conditions required for stable conduction welding are that the surface temperature exceeds the melting temperature but remains below the vaporization temperature. This upper limit of peak temperature avoids the formation of a keyhole and severe metal loss by vaporization and particle ejection, which can contaminate the microwelding environment with metal vapors and ejected metal particles.

In a laser scanning operation during welding as shown in Figure 10.2, transient heating is involved. In this case, the transient heat transfer equation in the Cartesian coordinates for laser welding is governed by (Yilbas et al., 2010; Seuzalec, 1987; He et al., 2005)

$$\rho(T)\frac{\partial(E)}{\partial t} - \rho(T)v\frac{\partial(E)}{\partial z} = \nabla \bullet (k(T)\,\nabla\,T) + S \tag{10.3}$$

where $\rho(T)$ is the temperature-dependent density, E is the energy gained by the substrate and expressed as $C\rho(T)\,T$, $k(T)$ is the temperature-dependent thermal conductivity, $C_p(T)$ is the temperature-dependent specific heat capacity, T is the temperature, and v is the velocity of the moving source along the z-axis. The source term given by S, which is proportional to the energy absorbed by the surface layers, is given by (Yilbas et al., 2010)

$$S = I_0\alpha(1-R)\exp\left[-\alpha x\right]\exp\left[-(y^2+z^2)\,/\,a^2\right] \tag{10.4}$$

where I_0 is the laser peak intensity, R is the surface reflectivity, α is the absorption coefficient whose value is wavelength and material dependent, a is the Gaussian parameter, and y and z are laser beam spot sizes along the respective axis while the laser beam scans the surface along the z-axis and the laser beam axis is x-axis (along the depth direction). For the intensity regime of interest to laser microwelding of metals, the mechanism of energy

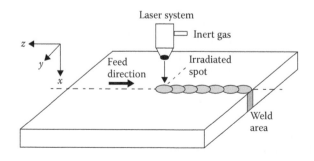

FIGURE 10.2
Laser welding model.

absorption involves damping of the electromagnetic radiation field of the incoming photons by the electron cloud associated with the lattice, which eventually converts the energy of photons to increased lattice vibrations indicative of localized heating. For a metallic target, the absorption occurs within a fraction of a wavelength distance from the surface (~100 nm) and is described by an exponential decrease in the electric field strength of the laser radiation with penetration distance below the surface. For nonmetals, the laser penetration distance is typically at least 10 times greater (or even >> 10× for transparent materials). Further, the laser absorptivity has a tendency to decrease with increasing wavelengths.

At the irradiated surface of the welded workpiece, the convective boundary can be assumed, and at the rear side of the workpiece, the convective and radiative boundary conditions can be considered. Therefore, the corresponding boundary condition at the irradiated surface and at the rear surface is

$$\frac{\partial T}{\partial z} = \frac{h}{k}\left(T_s - T_{ambient}\right)$$

(10.5)

$$\frac{\partial T}{\partial y} = \frac{h}{k}\left(T_s - T_{ambient}\right) + \frac{\varepsilon\sigma}{k}\left(T_s^4 - T_{ambient}^4\right)$$

(10.6)

respectively, where h is the heat transfer coefficient due to natural convection, T_s and $T_{ambient}$ are the surface and ambient temperatures, respectively, ϵ is the emissivity, and σ is the Stefan–Boltzmann constant ($\sigma = 5.67 \times 10^{-8}$ W/m^2 K). At the farthest boundary (at edges of the solution domain) constant temperature (=293 K) can be assumed.

Equation 10.3 can be solved numerically with the appropriate boundary conditions to predict the temperature field in the substrate. To analyze the phase change problem, a nonlinear transient thermal analysis is performed employing the enthalpy. To account for latent heat evolution during phase change, the enthalpy of the material as a function of temperature is incorporated in the energy equation.

In contrast, keyhole welding requires a high power density to form a narrow deeply penetrating vapor cavity, but the interaction time is short (i.e., welding speed is high) to maximize productivity. Data cover a range of interaction time from 10^{-4} s, which approaches the minimum for noticeable heat transfer by thermal diffusion, to 10 s, beyond which the rate of laser processing normally becomes uneconomically low.

In laser microwelding, laser energy is introduced at the surface, or at least within an absorption length. Thermal conduction into the material from this surface heating is well understood and easily modeled both analytically and numerically. However, this simple behavior becomes complex when the energy transport occurs at a high enough rate to cause melting and evaporation as it occurs in keyhole welding. Generation of plasma above the laser irradiation point can cause scattering and absorption of the laser at long wavelength. This may alter the energy distribution of the laser beam reaching the surface. Since these effects tend to be transient and somewhat chaotic, the thermal modeling approach becomes complex. Further, as the solid material melts, convective transport within the fusion zone becomes important, even predominant, with forces due to buoyancy, aerodynamic drag, surface tension, and Marangoni force due to the temperature dependence of the liquid–vapor surface energy. The surface to volume ratio of microparts is typically much larger than that of macroparts, and hence the effect of part dimensionality should also be considered in the analysis. Further, in laser–material interaction for ultrashort (picosecond or femtosecond) pulse energy input, equilibrium may not be established between the electrons and the lattice. As the conventional thermal diffusion models are not adequate for such nonequilibrium conditions, a two-temperature model is used to account for this phenomenon (Lee and Asibu, 2009).

10.3.2 Laser Microwelding System

Figure 10.3 shows a typical laser microwelding setup with a camera-based weld-imaging system for microweld monitoring. Various process parameters used for controlled microwelding are also described in Figure 10.3. On the basis of microwelding application requirements, a near-diffraction limited laser and beam delivery system capable of producing focused laser spots of <100 μm diameter can be considered. In order to take advantage of the small spot size, a motion control system, whether it moves the laser spot, the part, or both, should be of at least 1 order of magnitude better positional accuracy than the spot diameter. Motion control will need to be computer controlled unless only spot welding is contemplated. Further, it is difficult to make reliable microwelds at such size scales unless a high-quality vision system of appropriate magnification is incorporated into the system. Finally, depending on the speed capability, an average laser power requirement may be estimated. The faster the motion system, the higher may be the power employed, which can be estimated on the basis of the thermal model.

CO_2 ($\lambda = 10.6$ μm), Nd^{3+}:YAG ($\lambda = 1.06$ μm), and diode lasers can be used for direct energy conduction welding of metals and alloys. The absorptivity of metals increases as the wavelength decreases. Figure 10.4 shows the minimum laser power needed to melt the selected metals with an Nd^{3+}:YAG laser at 1.06 μm and 0.532 μm, and with a focused spot diameter of 50 μm, and it can be estimated from Equation 10.2. Silver and copper have

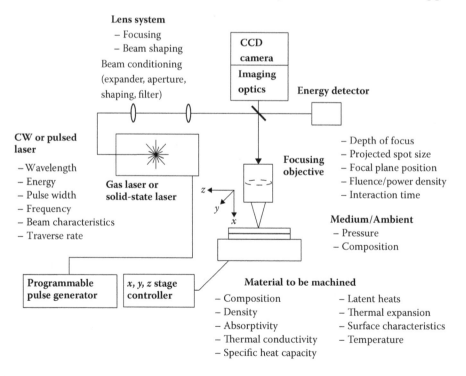

FIGURE 10.3
Laser microwelding system and process parameters.

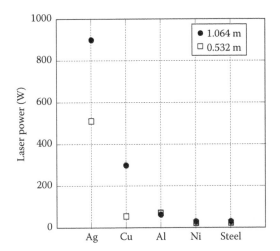

FIGURE 10.4
Minimum laser power required for melting different metals with Nd^{3+}:YAG laser at wavelengths of 1.064 μm and 0.532 μm. Spot diameter 50 μm at $1/e$ of maximum power.

low absorptivity for CO_2 laser and the fundamental wavelength of the Nd^{3+}:YAG laser. Therefore silver and copper require considerably higher laser power as compared to other metals. Further, as a typical example, the absorptivity of copper is ≈10% at the fundamental wavelength of the Nd^{3+}:YAG laser, whereas it increases to ≈50% at the second harmonics (λ = 0.532 μm) generation. Therefore copper can be welded at considerably lower power with the Nd^{3+}:YAG laser at λ = 0.532 μm. Surface condition of components to be welded can also influence the absorptivity.

The absorption bands of most polymers lie in the wavelength range of 2–10 μm. CO_2 lasers and those diode lasers that produce an output wavelength >2 μm are therefore suitable for polymer (plastic) welding. Laser output with a wavelength in the transmission range of most polymers (0.4–1.5 μm) is suitable for transmission welding; Nd^{3+}:YAG laser, fiber lasers, and many diode lasers are thus suitable sources of energy. Since conduction welding is normally used with relatively small components, the beam is delivered to the workpiece via optics with a simple beam defocusing to a projected diameter that corresponds to the size of the weld to be made. Spot welds are made by using pulsed lasers to melt sufficient material to form the joint. Continuous welds may be made by overlapping pulsed spot welds or by using a CW laser. Processing is normally carried out at room temperature in a clean environment. Molten weld metal is protected from environmental contamination by a quiescent blanket of inert shielding gas such as argon.

Along a manufacturing line, it is not easy to maintain the output power and the required characteristics of the laser beam in long-term operation in order to produce good quality weld beads. Weld quality is also affected by variation in materials, gap in butt or lap welding, surface contamination, fixturing, and so on. Therefore, it is desirable to evaluate the weld via *in situ* monitoring. The quality of microbeads can be ensured by real-time control of the laser parameters based on an *in situ* monitoring technique. In recent years, a variety of in-process monitoring techniques have been developed such as measuring the emission spectrum from the plume, measuring thermal emission, laser beam reflection analysis, charge coupled device (CCD) camera-based measurements, and measurement of sound pressure and acoustic waves. In the CCD camera-based system, the weld bead is illuminated by a line-shaped laser beam, and dimensions such as bead width and height, undercut, workpiece misalignment can be evaluated.

10.3.3 Defects in Laser Microwelding

Various defects, associated with material properties and discontinuity of the joint, are encountered in laser microwelding. The material-related defect is related to the changes in microstructure during melting and resolidification. Owing to high surface to volume ratio of micro fusion areas, sensitivity to contamination increases. Before joining, materials should be cleaned and baked to remove moisture and contaminants from the surface.

Defects are also encountered when welding two dissimilar materials with different metallurgical and physical properties. The formation of certain metallurgical phases in these materials could result in brittleness. To solve these problems and to obtain welds with adequate properties, it is essential to precisely control the process and the process parameters. Modification in the temporal shape of the pulse laser and hence controlling of the heating or cooling rates of the weld pool or variation of the position of the laser beam with respect to the joint and supply of different energies to the two materials could compensate for the differences in absorption of the laser beam and in the thermal conductivities (Berretta et al., 2007).

Discontinuity-type defects, such as lack of fusion, lack of fill, disbonding, distortion, porosity, humped bead, and burn-through, are encountered in laser microwelding. Lack of fusion results from insufficient energy delivered to the joint. Lack of fill (underfilled bead) is encountered because of (a) pulsed laser welding with too steep a pulse shape resulting in a splashing out of molten metal out of the keyhole and (b) large gap between lap or butt joints, which can be overcome by reducing the gap.

Disbonding can be caused by overheating during processing or insufficient cleaning. Distortion arises from localized heating and can be controlled by correct joint design and fixturing. Porosity may occur when gas is entrapped, particularly in blind joints, and can be minimized by designing joints that are able to vent vaporized materials. When the ramp-down period in pulsed welding is too short, the smooth metal flow needed to refill the keyhole cavity is prevented, the keyhole sides collapse, and bubbles are retained in the molten pool. The rapidly solidifying pool traps bubbles/pores before they can float to the surface. Humped beads may result owing to uncontrolled welding speed. The molten metal ahead of the keyhole is transferred around the perimeter of the keyhole as it moves forward. The metal flow produces a pressure drop at the keyhole perimeter, and a vortex flow takes place behind the keyhole. Both phenomena combine to make the molten metal flow unstable, resulting in a bead that exhibits regularly arranged humps after solidification. The flow velocity decreases with decreasing keyhole diameter. Hence, humping can be suppressed by decreasing the welding speed and the keyhole diameter.

10.4 Laser Microwelding Applications

Laser-assisted microjoining techniques are used in the microelectronics and biomedical industries. Some combinations of dissimilar materials can be joined using conduction welding by positioning the beam within the material of highest melting temperature, such that heat is conducted into the other material to cause limited interfacial melting with only a thin intermetallic interface. Pulsed lasers are used for a variety of welding applications in the electronics industries, including the attachment of wires to contacts and sealing of electronic packages. Many hermetic semiconductor packages are made from aluminum alloys with a wall thickness of about 1 mm and they contain delicate components that are easily damaged by heating. The increasing use of laser soldering in the microelectronics industry is being driven by the need for a precision joining process capable of handling high densities of fine joints. Possible joint geometries are limited by access of the laser beam and by fixturing considerations. Thus, butt, lap, corner, edge, fillet, T-joints, and the cross-wire geometry for wires are all possible. Figures 10.5 and 10.6 show lap-welding and butt-welding approaches used in laser microwelding, respectively.

The biomedical sector is also examining the potential application of laser soldering as a means of joining tissue without the need for sutures. Biomedical solders are based on albumin proteins that coagulate on interaction with the laser beam to form the joint. A lower energy is used than with laser welding in such procedures, and so the risk of damage to the surrounding tissue is reduced. Its use in biomedical applications is driven by the desire for minimally invasive surgical joining procedures in which the surrounding tissue is undamaged. Various applications of laser microwelding have been described by different research groups (Zhou et al., 2009; Guo, 2009; Miyamoto and Knorovsky, 2008; von Bulow et al., 2009; Sari et al., 2008; Haberstroh et al., 2006).

10.4.1 Typical Laser Microwelding Applications

Some examples of laser microwelding include (a) welding of 0.2 mm thick aluminum alloy case to the 1-mm-thick end cap in lithium ion batteries by a pulsed Nd^{3+}:YAG laser

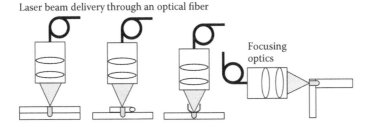

FIGURE 10.5
Laser-assisted micro-lap-welding techniques: (a) lap joint, (b) lap-on-terminal, (c) cup joint, and (d) side lap.

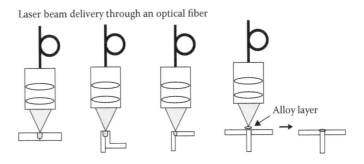

FIGURE 10.6
Laser-assisted micro-butt-welding techniques: (a) butt joint, (b) brim joint, (c) corner joint, and (d) hermetic seal welding.

(1.5 J/pulse, 200 Hz) at 30 mm/s; (b) seam welding of pressed metal sheets used in small size motors for electronic devices, including CD/DVD players and hard disk drives, using a pulsed Nd^{3+}:YAG laser with three split beams simultaneously; (c) spot welding of (120 μm diameter) hard disk suspension made from a stainless steel foil using a pulsed Nd^{3+}:YAG laser; (d) laser welding of optical fiber connectors producing a strong weld joint with a minimal thermal distortion. Multiple location welding is performed simultaneously by split pulsed Nd^{3+}:YAG laser beams to compensate for thermal deformation; (e) laser microwelding of dissimilar materials such as steel–copper, steel–aluminum, and steel–nickel; and (f) ultrashort laser microwelding of silicate glass substrates based on a localized heat accumulation effect using an amplified femtosecond Er^{3+}–fiber laser with a wavelength of 1558 nm and a repetition rate of 500 kHz.

10.4.2 Recent Advances in Laser Microwelding

10.4.2.1 Microwelding of Polymers

There is an increasing demand for more reliable and flexible methods of joining polymer parts. The NIR laser system, which is suitable for welding most metals, can also be considered for polymers and some glasses. Joining two polymer parts using transmission laser welding is usually achieved by one of two approaches (von Bulow et al., 2009; Sari et al., 2008; Haberstroh et al., 2006). One approach makes use of an intermediate absorbing layer at the interface between the joining parts. The absorbing layer is applied as a solvent using ink-jet printing technologies or other dispensing methods, but these methods are challenging additional process steps for high-throughput industrial production lines. In the other approach shown in Figure 10.7, the top part is transparent at the laser wavelength and the bottom part comprises absorbers, which can be pigments or dyes. The most widely used absorber is carbon black (Sari et al., 2008). As shown in Figure 10.7, the joining parts must be pressed on each other. The energy of the laser beam is transformed into heat in the contact zone. In laser transmission welding of plastics, melt is generated in the joining area.

For medical device applications, carbon black is not an acceptable material, and a better alternative is NIR absorbers, including Lumogen (IR1050) marketed by BASF and Pro-Jet marketed by Fujifilm Imaging Colorants. These dyes are highly absorbing in the NIR range of the electromagnetic spectrum and affect the visible range only to a moderate

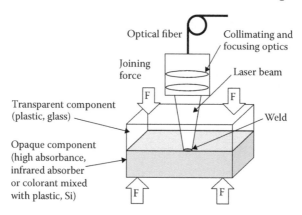

FIGURE 10.7
Laser microwelding of polymer or brittle–rigid materials, such as glass–Si.

extent. It is therefore possible to tailor a particular color and obtain sufficient absorbance at the laser welding wavelength to get a weld of sufficient tensile strength. Welds can be made in three-dimensional configurations by designing the polymer transparent to the laser wavelength at positions where no welds are desirable and designing NIR-absorbing polymer parts where welds are desirable. Transmission laser welding therefore opens up new fields of application and offers design possibilities for polymer products that can only be processed by this welding technique. Laser sources usually employed in this process are Nd^{3+}:YAG ($\lambda = 1064$ nm) and high-power diode lasers ($\lambda = 800$–1100 nm).

10.4.2.2 Laser Microwelding of Brittle–Rigid Materials

In laser-assisted bonding of brittle–rigid materials, such as a glass–silicon combination, one of the parts (glass) to be joined is transparent and the other part (silicon) is absorbent for the laser radiation used as shown in Figure 10.7. The main part of the energy of the laser radiation (Nd^{3+}:YAG laser, $\lambda = 1064$ nm) is converted into heat energy at the interface at absorption lengths of the order of several nanometers. The temperature increases at the interface and the joint is performed by thermochemical activation of the materials with a subsequent formation of covalent oxygen bonds (siloxane bonds). The heat energy assists the formation of chemical bonds at the interface and activates the adhesion forces. To achieve good bond results, exact physical contact between the sample surfaces must be assured.

In addition to bonding of dissimilar material combinations such as silicon and glass, the laser-assisted bonding technique can also be used to join similar material combinations such as silicon–silicon and glass–glass. The application of a suitable wavelength of the laser radiation is important for joining of similar materials such as silicon to silicon. A sufficient transmission of the laser radiation by the upper silicon sample must be guaranteed. In this case, the laser radiation is transformed into heat by the application of absorbent metallic interlayer in the joining zone. As an example, a CW thulium fiber laser ($\lambda = 1908$ nm, laser spot ~41–50 µm) with a maximum laser power of ~52 W was used for Si–Si microwelding (Sari et al., 2008). Silicon has a transmission capacity of ~55% for this wavelength and shows almost no absorption. Sputter-coated platinum/palladium (Pt/Pd, 80:20) with a thickness of 50 nm was used as an interlayer with an absorption of ~25% at the wavelength $\lambda = 1908$ nm.

Fusion welding of glass is very difficult as it requires high heat input, and the thermal gradient induces cracks. Fused silica glass, which possesses a very small thermal expansion coefficient, can be laser welded. Most glass joining involves adhesive bonding with relatively poor mechanical, thermal, and chemical durability. Recently, ultrashort laser pulses were used for lap welding of glass plates without any addition of colorant or absorbing material (Miyamoto and Knorovsky, 2008; Itoh et al., 2009; Watanabe et al., 2007; Tamaki et al., 2006; Miyamoto et al., 2007a,b). In this technique, application of picosecond (Miyamoto et al., 2007a) and shorter pulse length (few tens of femtosecond order) (Miyamoto et al., 2007b) lasers can be considered for controlled heating via a nonlinear absorption mechanism, and crack-free welding is possible even in glass with a large thermal expansion coefficient. Ultrafast laser pulses allow strong and localized nonlinear optical interactions with transparent materials at the focal point. The intensity of an ultrashort pulse laser is high enough to drive highly nonlinear absorption processes in materials that do not normally absorb the laser wavelength. When ultrashort laser pulses are focused onto bulk glass with a bandgap larger than the photon energy, free electrons are generated in the conduction band through multiphoton ionization. The free electrons subsequently gain kinetic energy from the electric field to produce more free electrons by avalanche ionization. Their energy is transferred to the lattice to provide a temperature rise in the bulk glass after the laser pulse. The absorbed laser energy is regarded as an instantaneous heat source since the energy transfer from the electrons to the lattice is much faster than thermal diffusion in the lattice (Miyamoto et al., 2010). Owing to localized properties of the nonlinear absorption, such as multiphoton and/or tunnel absorption and avalanche ionization, the femtosecond laser-assisted microwelding technique can be realized without using a light-absorbing intermediate layer. As multiphoton absorption does not depend on the presence of free electrons and on the linear absorption at the laser wavelength, any material can be processed. Thermal stress is significantly reduced because the molten pool is localized in the vicinity of the focus, which allows direct welding of similar and dissimilar glass substrates, and glass and transparent semiconductor substrates. In addition, welding of glass and metal is possible with reduction in unwanted thermal effects, as focused femtosecond pulses can directly melt not only metal but also glass through nonlinear absorption.

As shown in Figure 10.8a, when an ultrashort pulse laser is focused at the interface of two glass plates, laser energy absorbed by a multiphoton absorption process is transferred to the glass material after the laser pulse, and then the surrounding area is heated by thermal conduction. Only the interface of the two plates is heated and melted since the glass material is transparent at the laser wavelength except near the focused spot size. The liquid pool created at the interface fills up the gap between the two materials. The subsequent resolidification of the liquid pool joins the two materials. The focal region elongates along the optical axis due to nonlinear propagation such as filamentation. The filamentary propagation of femotosecond laser pulses bridges the two substrates along the laser propagation axis (Miyamoto et al., 2007b, 2010). Filamentary propagation is superior for laser welding because the elongated liquid pool means that it is not necessary to translate the focal spot along the laser beam axis.

The joining of dissimilar glasses, that is, borosilicate glass and fused silica, whose coefficients of thermal expansion are different, was demonstrated by the use of 1 kHz femtosecond laser pulses (Watanabe et al., 2007). Welding of nonalkali glass substrates was also demonstrated using an amplified femtosecond pulse at a wavelength of 1558 nm and a repletion rate of 500 kHz from an Er^{3+}–fiber laser system. Welding of a nonalkali glass substrate and a silicon substrate, which is transparent at 1558 nm, was also demonstrated successfully (Tamaki et al., 2006). The nonlinear absorptivity of the femtosecond laser pulses

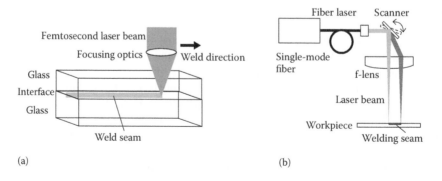

(a)   (b)

FIGURE 10.8
Laser microwelding: (a) ultrashort laser-assisted microwelding of glass plates and (b) single-mode fiber laser-assisted, high-speed scanning for welding seam.

is strongly affected by the repetition rate of the laser pulses. This is because the temperature just before the pulse arrival increases as the repetition rate is increased so that the
contribution of thermally excited electrons to the conduction band increases to enhance
the avalanche ionization. Femtosecond pulses are proved to be advantageous compared
with nanosecond pulses for joining of glass and metal with a reduced average power,
which effectively suppresses unwanted thermal effects (Ozeki et al., 2008). On the basis of
the comparison between picosecond and femtosecond ranges, the picosecond order range
is shown to be more attractive owing to not only the simplicity of the laser system but also
the higher melting and joining efficiencies as a heat source for fusion welding of glass,
because the picosecond range provides higher nonlinear absorptivity due to a larger contribution of avalanche ionization (Miyamoto et al., 2007b).

10.4.2.3 Microwelding via High-Speed Laser Scanning

The Fraunhofer Institute for Laser Technology (ILT) has developed the SHADOW® welding (Stepless High-Speed Accurate and Discrete One-Pulse Welding) technique, in which
high-speed joining with minimum distortion is possible using a pulsed Nd^{3+}:YAG laser
(pulse width 20 ms). A single pulse from a pulsed laser is combined with the rapid motion
of a workpiece or scanning of the laser beam. The processing length is determined by the
combination of pulse duration and velocity of the laser beam. The energy density can be
calculated as follows:

$$E = \frac{P}{Vd}$$

$$(10.7)$$

where E is the energy density, P is the laser power, V is the beam scanning velocity, and d is
the beam diameter at focus point. Initial application of this technique was in the mechanical watch industry where gear-to-shaft welds of extreme regularity and without any ripple
on the surface were demonstrated. Figure 10.8b shows the schematic of the setup of a
single-mode fiber for high-speed laser scanning application. The possibility of high-speed
welding with a single-mode fiber laser (1090 nm) has also been reported (Okamoto et al.,
2008). The combination of microbeam and high-speed laser scanning (600–2000 mm/s)
was possible for thin metal sheet welding. A small focus diameter of 22 μm was found in
general to provide good control of the penetration depth at the energy densities <1 J/mm².
The overlap welding of 25 μm thickness was successfully performed, regardless of the
small butt-gap distance between two sheets. At the velocity of 666 mm/s and 833 mm/s,
the excellent fine welding without humping was performed with bead width <30 μm. The
welded joint shows a smooth surface compared to the spaced spot welding.

10.5 Summary

Laser has the capability to focus down to a few micrometers, which facilitates high local
resolution, low heat deposition in the workpiece, and a high level of flexibility. Hence,
lasers can be used for various microwelding applications such as electronics, sensor systems, MEMS, and biomedical devices. CW or pulsed Nd^{3+}:YAG lasers, fiber lasers, and
diode lasers with high beam quality enable highly reproducible joining of metals with

weld seam widths below 100 μm with feed rates up to 1 m/s. Even combinations such as steel–copper or steel–aluminum can be joined where conventional welding processes used to be unsuccessful. Laser beam welding of thermoplastic polymers provides weld seams of high optical and mechanical quality. By adapting the laser wavelength to the absorption behavior of the polymers, even transparent materials can be welded without additional absorbers. Laser bonding of silicon and glass is a nonmelting solid-state joining process and is based on the formation of oxygen bridges. This process is particularly suited for bonding and encapsulation of microsystems with moving parts and thermally sensitive elements. By the use of lasers with wavelengths in the 1500–1900 nm range, silicon–silicon can be joined. Laser microwelding of transparent materials, such as glass–glass, based on a localized heat accumulation effect using an ultrashort laser is also possible. In this approach, an amplified femtosecond laser or a picosecond laser, with a high repletion rate is used. This is a useful technique for semiconductor materials, which may have various applications such as three-dimensional stacks, assembly of electronic devices, and fabrication of MEMS devices. Recently, a high-speed laser scanning technique (600–800 mm/s) with both single-mode fiber laser and pulsed Nd^{3+}:YAG laser has been studied for microwelding of thin stainless sheets.

References

Berretta, J.R., de Rossi, W., das Neves, M.D.M., de Almeida, I.A., and Junior, N.D.V. 2007. Pulsed Nd:YAG laser welding of AISI 304 to AISI 420 stainless steel. *Optics and Lasers in Engineering* 45: 960–966.

von Bulow, J.F., Bager, K., and Thirstrup, C. 2009. Utilization of light scattering in transmission laser welding of medical devices. *Applied Surface Science* 256: 900–908.

Chryssolouris, G. 1991. *Laser Machining: Theory and Practice*. New York: Springer-Verlag.

Dahotre, N.B. and Harimkar, S.P. 2008. *Laser Fabrication and Machining of Materials*. New York: Springer Science + Multi Media.

Guo, K.W. 2009. A review of micro/nano welding and its future developments. *Recent Patents on Nanotechnology* 3: 53–60.

Haberstroh, E., Hoffmann, W.M., Poprawe, R., and Sari, F. 2006. Laser transmission joining in micro-technology. *Microsystem Technologies* 12: 632–639.

He, X., Elmer, J.W., and DebRoy, I. 2005. Heat transfer and fluid flow in laser microwelding. *Journal of Applied Physics* 97: 0849091–0849099.

Ion, J.C. 2005. *Laser Processing of Engineering Materials*. Oxford, MA: Elsevier Butterworth-Heinemann.

Itoh, K., Watanabe, W., and Ozeki, Y. 2009. Nonlinear ultrafast focal-point optics for microscopic imaging, manipulation, and machining. *Proceedings of the IEEE* 97: 1011–1030.

Lee, D. and Asibu, E.K. Jr. 2009. Numerical analysis of ultrashort pulse laser-material interaction using ABAQUS. *ASME Journal of Manufacturing Science and Engineering* 131: 021005–021015.

Miyamoto, I. and Knorovsky, G.A. 2008. Laser microwelding. *Microjoining and Nanojoining*, ed. Y. Zhou. Cambridge: Woodhead Publishing Ltd. pp. 345–417.

Miyamoto, I., Cvecek, K., Okamoto, Y., and Schmidt, M. 2010. Novel fusion welding technology of glass using ultrashort pulse lasers. *Physics Procedia* 5: 483–493.

Miyamoto, I., Horn, A., and Gottmann, J. 2007a. Local melting of glass material and its application to direct fusion welding by ps-laser pulses. *JLMN-Journal of Laser Micro/Nanoengineering* 2: 7–14.

Miyamoto, I., Horn, A., Gottmann, J., Wortmann, D., and Yoshino, F. 2007b. Fusion welding of glass using femtosecond laser pulses with high-repetition rates. *JLMN-Journal of Laser Micro/Nanoengineering* 2: 57–63.

Okamoto, Y., Gillner, A., Olowinsky, A., Gedicke, J., and Uno, Y. 2008. Fine micro-welding of thin stainless steel sheet by high speed laser scanning. *JLMN-Journal of Laser Micro/Nanoengineering* 3: 95–99.

Ozeki, Y., Inoue, T., Tamaki, T., Yamaguchi, H., Onda, S., Watanabe, W., Sano, T., Nishiuchi, S., Hirose, A., and Itoh., K. 2008. Direct welding between copper and glass substrates with femtosecond laser pulses. *Applied Physics Express* 1: 082601.

Ready, J.F. 1971. *Effects of High-Power Laser Radiation*. New York: Academic Press.

Sari, F., Hoffmann, W.M., Haberstroh, E., and Poprawe, R. 2008. Applications of laser transmission processes for the joining of plastics, silicon and glass micro parts. *Microsystem Technologies* 14: 1879–1886.

Seuzalec, A. 1987. Thermal effects in laser microwelding. *Computers and Structures* 25: 29–34.

Tamaki, T., Watanabe, W., and Itoh, K. 2006. Laser micro-welding of transparent materials by a localized heat accumulation effect using a femtosecond fiber laser at 1558 nm. *Optics Express* 14: 10460–10468.

Watanabe, W., Onda, S., Tamaki, T., and Itoh, K. 2007. Direct joining of glass substrates by 1 kHz femtosecond laser pulses. *Applied Physics B* 87: 85–89.

Yilbas, B.S., Arif, A.F.M., and Abdul Aleem, B.J. 2010. Laser welding of low carbon steel and thermal stress analysis. *Optics and Laser Technology* 42: 760–768.

Zhou, W., Hu, A., Khan, M.I., Wu, W., Tam, B., and Yavuz, M. 2009. Recent progress in micro and nano-joining. *Journal of Physics: Conference Series* 165: 012012.

11

Electron Beams for Macro- and Microwelding Applications

D.K. Pratihar and V. Dey

Indian Institute of Technology Kharagpur

A.V. Bapat and K. Easwaramoorthy

Bhabha Atomic Research Centre

CONTENTS

11.1 Introduction

Electron beam welding (EBW) is a metal-joining technology that is about 60 years old [1]. Here, a high-energy focused electron beam (spot diameter of a few microns to a few millimeters) is used as an intense heat source for joining two metals. This intense heat source has an exceptional capability to raise the material to virtually any temperature in an extremely short duration. Because of this unique property, EBW offers the following advantages:

- Joining dissimilar metals of largely varying melting points
- Narrow and deep welds
- Ability to join both thick and thin sections
- Low heat-affected zone
- Joining highly reactive metals
- Filler-material-free weld

This chapter introduces the principle of EBW. The applicability of electron beams to both macro- and microwelding has also been discussed. Macro EBW technology was initiated mainly by the requirements of the nuclear and aeronautical industries and micro EBW by the micro electro mechanical systems (MEMS) and semiconductor industries. Assembly of miniature chemical test kits (micro total analysis systems) and joining of carbon nanotubes are some promising areas in which electron beam microwelding (EBμW) can be successfully applied. EBμW can be thought of as a scaled-down, high-precision version of the conventional EBW. EBμW machine development is expected to play a major role in realizing modern, cutting-edge technologies. However, the development of the EBμW machine has not yet fully matured, although the working principles of macro and micro EBW are the same. The following section gives a brief idea of conventional EBW machines before introducing some concepts of EBμW machines.

11.2 Description of an EBW Setup

An EBW machine is an engineered ensemble of various components. Multidisciplinary expertise in electrical, electronics, vacuum, and mechanical engineering is required to realize the hi-tech metal-joining process on the shop floor. An EBW machine consists of a vacuum chamber fitted with an electron gun column. The main components of an EBW gun column are as follows:

- An electron gun assembly consisting of a bias, anode, and a filament
- An electromagnetic (EM) beam focusing lens
- An EM beam deflection and oscillation lens

The electron source is energized with an externally connected electrical power source through high-voltage cables. The system is also provided with control systems, safety features, and gadgets. A job-handling system facilitates the holding and manipulation of the weld job inside the vacuum chamber. A schematic view of an EBW machine is shown in Figure 11.1.

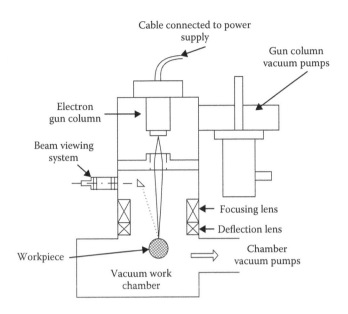

FIGURE 11.1
Schematic of an EBW machine.

11.3 Design Considerations of the Electron Gun Column

A well-engineered EBW machine consistently produces a superior metallurgical and aesthetic weld quality. EBW machine development has matured to a great extent, contributing toward the ever-growing field of materials science and technology. However, the following issues are receiving the attention of EBW machine developers from all over the world:

- Miniaturization
- Operator friendliness
- Machine intelligence
- Low production and maintenance time
- Standards for beam quality

An electron gun is a device capable of producing a continuous, high-velocity, directed stream of electrons. The electron gun assembly consists of an electron source and the electrodes necessary to produce an electrostatic field of the required magnitude and distribution.

11.3.1 Electron Source

An electron source is capable of supplying an abundance of free electrons continuously with time. This component is also called a *cathode*. A wide variety of cathode shapes is used in electron guns, such as simple hairpin, helical, spiral, or strip. The choice of the emitter shape is constrained by the required beam shape at the target. A comparison of these shapes can be seen in Reference 2.

Emission of electrons from incandescent bodies (i.e., thermionic emission) is the most preferred method of electron generation for welding applications among the many emission methods available [3]. The reason is that thermionic emission is a stable, long-duration, continuous source of high-electron current. In this method, electrons are emitted from a material heated to incandescence in vacuum. Tantalum and tungsten are widely used as emitters in such electron guns. Tungsten–rhenium and lanthanum hexaborate are also reported as cathodes [4]. Conventionally, an electric current is passed through the emitting material to raise its temperature. Other methods such as bombarding the cathode with low-velocity electrons or laser beams have also been reported. In addition to the above, the emitter should have the following characteristics:

- Shape stability at high (more than 2000°C) temperatures
- High ratio of extraction to radiation area (emitter efficiency)
- Low lead losses
- High emitter life
- Dimensional tolerances achievable during manufacture

Moreover, a good filament should have high shape stability, emitter efficiency, and life.

11.3.2 Electrostatic Lens

The electron cloud generated by thermionic emission explained earlier is required to be shaped and accelerated suitably. The kinetic energy of an electron increases when subjected to a time-independent electric field in vacuum. An electron appearing with negligible initial velocity near the cathode surface is accelerated toward the anode by the electric field between the electrodes. The electric field is produced between suitably shaped (contoured) electrodes (Figure 11.2), which are connected to various direct current power sources. When a fairly large number of electrons (e.g., $\sim10^{18}$) are accelerated, the resultant swarm is called an *electron beam*.

The technique for contouring the electrodes is fairly complex. Conventionally, this problem is solved using electron optics, where an analogy is drawn between the electron motion in electric (and magnetic) field and that of light rays in a refracting medium. This analogy is fully explored in *electron optics*. The desired electrode shapes are arrived at using the principle of ray tracing (a classical method in light optics) by either analytical or numerical methods [5]. The exquisite properties of electrons such as relativistic effects, mutual coulomb repulsion, self-generated magnetic fields, and random velocities at the emission point are included for more accurate calculations. The aberrations in light optics are equally valid for electron optics as well [3,6,7]. Alternatively, for complex electrode geometries, the EM fields are computed numerically by the finite difference, finite element, boundary element, finite integration, or hybrid method. The trajectories are then computed by solving the relativistic analogs of Newton's equations of motion. The space charge and self-generated magnetic field effects are included by superimposing the EM fields produced by the electrodes alone on the beam-generated EM fields using iterative methods until a self-consistent solution is obtained. Figure 11.3 shows the photograph of an 80-kV, 12-kW EBW gun (developed by BARC) in open condition.

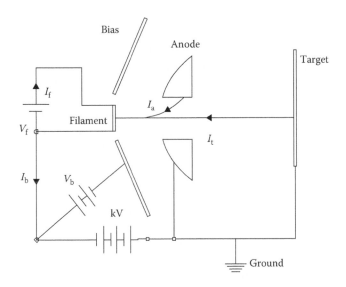

FIGURE 11.2
Line diagram of an electron gun and its power supplies. (V_f, filament heating voltage; I_f, filament heating current; V_b, bias voltage; kV, acceleration voltage; I_b, beam current; I_t, target current; I_a anode current.) The currents indicated in the circuit are the conventional currents whose direction is opposite to that of the electron flow.

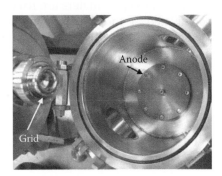

FIGURE 11.3
Photograph of the fabricated 80-kV, 12-kW EBW gun in open condition (filament is not seen).

11.3.3 Electromagnetic Lens and Its Basic Design

The electron beam formed by the electron gun is transported to the target with the help of lenses. These devices, which work on EM forces, bring back the diverging electron beam emerging out of the electron gun onto a target (usually placed a few hundred millimeters away). A diverging beam from an electron gun is caused by:

- Diverging lens effect of the anode aperture
- Coulomb repulsion between electrons (space charge effect)
- Thermal velocity distribution of electrons at the cathode (in thermionic emission guns)

EM lenses are used for focusing and transporting the electron beams onto the target. When an electron enters the magnetic field produced within an axisymmetric solenoid, it experiences a force that always acts toward the axis irrespective of the direction of the axial field component and is proportional to the radial distance of the electron from the axis. These are the conditions for image formation by a lens. Since the magnetic field strength is proportional to the coil current, the focal distance of the lens, which is a function of the magnetic field strength, can be varied by changing the coil current. The image obtained by an EM lens will be a rotated image rather than an upright one.

For practical purposes, a simple air core coil is insufficient owing to the large consumption of electrical power and the weight of copper to obtain a given magnetic field at the center. It is hence essential to concentrate the magnetic field in a small volume. A great economy in electrical energy is gained by shrouding the coil with a high-permeability magnetic material. The thickness of the magnetic material is chosen in such a way that it does not run to magnetic saturation [8]. Conventionally, pure soft iron is recommended. Magnetic stainless steel (AISI 400 series) is an alternate choice. The number of turns in the EM lens (axisymmetric solenoid) does not affect the properties of the lens. However, fewer turns will require a large current, increasing the electrical I^2R loss. On the other hand, a large number of turns makes the coil bulky and electrically sluggish. The number of turns should hence be judiciously chosen. The coil should be separated from the vacuum by a proper mechanical design. There are cases where hermetically sealed coils are used. A coil outside the vacuum is much easier to design and fabricate and is recommended for EBW gun columns.

Bore diameter of the EM lens (D), air gap (S), and number of turns are the geometric parameters characterizing the lens. These parameters are fixed at the time of design. The lens excitation current can be adjusted by the operator for varying the focal length, and hence, the beam can be focused at various job distances. The beam cannot be focused by an EM lens to a point on the target, and is limited by some defects (called *aberrations*) listed below.

- Spherical aberration
- Space charge aberration
- Thermal velocity aberration
- Chromatic aberration
- Lens machining tolerances
- Lens alignment (tilt and axial shift)

Spherical aberration is due to the different focal distances for paraxial (near-axis rays) and marginal rays (rays far away from the axis). The aberration is positive (longer focal length for paraxial rays and shorter focal length for marginal rays) for magnetic lenses. This causes a point source to be imaged as a disc. Space charge aberration occurs because of the mutual electrical repulsion of electrons. The electrons are hence brought to a minimum spot of finite diameter rather than a geometrical point, for a given radial force (in case of axisymmetry). Interested readers may refer to References 6 and 7 for a more elaborate discussion on other defects.

An important component for the operation of an electron gun is its power source. Any variation in the acceleration, filament grid, and focusing coil supply adversely affects the gun performance. Hence, great care has to be exercised in the design and fabrication of these power supplies. The deflection and oscillation lens is another important gadget in the EBW gun column. This is used to obtain a precise beam alignment or to steer the beam on difficult-to-access joints and to overcome voids appearing in the weld root. The concepts run along the same lines as those of an EM focusing lens, and hence are not dealt with in this chapter.

The engineering design of an electron gun column in accordance with the structural, vacuum, and high-voltage design principles requires rich design and practical experience. The engineering design incorporates the following points:

- Material selection and qualification
- Tolerance on the components
- Welding qualification
- Choice of vacuum pumping, vacuum plumbing, and instrumentation
- Assembly procedure and quality assurance
- Water cooling requirements for the critical components (e.g., anode)
- Operator friendliness in servicing and maintenance

The complete 12-kW electron gun column of an EBW machine, designed according to the above design inputs by the BARC, Mumbai, is given in Figure 11.4a. The electron gun for this column is numerically designed (Figure 11.4b) with a commercial charged particle tracing code. The design objectives are to obtain the required beam current, the desired spot size on the target, and low beam losses. The electron optical ray diagram that forms the base for the mechanical design of the gun column is given in Figure 11.4c.

The authors and their team, with their experience in developing macro EBW machines, are venturing into the development of EBW machines for microwelding. The perspectives on micro electron beams are now addressed.

11.4 Electron Beams for Micro Operations

Micro operations are used for the fabrication of microcomponents of thickness ranging from a few to a few hundreds of micrometers. An excellent review of electron beams and lasers, especially on micro/nanowelding, can be found in Reference 9. The following section deals with the fundamentals of qualifying electron beams as a microwelding tool. The challenges and unsolved issues are discussed in the last section.

11.4.1 Electron Beam as Microwelding Tool

In *macro* electron beams, the spot size (1 to few tens of millimeters) is dominated by coulombic repulsion owing to large beam powers, whereas the beams in electron microscopes (less than a micron) are dominated by aberrations and thermal velocities. *Micro* electron beams are between these two classes of beams. Hence, the first challenge is to handle both the space charge and aberrations simultaneously and optimize the beam spot size. Successful attempts at microwelding have been made with beams having spot size of 40 μm at 5-W beam power [10] and a beam power density of 1×10^6 W/cm^2. This power density is adequate for carrying out welding operations. Electrons are virtually massless ($m_e = 9 \times 10^{-31}$ kg), which enormously eases the steering of electron beams. Hence, the heat input to the job can be precisely controlled. Literature is available on rapid beam sweeping frequencies up to 10 kHz. Additionally, electrons do not contaminate the target material during the interaction. They simply deposit their kinetic energy on the target and mix with the ocean of conduction band electrons present in the target. Thus, all the electron beam processes are contamination free. The generation and transport of an electron beam needs a vacuum of 10^{-6} mbar. All electron beam processes are carried out in a highly inert environment. This facilitates operations on highly reactive metals.

All the points discussed above favor qualifying electron beams for micro operations. However, in the literature, there are very few reports on the development of dedicated industrial micro EBW machines. The first design step is to choose the parameters for electron beam microwelding machines, which is dealt with in the following section.

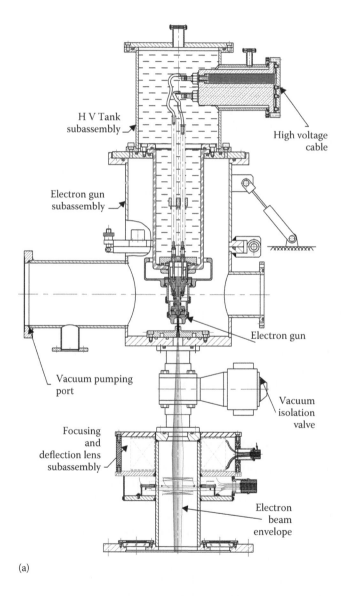

(a)

FIGURE 11.4

(a) Electron gun column of an 80 kV, 12 kW EBW machine developed at BARC, Mumbai.

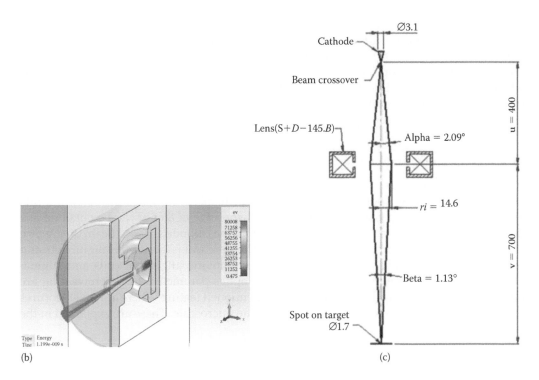

FIGURE 11.4 (continued)
(b) electron trajectories of the 80 kV, 12 kW EBW gun design, and (c) electron optical ray diagram of the gun shown in (a).

11.4.2 Choice of Parameters for Electron Beam Microwelding

The major parameters of electron beam machines are acceleration voltage, beam current, spot size on the target, welding speed, and pressure in the vacuum chamber. Accelerating voltage is the prime factor for achieving beam quality. Let us take the typical example of a 1-W electron beam at accelerating voltages of 10 and 100 kV. The former requires 100 mA of beam current as compared to 10 mA for the latter. The lesser beam current needs a smaller cathode size and hence, a smaller source size. A larger beam throw distance with a reasonably good beam quality can be achieved with a larger accelerating voltage. On the other hand, the electron gun is discharge prone at high voltage, which is detrimental. X-ray generation is more severe at higher voltages, which requires an additional lead shielding of the working chamber [1]. Considering the above pros and cons, one has to take a judicial decision on the choice of the accelerating voltage. Reports of successful welding of microcomponents with 30 kV accelerating voltage are available in the literature. This limits the beam throw distance to about a few tens of millimeters. Attempts are being made to establish different classes of micro electron beam machines with 30–50 kV accelerating voltage and 5–500 W beam power [13].

The spot size for microwelding lies in the range 10–100 μm depending on the beam power. The operating power density is in the range (1×10^6)–(1×10^7) W/cm². With these parameters, jobs of 10–1000 μm thick stainless steel can be carried out. These values are extrapolations of the weld depths obtained with macrobeams. The welding speed reported in the literature is from 25 to 100 μm/s [14] for joining stainless steel wires. Information on welding speed for other materials is not available in the literature. All the micro operations reported in the literature are carried out at high vacuum, that is, 1×10^{-6} mbar. Considering the volume of the working chamber, which is typically about 200 mm × 200 mm × 200 mm, the pumping of the working chamber could be done through a turbomolecular pump, which ensures a dry and clean atmosphere for the process. The broad operating regime for electron beam microwelding is understood from the above discussion. This information leads to the design of electron optics for microwelding.

11.4.3 Electron Optics for Microwelding

Electron optics for microwelding is similar to that of a macro-EBW machine with major emphasis on the beam spot diameter. The basic principle is to image the beam minima obtained in the electron gun using EM lenses. A spot size of 86 μm can be obtained at a throw distance of 120 mm. One can understand that the cathode size and beam divergence angles are smaller than those in macrobeams. The practical issues of misalignments and ellipticity of the beam can be compensated by employing alignment coils and stigmators at suitable locations. Some typical electron optics parameters for microwelding are given in Reference 10. In addition to the above-mentioned features, there are other features reported in the literature to assist operations on the microscale. These are explained in the following section.

11.4.4 Special Features for Micro Electron Beam Machines

The following special features are important for micro electron beam machines:

- Pulsed beam

  Pulsed electron beams are basically discontinuous beam current pulses with time. A pulsed beam is achieved by applying beam-blanking pulses to the bias electrode of the electron gun. The pulsed electron beam enhances the localized heat input, thus increasing the width-to-depth ratio of the weld bead [12]. Pulse frequencies less than 10 kHz can be preset by the operator.

- Multibeam

  Multibeam is a technique by which the beam is skipped between several positions with a speed so high that the thermal influence takes place on the structure at different points simultaneously [14]. Rapid beam movement is achieved with the help of low-current EM coils.

- Imaging facility

  Taking advantage of the available fine electron beams for microwelding, one can fit in the facility of using the same beam for low-magnification imaging (1-μm resolution). Inspection of the welded surface with the help of conventional optics, which is very difficult in the case of microdimensional jobs, is made possible with the imaging facility. This dual mode has been demonstrated for low-power microbeams [10,14], and it is yet to be proved for high-power beams.

- Preloading

  A problem unique to micro electron beam operations is to handle the huge surface tension and the capillary forces in the jobs. This leads to a failure of welded joints due to thermal contraction and deformation. Application of a critically set preload [14] is a promising solution to overcome these problems, especially while joining 30-μm thick foils.

11.4.5 Challenges and Unsolved Issues in Electron Beam Microwelding

Electron beam microwelding has been emerging rapidly during the last few years. Attempts have been successfully made to prove the capabilities of electron beams in carrying out operations on a microscale by modifying the existing scanning electron microscopes. However, no such dedicated industrial machines are available. The following challenges are anticipated and are to be solved for taking this technology to maturity.

- Electron beam characterization

  Characterizing macro electron beams in terms of beam spot size and current density distribution are well known. However, resolving the beam characterization problems for a microbeam is still under investigation [15]. The effect of beam impingement at beam minima or at the image point is yet to be resolved.

- Precise job-maneuvering device

  Job maneuvering at high resolution has been addressed in Reference 16. The DIN standard 32561 (2003) gives the dimensions and tolerances of production equipment for microsystems. An alternate approach to beam maneuvering using an ultrafast beam deflection system [17], rather than job movement, seems to be a feasible solution.

- Weld parametric studies for microjobs

  A database for fixing the parameters mentioned earlier in this section is not available for different materials and thicknesses as in the case of macrowelding [18]. A heat-transfer model for pulsed electron beams for micromilling is proposed in Reference 11. Further generation of data will be useful to the operator for carrying out the actual jobs successfully.

- Quality assurance of welds

  Macrowelds are qualified by either destructive or nondestructive methods. Destructive methods are usually recommended for trial samples before the actual job is carried out. The surface finish is, however, inspected *in situ* with the help of viewing optics. The validity of the quality assurance program for micro operations is yet to be established.

11.5 Statistical Analysis and Constrained Optimization on the Results of Macro Electron Beam Welding Process on ASS 304 Steel

The success of a weld depends on the peak temperature that is achieved by the process, the way heat travels in different directions during welding, and the way cooling takes place in the welded parts. All the above physical phenomena are coded into the solidified shape of the weld pool. A proper understanding of the development of weld-pool shape in EBW has always been a difficult task because of the inherent complexity of the process. Direct experimental investigations are only a few, as they are expensive. Petrov et al. [19] filmed the formation of the shape and size of the weld pool along with the keyhole with a charge-coupled device camera during an experimental investigation.

11.5.1 Analytical Approaches

There have been many theoretical approaches to study the effect of various physical phenomena on weld-pool shape. Modeling the weld pool in the case of EBW, as in any conventional welding process, has always been a challenging task, as there are too many physical phenomena involved in the process.

Hashimoto and Matsuda [20], in 1965, developed an analytical model to relate the depth of penetration, beam parameters, and material characteristics. Here, Hashimoto assumed the electron beam to have a square cross section at a constant power density. It was supposed that there was no interaction between the electron beam and the metal vapor in the capillary and that the fused peripheral zone was at the melting temperature. It was further considered that heat dissipation to the atmosphere was isotropic and that no heat was transported by convection within the fused peripheral zone. Although the last consideration was improper, nevertheless, the most difficult task in this model was to measure the focal spot diameter. Klemens [21] considered energy balance and energy loss mechanisms to predict the penetration of the electron beam. He assumed that the penetration of the beam was mostly governed by conduction heat transfer. The cross section of the electron beam was supposed to be sufficiently narrow with high energy content. The model predicted bead penetration (BP) in both the static and moving states of the beam. During EBW, the shape of the moving molten pool of metal is generally seen to be elliptical. Miyazaki and Giedt [22] attributed the elliptical shape to a uniformly concentrated heat source. They also derived a relationship between weld power and weld penetration.

Vijayan and Rohatgi [23] also assumed the heat source to be linear, and described the physical nature of transient fusion and deep penetration by an electron beam in semi-infinite metal targets. They studied both the formation of the transient fusion zone and the formation of the transient keyhole. In their model, they had supposed that the melting started at a point beneath the surface of the workpiece, and then the melt penetrated deeper into the workpiece. It was also supposed in the model that the line-heat source not only penetrated deep into the workpiece but also spread across the surface owing to conduction. Because of this, the workpiece also melted laterally. The model developed gave a good description of both transient and steady-state fusion. The model helped to distinguish between the melting point isothermal boundary and the fusion boundary of high thermal conductors in general. The model further justified the spreading of a fusion layer laterally by thermal conduction, and thereby, the weld profiles were predicted.

A more realistic model was developed by Elmer et al. [24]. Their model agreed with the point-source solution predicted by Rosenthal [25]. Both the point-source and the distributed-source models were developed on the basis of the assumption that the heating effect was limited to the surface of the workpiece. The authors compared the geometric shapes of the electron beam weld pools with already established distributed-source, point-source, and line-source heat-conduction models. They showed that each model could represent certain EBW regimes and that none of them could represent the entire paradigm. The authors showed that the weld-pool shape depends on the average energy absorbed per unit area on the surface of the workpiece and on the ratio of the beam power to the beam area. For energy densities greater than the critical value, the welds could be simulated both by the point source and by the line source. This critical energy density, which separates the different heating modes, was found to be material dependent. In the case of ASS 304, this was 10 J/mm^2. Finally, the authors could develop an empirical relationship between the penetration depth and the EBW parameters for the distributed-, point-, and line-source heating models. In their model, the weld depth could be predicted only if the weld width was known. Since the weld width could not be known a priori, they substituted that with the focal spot diameter. The greatest problem associated with this model was that any error in the measurement of the focal spot diameter of the electron beam would have resulted in an erroneous relationship between the welding parameters and the weld depth.

Koleva et al. [26] established a correlation between the weld depth and the weld width with operating parameters such as welding velocity, beam power, and position of the beam focus in relation to the sample surface. The authors considered a steady-state model involving a linear, uniformly distributed heat source in a coordinate system that moved with respect to the sample coordinate system. All the analytical models discussed were based on the shape of the source, distribution of the energy of the source, and the effect of boundary conditions on the heat transmission from the source. Coüedel et al. [27] studied the sensitivity of the thermal field to the source size and the effects of boundary conditions. In their study, they considered both line and Gaussian distributed-type heat sources. In actual practice, before the welding of massive plates, generally, the working welding parameters are tried out on smaller plates to check the full penetration. The width of weldment on narrow trial plates was seen to widen because of overheating. The developed analytical heat-transfer model [27,28] made it possible to estimate the critical conditions of weld-bead widening. The model was equally capable of predicting both the full and partial penetration welds. The 2D analytical solutions used Gaussian cylindrical or line models of heat source. However, this model did not consider latent heat, convection heat transfer, and other properties of the welded metal that were independent of temperature.

Ho [29] developed an analytical model to predict the fusion zone of an electron beam weld with a Gaussian profile of the beam, where the focal point could be located either above or below the surface to be welded. The shape of the keyhole cavity was assumed to be a paraboloid of revolution. The coordinate system of the moving frame for this model was assumed to be parabolic. The effects of beam focusing characteristics, such as location of the beam focus relative to the workpiece surface, spot size at the focus, and beam-convergence angle on the fusion zone were investigated. The predicted depths of the fusion zone varied with the location of the focus and the focal spot size. The penetration increased and reached the maximum when the position was found to be slightly below the top surface of the workpiece, and then decreased with further descent. The penetration was found to increase with low focal spot diameter. The transverse sections of the fusion zones, as predicted by the model, were conical with a spherical cap on the top for deep penetration, and were similar to a paraboloid of revolution in the case of shallow penetration.

Rai et al. [30] developed a 3D numerical model of heat transfer and fluid flow in a keyhole mode of the EBW. The model took into account the variation of wall temperature with depth and the effect of Marangoni convection on keyhole walls. Convection was the dominant mechanism of heat transfer in the weld pool, and the gradient of surface tension played an important role in fluid flow. The effect of the Lorentz force was found to be insignificant compared to that of the Marangoni force in their model. Welding parameters, such as beam radius, input power, and welding speed, were seen to have significant contributions on the weld-pool geometry.

11.5.2 Soft-Computing-Based Approaches

Dey et al. [31,32] performed bead-on-plate welding on stainless steel (ASS 304) and aluminum plates (Al 1100) with EBW. Weld runs were performed in accordance with a central composite design (CCD). A detailed study was performed for both the materials mentioned above to find the significance of the weld parameters in the weld-bead geometry. They used a genetic algorithm (GA) with penalty approach to look for welding parameters that would minimize weldment area, while maintaining the maximum BP. They could also replicate the complicated dagger shape of the weld cross section after predicting the weld-bead geometry for a particular combination of inputs. The inputs and outputs of the process were also successfully predicted in both forward and reverse directions using a radial basis function neural network.

11.5.3 Description of Experimental Setup and Data Collection

In the first phase of the study, actual bead-on-plate experiments on stainless steel plates were performed on an EBW machine in a small-scale industry, *M/s. Siddhi Engineering Company*, Mumbai, India. The 6-kW, 150-kV machine, shown in Figure 11.5 was developed in-house at BARC, Mumbai, India, in 1980 [33]. The other details of the machine are given in Table 11.1.

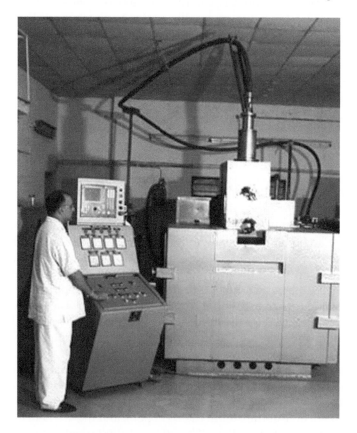

FIGURE 11.5
Photograph of the electron beam welding machine at M/s Siddhi Engineering Company, Mumbai, India. (From Saha, T.K. and Ray, A.K. *International Symposium on Vacuum Science and Technology, Journal of Physics: Conference Series*, 114, 2008, DOI: 10.1088/1742-6596/114/1/012047.)

TABLE 11.1

General Specifications of the 6-kW EBW Machine

Machine capacity	6 kW
Work chamber size	1500 mm × 1000 mm × 1000 mm
Work table size	800 mm × 500 mm
Table speed along X and Y directions	20–2000 mm/min
Gun vacuum	2×10^{-6} mbar
Chamber vacuum	5×10^{-5} mbar

In the studies conducted, three working parameters, namely, accelerating voltage, beam current, and welding speed, were varied to see their effects on bead geometry. The experiment was conducted in accordance with the CCD technique. The number of center points taken was three in this case. This resulted in 17 different combinations, which were again repeated three times to generate 51 trial runs. Test runs were also conducted to verify the models.

11.5.3.1 ASS-304 Welded Samples

The experiments on ASS-304 were done on plates of size 150 mm × 100 mm × 5 mm. The working ranges for accelerating voltage (V), beam current (I), and welding speed (S) were kept fixed at 60–90 kV, 7 mA–9 mA, and 60–90 cm/min, respectively. On average, six bead-on-plate runs were taken on each plate, as shown in Figure 11.6.

The polished specimens were electrochemically etched to clearly reveal the fused metal zone. The photographs of the etched sections are shown in Figure 11.7, and a schematic view of the same is shown in Figure 11.8.

11.5.4 Statistical Regression Analysis of Stainless Steel (ASS-304) Data

Bead geometries such as BP, bead width (BW), bead height (BH), and the coordinates of two more points P_1 (a_1, b_1) and P_2 (a_2, b_2), as shown in Figure 11.8, were measured for all 51 trial cases. The measured data were used to find the significance of welding parameters on the bead geometry (shown in Table 11.2). An analysis of variance (ANOVA) test further established the coherency of the welding process. The prediction of the regression equations developed was then evaluated for the test cases.

The results of the significance test help us investigate the contribution of process parameters to the said response. The terms X_1, X_2, and X_3 represent accelerating voltage, beam

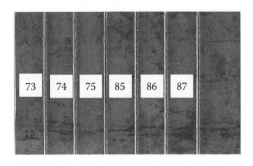

FIGURE 11.6
A stainless steel plate showing six weld runs.

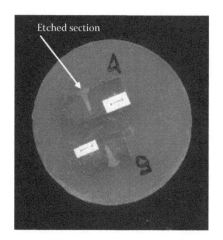

FIGURE 11.7
Photograph of the etched section of the weldment of an ASS-304 specimen.

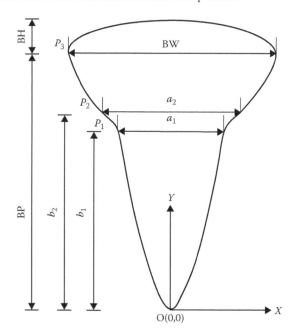

FIGURE 11.8
Schematic view of the fusion zone of ASS-304 welded specimens.

current, and welding speed, in their coded forms, respectively. The parameters X_1, X_2, X_3, X_1^2, X_2^2, X_3^2, and $X_1 X_2$ made significant contributions to BP. The relationship of BP with the process parameters was nonlinear. The correlation coefficient was found to be equal to 0.9469 for this response. Table 11.3 shows the results of ANOVA. The regression equation for BP was found to be as follows:

$$BP = 11.2643 - 0.304687\ V + 4.97343\ I + 0.0697153\ S + 0.00282728\ V^2 - 0.220125\ I^2$$
$$- 0.000389617\ S^2 - 0.00897604\ VI - 0.0000625632\ VS - 0.00212293\ IS$$

TABLE 11.2

Significance Test for BP

Sl No.	Term	Coefficient	SE Coefficient	T	p
1	Constant	3.45098	0.04971	69.428	0.000
2	X_1	0.63887	0.03673	17.392	0.000
3	X_2	0.60839	0.03673	16.562	0.009
4	X_3	−0.28598	0.03673	−7.785	0.000
5	X_1^2	0.63614	0.07097	8.964	0.000
6	X_2^2	−0.22013	0.07097	−3.102	0.003
7	X_3^2	−0.15585	0.07097	−2.196	0.034
8	X_1X_2	−0.13464	0.04107	−3.278	0.002
9	X_1X_3	−0.01877	0.04107	−0.457	0.650
10	X_2X_3	−0.04246	0.04107	−1.034	0.307

Standard error of the estimate $(S) = 0.201199$; Regression coefficient $(R^2) = 94.69\%$; Regression coefficient (adjusted) $(R^2[\text{adj}]) = 91.14\%$.

TABLE 11.3

ANOVA Test for BP

Source	DF	Seq SS	Adj SS	Adj MS	F	p
Regression	9	29.5836	29.5836	3.28707	81.20	0.000
Linear	3	25.8023	25.8023	8.60078	212.46	0.000
Square	3	3.2945	3.2945	1.09817	27.13	0.000
Interaction	3	0.4868	0.4868	0.16226	4.01	0.014
Residual error	41	1.6597	1.6597	0.04048	—	—
Lack-of-fit	5	0.2715	0.2715	0.05430	1.41	0.245
Pure error	36	1.3882	1.3882	0.03856	—	—
Total	50	31.2434	—	—	—	—

BP was found to increase with increase in accelerating voltage and beam current, whereas it decreased with increase in welding speed, as shown in Figure 11.9. Thus, it may be concluded that maximum BP was obtained when the welding was carried out at a lower welding speed and higher accelerating voltage and beam current. The performance of the developed model was tested on six cases. The values of BP predicted by the model were compared with the experimental ones for the test cases, and the percentage deviations in prediction shown in Figure 11.9 were computed. The values of percentage deviation in the prediction of BP lay between −1.51% and 15.18% (Figure 11.10). The mean squared deviation in the prediction of BP was found to be equal to 0.0789.

A similar analysis was carried out for BW and BH as well. The correlation coefficients were seen to be equal to 0.7676 and 0.9067, respectively. The mean squared deviation in predictions of BW and BH were obtained as 0.0124 and 0.0004, respectively.

11.5.5 Forward and Reverse Modeling of Electron Beam Welding Process Using Radial Basis Function Neural Networks

In order to automate a process, its input–output relationships are to be determined in both forward and reverse directions. An attempt was made to model input–output relationships of an EBW process in both forward and reverse directions using the radial basis functions neural network (RBFNN) [34]. The performance of this network depends significantly on

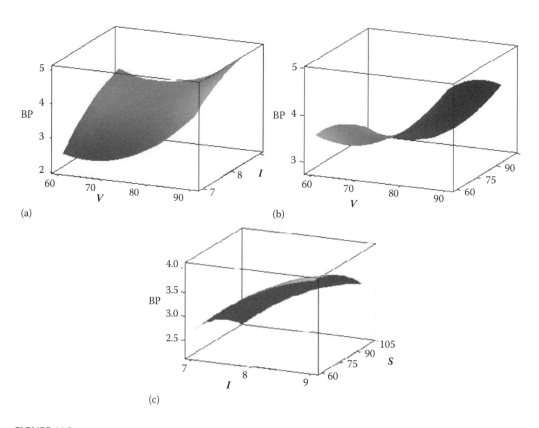

FIGURE 11.9
Surface plots of bead penetration (BP) for ASS-304 with varying input parameters: (a) accelerating voltage and beam current, (b) accelerating voltage and weld speed, and (c) weld speed and beam current.

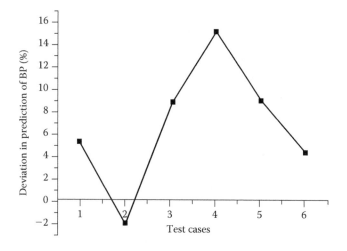

FIGURE 11.10
Percentage deviation in the prediction of BP for ASS-304 from the experimental values.

its architecture, which again is dependent on the number of neurons of the hidden layer. The number of hidden neurons of this network was kept equal to that of the clusters formed from the dataset. Three approaches were developed. The first approach used a fuzzy c-means algorithm to generate the number of clusters, while the second and third approaches used an entropy-based fuzzy clustering algorithm and a modified algorithm, respectively. The third approach was seen to outperform the other two in both forward and reverse mappings. This was because the modified clustering algorithm performs better than the other two approaches.

11.6 Summary

The importance of electron beams in both micro- and macrowelding has been felt. A considerable amount of literature is available on experimental studies related to macrowelding. The experimental results obtained have been analyzed. Electron beam macrowelding is used nowadays in modern industry. However, electron beam microwelding is still in the research phase. Research issues, as mentioned above, are to be solved first before electron beam microwelding finds its place in industries.

References

1. Bakish, R.A. 1962. *Introduction to Electron Beam Technology*. New York: John Wiley & Sons.
2. Schiller, S., Heisig, U., and Panzer, S. 1982. *Electron Beam Technology*. New York: John Wiley & Sons.
3. Spangenberg, K.R. 1948. *Vacuum Tubes*. New York: McGraw Hill.
4. Humphries, S. 1990. *Charged Particle Beams*. New York: John Wiley & Sons.
5. Pierce, J.R. 1954. *Theory and Design of Electron Beams*. New York: Van Nostrand.
6. Klemperer, O. and Barnett, M.E. 1971. *Electron Optics*. Cambridge: Cambridge University Press.
7. Grivet, P. 1965. *Electron Optics*. London: Pergamon Press.
8. Maleka, A.H. 1971. *Electron Beam Welding—Principles and Practices*. New York: McGraw-Hill.
9. Guo, K.W. 2009. A review of micro/nano welding and its future developments. *Recent Patents on Nanotechnology* 3(1): 53–60.
10. Baertle, J., Lower, T. and Dobeneck, D.V. 2006. Electron beam welding beyond the ordinary scale. *8th International Conference on Electron Beam Technologies*. Bulgaria: Varna.
11. Brown, G. and Nichols, K.G. 1966. A review of the use of electron beam machines for thermal milling. *Journal of Materials Science* 1(1), pp. 96–111.
12. McGeough, J. 2002. Chapter 11. *Electron Beam Machining, Micromachining of Engineering Materials*. New York: Marcel Decker Inc.
13. Tanasie, G., Bohm, S., Bartle, J., Lower, T., Reiter, A., Frazkowiak, M., and Reinhart, G. Electron beam machining of microsystem products, www.mikroprodz.de.
14. Reisgen, U. and Dorfmuller, T. Microelectron beam welding with a modified scanning electron microscopy: Findings and prospects. *Journal of Vacuum Science & Technology B* 27(3), 1310.
15. Teruya, A.T. 2008. Miniature modified faraday cup for micro electron beams, U.S. Patent Application No 60/582166, through Google patents.

16. Zah, M.F. and Franzkowiak, M. 2006. Flexible handling system for an electron beam-based microsystem production centre. *Proceedings of the 6th euspen International Conference.* Vol. 2. pp. 24–27, Vienna.

17. Dilthey, U. and Dorfmuller, T. 2006. Micro electron beam welding. *Microsystems Technology* 12: 626–631.

18. Koleva, E. 2005. Electron beam weld parameters and thermal efficiency improvement. *Vacuum* 77: 413–421.

19. Petrov, P., Georgiev, C., and Petrov, G. 1998. Experimental investigation of weld pool formation in electron beam welding. *Vacuum* 51: 339–343.

20. Hashimoto, T. and Matsuda, F. 1965. Effect of welding variables and materials upon bead shape in electron beam welding. *Transactions of NIRM* 7(3): 96–109.

21. Klemens, P.G. 1969. Energy considerations in electron beam welding. *Journal of Electrochemical Society: Electrochemical Science* 116(2): 196–198.

22. Miyazaki, T. and Giedt, W.H. 1982. Heat transfer from an elliptical cylinder moving through an infinite plate applied to electron beam welding. *International Journal of Heat Mass Transfer* 25(6): 807–814.

23. Vijayan, T. and Rohatgi, V.K. 1984. Physical behavior of electron-beam fusion heat transfer and deep penetration in metals. *International Journal of Heat Mass Transfer* 27(11): 1985–1998.

24. Elmer, J.W., Giedt, W.H., and Eagar, T.W. 1990. The transition from shallow to deep penetration during electron beam welding. *Welding Journal* 69: 167s–176s.

25. Rosenthal, D. 1946. The theory of moving sources of heat and its application to metal treatments. *Transactions ASME* 43(11): 849–866.

26. Koleva, E., Mladenov, G., and Vutova, K. 1999. Calculation of weld parameters and thermal efficiency in electron beam welding. *Vacuum* 53: 67–70.

27. Couëdel, D., Rogeon, P., Lemasson, P., Carin, M., Parpillon, J.C., and Berthet, R. 2003. 2D heat transfer modelling within limited regions using moving sources: Application to electron beam welding. *International Journal of Heat and Mass Transfer* 46: 4553–4559.

28. Rogeon, P., Couëdel, D., Le Masson, P., Carbon, D., and Quemener, J.J. 2004. Determination of critical sample width for electron beam welding process using analytical modeling. *Heat Transfer Engineering* 25(2): 52–62.

29. Ho, C.Y. 2005. Fusion zone during focused electron-beam welding. *Journal of Materials Processing Technology* 167: 265–272.

30. Rai, R., Burgardt, P., Milewski, J.O., Lienert, T.J., and Debroy, T. 2009. Heat transfer and fluid flow during electron beam welding of 21Cr–6Ni–9Mn steel and Ti–6Al–4V alloy. *Journal of Physics D: Applied Physics* 42: 1–12.

31. Dey, V., Pratihar, D.K., Datta, G.L., Jha, M.N., Saha, T.K., and Bapat, A.V. 2009. Optimization of bead geometry in electron beam welding using a genetic algorithm. *Journal of Materials Processing Technology* 209: 1151–1157.

32. Dey, V., Pratihar, D.K., Datta, G.L., Jha, M.N., Saha, T.K., and Bapat, A.V. 2010. Optimization and prediction of weldment profile in bead-on-plate welding of Al-1100 plates using electron beam. *International Journal of Advanced Manufacturing Technology* 48: 513–528.

33. Saha, T.K. and Ray, A.K. 2008. Vacuum the ideal environment for welding of reactive materials. *International Symposium on Vacuum Science and Technology, Journal of Physics: Conference Series* 114: DOI:10.1088/1742-6596/114/1/012047.

34. Dey, V., Pratihar, D.K., and Datta, G.L. 2010. Forward and Reverse modeling of electron beam welding process using radial basis function neural networks. *International Journal of Knowledge-based and Intelligent Engineering Systems* 14(4): 201–215.

Section V

Microforming

12

Micro- and Nanostructured Surface Development by Nano Plastic Forming and Roller Imprinting

M. Yoshino, K. Willy, and A. Yamanaka
Tokyo Institute of Technology

T. Matsumura
Tokyo Denki University

S. Aravindan and P.V. Rao
Indian Institute of Technology Delhi

CONTENTS

12.1 Introduction

Micromanufacturing is the bridge between nano- and macromanufacturing. There is an increasing demand for miniaturized products such as microsystems, microreactors, fuel cells, and micromedical devices. Micromanufacturing involves techniques such as photolithography, chemical etching, plating, LIGA, and laser ablation. It also involves non-silicon applications through mechanical fabrication methods such as micromechanical cutting, microembossing, injection molding, and microextrusion. It has been realized that no single process can satisfy the increasing demand for miniaturized components and devices. A single process such as micromachining is suitable for materials fabrication in the wider sense; however, it is not suitable when the number of components is large. Since a combination of processes increases the output dramatically, the thrust is increasingly toward the development of hybrid processes or integration of processes. A confluence of processing technologies is needed to address demands such as surfaces having high aspect ratio structures, three-dimensional complex/intricate structures, the method not being restricted to silicon, the process not being limited to clean room environment, rapid production at low cost, and environment friendliness.

An engineering surface can be defined as a surface that has been modified using nano/micromanufacturing technologies in such a manner that there now exists a strong affinity to a particular function that was lacking in the unmodified surface. Modification capability of engineering surfaces in terms of friction, optical, acoustic, thermal, electrical, chemical, and fluid properties are potentially useful in creating new cutting-edge technologies [1]. Recent studies reveal that these surface characteristics depend not only on the chemical components but also on the topography and microscopic structure of the surface. Therefore, controllability and function integration capability of these characteristics at an arbitrary location on a material surface are important to realize unlimited application possibilities in the fields of energy, heat transfer, fluid dynamics, optics, bioengineering, cleaning, and manufacturing [2].

Various research studies on engineering surfaces in terms of surface optical and wettability properties have been reported. For optical properties, one of the most important functions is the reflectance. It is possible to modify surface reflectance by creating a material with gradual change in refractive index over a range of optical spectra, that is, multilayered thin films [3]. Alternatively, a micro/nanomanufacturing process is utilized to fabricate subwavelength structures having a period and height shorter than the wavelength of light. Many researchers focus on photolithography and wet-etching processes to attain these kinds of structures [4,5]. Gombert et al. combine a holographic exposure technique with embossing to fabricate linear and crossed grating with periods from 205 to 301 nm on PMMA and polycarbonate [6]. Recently, Saarikoski et al. [7] reported the use of nanoporous anodized aluminum oxide (AAO) as mold inserts in injection molding to pattern polycarbonate surfaces with nanopillars, microbumps, or nanopillars superimposed on microbumps. The products of micro/nanomanufacturing processes are found to be more advantageous than multilayered thin films since they are stable and the transmitted waveform is not degraded.

Micro/nanostructured surfaces are finding applications in the reduction of friction in fluid flow and the development of antifriction surfaces and superhydrophilic and superhydrophobic surfaces. This topography is useful in devices such as lab on a chip, microfluidic devices, and intelligent biosensors. Such structured surfaces are useful for studying cell behavior (cell orientation, migration, and morphology), stem cell differentiation, and development of implants. Wettability is considered to be the most important phenomenon governed by surface energy. This property is characterized by Young's contact angle (θ) between a liquid droplet and a solid surface at the liquid/vapor interface. Wenzel [8], followed by Cassie and Baxter [9], has reported that surface topography plays a major role in altering the interfacial tension balance, resulting in a lower or higher contact angle. The impact of surface topography effects on the contact angle reveals the possibility of employing micro/nanomanufacturing technologies to create artificially controlled surface topography by micro/nanostructures. Since it has the capability of creating well-defined patterns with good controllability, the photolithography process is commonly selected as the fabrication technique [10–12]. In order to increase fabrication speed, nanoimprint lithography and wet-etching processes have been utilized to fabricate an array of structures on the surface of silicon [13]. Moreover, roller imprinting of a porous anodic alumina template to fabricate nanopillar arrays on polycarbonate has also been reported [14]. Aside from the photolithography-based process, other techniques such as micromachining [15,16], plasma treatment [17], crystal growth [18], and aggregation [19] have been used. However, limitations on the fabrication speed, flexibility in materials, and controllability have made these approaches less popular. In spite of their capability of modifying the surface optical and wettability properties, the above-mentioned methods are not always suitable for commercial purposes because of limitations on work materials, stringency in methods, high cost, and slow fabrication time. In addition, the use of dangerous chemicals in the photolithography-based process leads to fundamental environmental problems. Little work has been done on the use of metals as the work material despite their extensive usage in industry. High electrical and thermal conductivities, high strength-to-density ratio, and an ability to be deformed without cleaving have made metals an interesting target for engineering surface application.

A combination of nano plastic forming, coating, and roller imprinting processes, abbreviated as NPF-CRI [20], is used to rapidly modify the engineering surface of a metallic material. This technique is useful in fabricating micro/nanostructures with good controllability, high throughput, low cost, low emission, flexibility in geometries, and flexibility in work materials. In addition, this technique eliminates the need for the lithography process. As a result, both fabrication cost and chemical usage are significantly reduced. The structure shapes and geometries that can be fabricated rely solely on the shape and geometry of the diamond tool used in the nano plastic forming (NPF) process, which makes the fabrication of various three-dimensional shapes possible.

Surface optical and wettability properties are considered in showing the capability of the NPF-CRI technique to modify the engineering surface of metals. Line, cross, diagonal, and interval patterns are fabricated with variation in pitch settings on the surface of mirror-finished aluminum plates. Effects of these patterns and pitch settings on the engineering surface in terms of surface optical and wettability properties are also discussed.

12.2 Nano Plastic Forming

The NPF process was first developed by Yoshino et al. [21] to fabricate fine patterns on hard, brittle material surfaces. The principle of the NPF process is shown in Figure 12.1. It is similar to the indentation process. Structured tools are fabricated by either focused ion beam machining or electron beam/laser machining. Focused ion beam machining has found increased application in the production of precisely located arrays of structures on any material. The working principle of focused ion beam machining and methods for fabrication of diamond tools are explained in detail by the same authors [22]. Diamond is the strongest known material, having extreme mechanical hardness (~90 GPa), bulk modulus, good thermal conductivity at room temperature, broad optical transparency from the deep UV to the far IR region of the electromagnetic spectrum, and good electrical insulating properties. They are very resistant to chemical corrosion and are biologically compatible. Since diamond can retain its atomic sharpness, micro- and nanostructures can be fabricated on diamond. Various types of diamond tools such as point tool, knife-edge tool, I-type, T-type, and multi-nanostructures are available for fabricating different

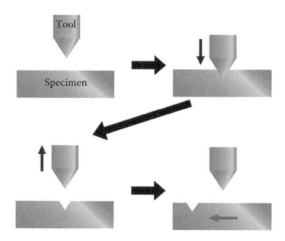

FIGURE 12.1
Schematic representation of the nano plastic forming process.

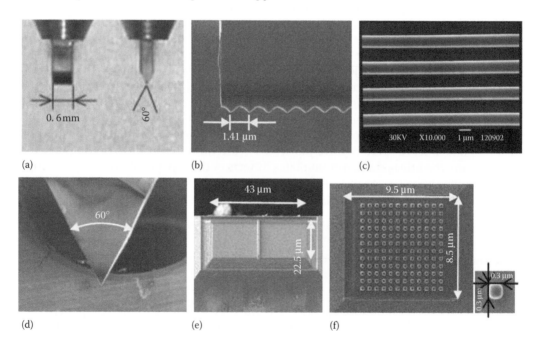

FIGURE 12.2
Typical diamond tools: (a) knife-edge tool, (b) multipoint tool, (c) line and space tool, (d) point tool, (e) T-type tool, and (f) nanostructure tool.

kinds of micro/nanostructures [23]. Figure 12.2 shows some typical diamond tools. The structures on a tool are imprinted on the work material surface by accelerating the tool with a piezoactuator in the vertical direction toward the specimen surface, and the load is applied on the surface of the specimen according to specified loading conditions such as the imprinting load and holding time. The structure of the tool can be imprinted on the surface. After the specified period, the tool is withdrawn and the stage is moved to the next position dictated by the controlling system and software. This facilitates the fabrication

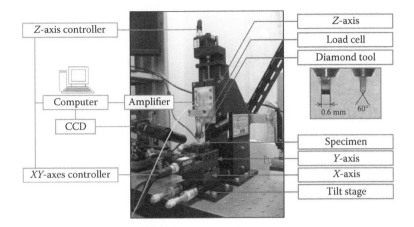

FIGURE 12.3
Nano plastic forming equipment.

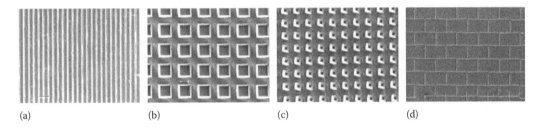

(a)   (b)   (c)   (d)

FIGURE 12.4
Examples of micro- and nanostructures: (a) line structure, (b) stamp structure, (c) square with dot structure, and (d) bricklike structure.

of micro/nanostructures at meso/macrolevels at the surface. By controlling the process parameters such as imprinting load, imprinting speed, and holding time, structured surfaces can be created on all materials, for example, ceramic, polymer, and metal. This process is more advantageous compared to lithographic processes owing to its lower cost, nonusage of dangerous chemicals, less stringency in the fabrication process, and flexibility in work materials. The NPF setup (Figure 12.3) consists of a linear motion controller, a data acquisition system, and a diamond tool. The linear motion controller is controlled by the computer for its xy-axes and z-axis. Resolution of the stages is 10 nm, and the strokes are 20 and 40 mm for the xy-axes and z-axis, respectively. The xy-axes are mounted on a tilted stage that has a resolution of 0.015°. A high-resolution load cell (Kistler 9205) is mounted on the z-axis to measure the indentation load. Typical micro- and nanostructures fabricated using NPF alone are presented in Figure 12.4.

12.3 NPF-CRI Technique

The recent trend is toward combining multiple processes to realize rapid production. High-precision microarray holes of nickel–diamond composites are fabricated by combining microelectric discharge machining with electroforming [24]. Structured surfaces

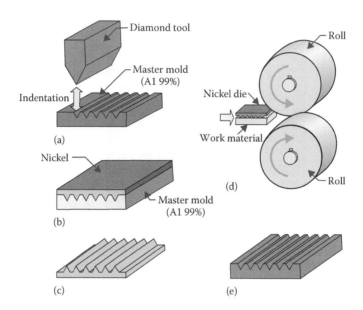

FIGURE 12.5
Schematic diagram of the NPF-CRI method: (a) nano plastic forming, (b) nickel coating, (c) nickel die, (d) roller imprinting, and (e) end product.

can be produced on the metal mold by NPF, and such structured metallic molds can be used to produce a larger number of structured surface devices by integrating the mold with the microinjection molding process. The NPF process is integrated with coating and roller imprinting. The NPF-CRI technique has three steps. Figure 12.5 shows the schematic diagram of the whole processes. First, the NPF process is used to fabricate micro/nanopatterns using a single-crystal diamond knife-edge tool on the surface of a mirror-finished aluminum plate (the master mold). Aluminum is used owing to its high formability. Second, the master mold is coated with nickel and the aluminum is peeled off to obtain a nickel die. Third, the nickel die is used for double plate roller imprinting to transfer the patterns onto the surface of the rolling material. For the rolling work material, aluminum is selected as representative of metallic materials because of its remarkable combination of characteristics such as low density, high corrosion resistance, high electrical conductivity, high thermal conductivity, and high formability.

12.4 Micro- and Nanostructured Surface Development

In order to study the modification of an engineering surface, master molds with four kinds of patterns, namely, line, cross, diagonal, and interval patterns, are fabricated using the NPF process. Figure 12.6 illustrates the schematic diagram of these patterns, and Table 12.1 shows the experimental parameters. During the fabrication of a line pattern, a knife-edge tool is made to traverse over the surface of a specimen until contact occurs. Then, indentation is performed while maintaining the indentation load according to the load parameter (0.3 N). When the required indentation load is achieved, the diamond tool is retracted from the specimen's surface. The process is then continued to the next indentation point

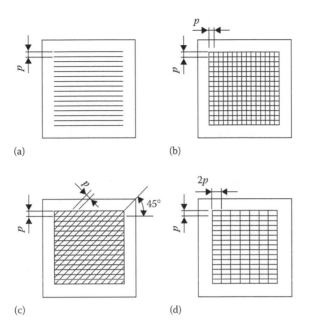

FIGURE 12.6
Schematic diagram of fabricated patterns: (a) line pattern, (b) cross pattern, (c) diagonal pattern, and (d) interval pattern.

TABLE 12.1

Experimental Parameters for the Nano Plastic Forming Process

Work material	15 mm × 15 mm × 0.5 mm aluminum (99%, mirror finished)
Diamond tool	Knife-edge tool (edge angle: 60°, width: 600 mm, tip radius: ~20 nm)
Indentation load	0.3 N
Pitch	1–8 mm
Machining area	4 mm × 4 mm
Type of patterns	Line, cross, diagonal, and interval patterns

or stopped when the number of indentations required has been reached. The cross and diagonal patterns are fabricated by first creating the line pattern and then subsequently rotating the workpiece through 90° and 45°, respectively, to create different orientations of grooves. In the case of the interval pattern, the pitch setting of the vertical orientation grooves is twice that of the horizontal orientation. The whole process is carried out inside a clean chamber to control the machining environment. A CCD camera is used to observe the NPF process in real time.

Nickel Electroplating: Nickel electroplating is selected as the coating process to ensure smoothness and hardness of the coating. Before the electroplating process, the master mold is cleaned using an ultrasonic cleaner. The master mold is then subjected to sputter etching to remove contamination layers from its surface. Then, a thin layer of gold (5 nm) is sputtered on the master mold surface as the electroplating base. The process is then continued by nickel electroplating, using a commercially available electroplating solution (N-100ES) to obtain a nickel die of 200 μm thickness. The parameters used for coating are presented in Table 12.2.

Roller Imprinting: The rolling mill used in this experiment is shown in Figure 12.7. Rolls having a diameter of 27 mm and width of 30 mm are used. The mill has a provision to control the roll gap, and it can be operated at a rolling speed of 850 mm/min. It has strain gauges, an amplifier, and a data recorder to measure the rolling forces. The nickel die obtained from the coating process is used in the flat roller imprinting (dry) process, where the die and work material are butted and rolled together. Using this method, the micro/nanopatterns are replicated onto the surface of a 15 mm × 15 mm × 0.5 mm mirror-finished aluminum sheet. The roller imprinting process is carried out with a rolling load of 120 N. Table 12.3 shows the experimental parameters for the roller imprinting process.

Measurements: Control of structure geometries during the NPF process is performed by controlling the indentation load using a high-resolution load cell (Kistler 9205). In the roller imprinting process, strain gauges are installed to measure and control the rolling load. This is important since the rolling load affects the transferability, so it has to be maintained within the range 100–120 N to facilitate optimum roller imprinting results [20].

The fabrication results are observed using a scanning electron microscope (SEM). Quantitative measurements of structure geometries are performed using atomic force microscopy (AFM). Five replications are used to ensure reliability. In order to analyze the surface optical property, a spectrometer (SEC2000-VIS/NIR) is used to measure the reflectance of the specimen in the wavelength range 400–850 nm. A mirror-finished aluminum plate is used as reference. For the surface wettability property, contact angle measurements are conducted using a contact angle meter with 1 mL deionized water as the test liquid. Figure 12.8 is a schematic diagram of the contact angle measurement setup. Five measurements are taken for each specimen, and the average contact angle value is calculated.

TABLE 12.2

Experimental Parameters for the Coating Process

No.	Process	Remarks	Current	Time
1.	Sputter etching	—	4 mA	4 min
2.	Sputter coating	Target: gold; thickness: 5 nm	6 mA	1 min
3.	Nickel electroplating	N-100ES:H_2O = 2:3 (room temperature)	0.01 A	24 h

TABLE 12.3

Experimental Parameters for the Roller Imprinting Process

Roll diameter	27 mm
Roll width	30 mm
Rolling speed	850 mm/min
Maximum rolling force	46.67 kN
Maximum torque	123.48 Nm
Applied rolling force	120 N

FIGURE 12.7
Roller imprinting equipment.

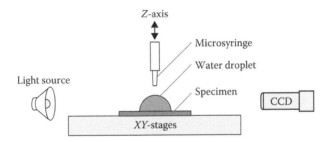

FIGURE 12.8
Schematic diagram of contact angle measurement.

12.4.1 Application Potentials of Micro- and Nanostructured Surfaces

By way of controlling the structures on the surfaces, special functionalities such as friction, optical, fluid flow, and heat transfer. This finds applications in micro-optical devices, antireflection films, photonic crystals, and hydrophobic and hydrophilic surfaces. Micro- and nanostructures especially at metal surfaces are finding applications in optical devices used in extreme operating conditions. Antireflection surfaces are finding applications as light absorbers, stray-light management components in telescopic systems, radiometers for space applications, and solar thermal converters. Line, cross, diagonal, and interval patterns are successfully fabricated on the surface of aluminum plates using the NPF-CRI technique. Figure 12.9 shows typical SEM observation results for each pattern. The patterns are uniformly transferred onto the surface of aluminum plates during the roller imprinting process. A comparison between the master mold and the rolling results reveals a good transferability of pattern geometries with an average transfer ratio of 91%.

12.4.1.1 Surface Optical Property

Reflectance measurements are carried out on the micro/nanostructured surfaces and on the polished surface of an aluminum plate. Figure 12.10 shows the reflectance measurement results of the fabricated structures. It can be seen from the figure that reflectance

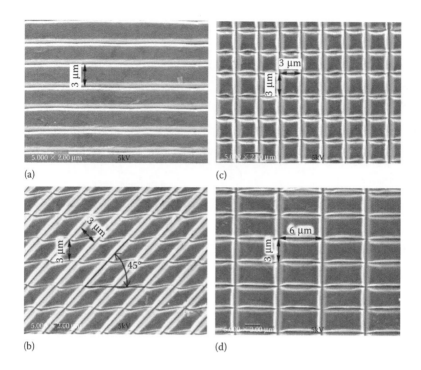

FIGURE 12.9
Typical line, cross, diagonal, and interval patterns fabricated using the NPF-CRI technique: (a) line pattern, (b) cross pattern, (c) diagonal pattern, and (d) interval pattern.

of the aluminum plate is greatly affected by the fabricated structures on its surface. In Figure 12.10a, the reflectance values are significantly decreased by approximately 60% relative to those of the mirror-finished aluminum when 8 µm pitch structures are fabricated on the aluminum surface. This is because its structures are inclined at 30° with respect to the normal direction. Hence, the light is dispersed to the side, and only the part of it that strikes the flat surface is reflected. The ratio between the flat area and the overall surface area is 83%, assuming that there is no pileup on the material surfaces. Hence, when the effect of pileup is taken into account, a 40% reduction in the reflectance value is reasonable. These values decrease with smaller pitch settings. At 1- and 2-mm pitch settings, a peculiar optical property is observed at wavelengths between 500 and 700 nm, which correspond to the green, yellow, and red regions. A maximum reflectance peak of 70.1% is recorded at 610 nm. A similar behavior is observed in diagonal patterns (Figure 12.10c). However, since the ratio between the flat and overall surface areas for 8-mm pitch setting is 70%, the reflectance value is further reduced to less than 40%. In addition, the peculiar optical property is shifted to the wavelength range 570–800 nm with a peak of 47% at 680 nm. A similar reduction in reflectance values is observed for the cross and interval patterns (Figure 12.10b and d), but the peculiar optical property is less pronounced. This peculiar optical reflectance can only be observed when the depth and width of the V-grooves are less than the wavelength of the light source. During illumination in a broad wavelength, standing-wave patterns, with higher maximum intensity compared to that of the incoming plasmons, are formed inside the V-grooves. Under this condition, resonant field enhancement occurs owing to the interference of counter-propagating gap surface plasmon modes, bouncing between the V-groove bottom and opening [25]. The origin of the reflectivity maxima on a deep linear groove has also been numerically explained by Kuttge et al. [26]. By using two-dimensional finite-difference time-domain and boundary-element method calculations, the local magnetic field inside the groove was calculated. Their results showed that on resonance, surface plasmons are coupled in the groove to excite a localized groove mode, resulting in an increased reflectivity. Measurement results show that fabricated patterns with 1-mm pitch setting have the capability of filtering the reflectance of particular wavelengths. On the basis of these results, it can be concluded that fabricated patterns using the NPF-CRI technique successfully modified the surface optical property of an aluminum plate. Such structured surfaces can be fabricated on other metallic materials such as titanium alloy, stainless steel, and copper to realize optical devices.

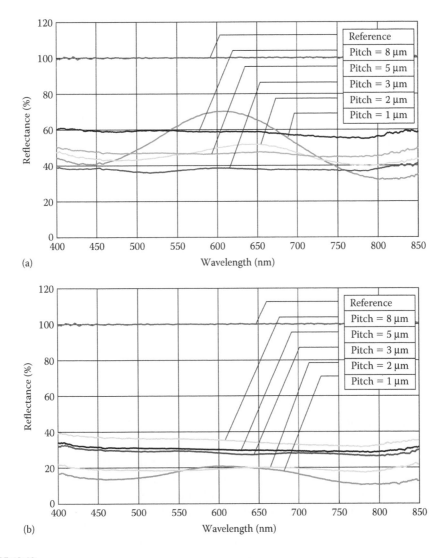

FIGURE 12.10
Reflectance measurement results.

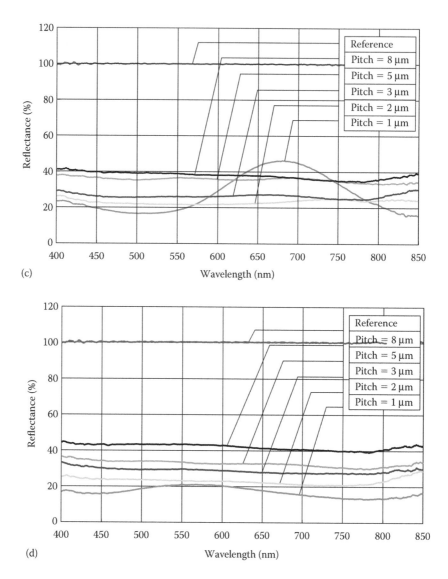

FIGURE 12.10   (continued)

12.4.1.2 Surface Wettability

Biomimicking of structures on metallic surfaces results in hydrophobic and self-cleaning surfaces. Micro/nanostructured surfaces exhibit different wettability characteristics. Figure 12.11 shows typical changes in the contact angle of a water droplet on an aluminum surface after micro/nanostructures were patterned using the NPF-CRI technique. Young's static contact angle (θ) is found from the measurement result on a mirror-finished aluminum plate (Figure 12.11a) as 82.6°, showing that the aluminum plate is slightly hydrophilic. Figure 12.11b–e clearly shows that micro- and nanostructures fabricated using the NPF-CRI technique are effective in modifying the contact angles of a water droplet on an aluminum surface, changing the surface from hydrophilic to hydrophobic.

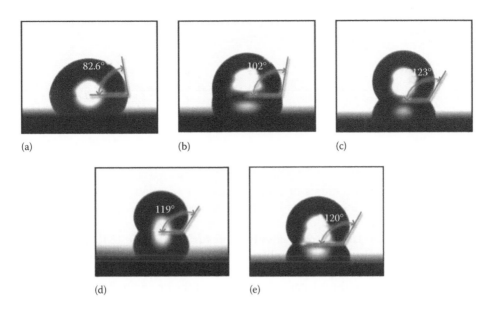

FIGURE 12.11
Modification of contact angles by nano/microstructures (pitch setting = 3 µm): (a) aluminum plate, (b) with line pattern, (c) with cross pattern, (d) with diagonal pattern, and (e) with interval pattern.

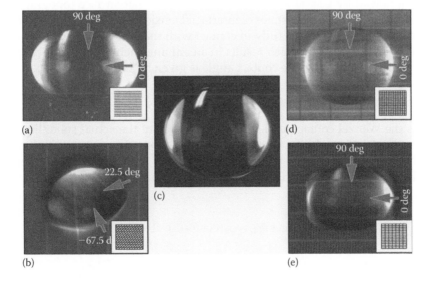

FIGURE 12.12
Droplets of water on aluminum and patterned surfaces. (a) Line pattern, (b) diagonal pattern, (c) aluminum plate, (d) cross pattern, and (e) interval pattern.

Figure 12.12 shows the top view of water droplets on the mirror-finished aluminum plate and patterned surfaces. Water droplets were gently deposited, and the pictures were taken from the top using a digital camera. From these figures, it can be seen that other than the cross pattern, water droplets in line, diagonal, and interval patterns are asymmetrical. This phenomenon is attributed to the shape of the structures, which results in different interfacial tensions for different directions. In order to take the asymmetrical

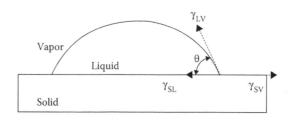

FIGURE 12.13
Contact angle of a droplet on a smooth solid surface.

shape into account, measurements of contact angles are carried out from the two directions that correspond to the maximum and minimum contact angles, as shown by the arrows in Figure 12.12.

The behavior of water droplets on a smooth solid surface (Figure 12.13) can be explained by Young's theory, in which the wettability property is characterized by Young's contact angle (θ) at which a liquid/vapor interface meets the solid surface. This contact angle is governed by the balance of three interfacial tensions at the boundary between solid–vapor (γ_{SV}), solid–liquid (γ_{SL}), and liquid–vapor (γ_{LV}) interfaces; as given by Young's equation:

$$\gamma_{SV} = \gamma_{SL} + \gamma_{LV} \cos\theta \tag{12.1}$$

Wenzel [8], followed by Cassie and Baxter [9], reported that surface topography plays a major role in altering the balance states of interfacial tensions, resulting in a lower or higher contact angle. When the liquid is fully in contact with the surface topography, it falls into Wenzel's state (Figure 12.14a), which results in an enhancement of surface hydrophobicity or hydrophilicity. In this state, the contact angle is given by Equation 12.2 [8]:

$$\cos\theta_w = r \cos\theta \tag{12.2}$$

where θ_W is the Wenzel contact angle and r is the ratio of the actual solid/liquid contact area to its vertical projected area. On the other hand, when incomplete liquid penetration occurs so that the droplet sits on an air pocket between the liquid and solid interfaces, it falls into the Cassie–Baxter state (Figure 12.14b). As a result, the contact angle is given by Equation 12.3 [9]:

$$\cos\theta_{CB} = \varphi_s\left(\cos\theta + 1\right) - 1 \tag{12.3}$$

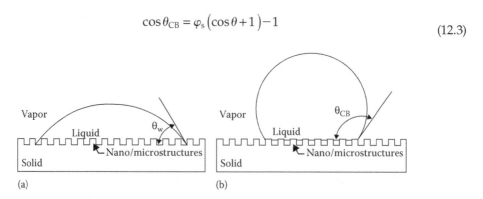

(a)   (b)

FIGURE 12.14
Contact angle of a droplet on a patterned solid surface: (a) Wenzel's state and (b) Cassie–Baxter state.

where θ_{CB} is the Cassie–Baxter contact angle and φ_s is the area fraction of the solid surface in direct contact with the liquid droplet. This equation indicates that the Cassie–Baxter state always results in an increase in contact angle. When φ_s approaches zero, the contact angle approaches 180°. The Wenzel and Cassie–Baxter states are usually considered to be distinct energy minima with an energy barrier to change from one state to another [27]. The droplet generally falls into the state that provides the minimum energy. However, this also depends on how the droplet is formed [28].

To compare the contact angle measurement results, calculation of contact angles predicted by Wenzel and Cassie–Baxter equations are performed using pattern geometries measured by AFM. Contact angle measurement results from different views plotted together with the theoretical calculation results are shown in Figure 12.15. It can be seen from these plots that most of the measured contact angles are closer to the values predicted by Cassie–Baxter's equation as compared to the prediction results from Wenzel's equation. This suggests that these droplets are all in the Cassie–Baxter state and that there is an air pocket between the water and aluminum interfaces. The measurement results reveal that the contact angle of water droplets on cross patterns is symmetrical. On the contrary,

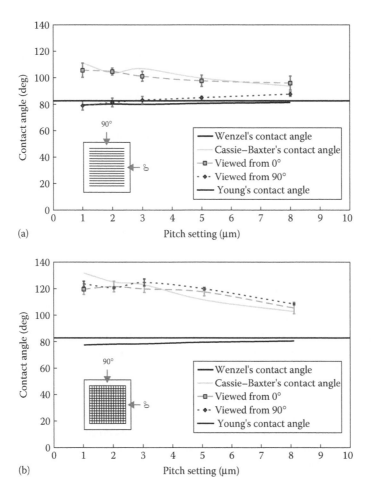

FIGURE 12.15
Contact angle measurement results.

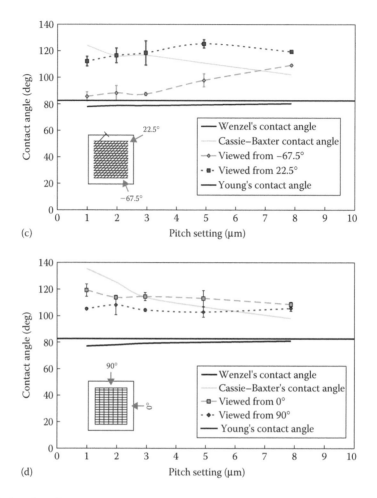

FIGURE 12.15 (continued)

contact angles of line, diagonal, and interval patterns vary with the viewing direction. For line patterns (Figure 12.15a), the results show that there is no apparent change in the contact angle viewed from the 90° direction compared to Young's contact angle. However, the contact angle viewed from the 0° direction increases and is in close agreement with the results obtained using Cassie–Baxter's equation. A maximum contact angle of 105° is recorded at 1-μm pitch setting. In the case of a cross pattern (Figure 12.15b), contact angles from the 0° and 90° directions are symmetrical and agree well with the Cassie–Baxter calculation results. The maximum contact angle is found to be 124°. For the diagonal pattern (Figure 12.15c), the droplet is twisted and the maximum contact angle observed from the 22.5° direction is 126° at 5-μm pitch setting. The behavior of water droplets from 1- to 3-μm pitch settings is almost similar to the line pattern. This is attributed to the fact that in the diagonal pattern, 0° orientation grooves are very small as compared to those in 45° orientation when the pitch setting is less than 5 μm (see Figure 12.9), making it very similar to the shape of a line pattern. The effect of grooves on the 0° orientation starts to show from a pitch setting of 5 μm where the contact angles viewed from a direction of −67.5° increase and are in good agreement with the results of calculation using Cassie–Baxter's equation. In the case of interval pattern (Figure 12.15d), contact angles from the 0° direction are constantly larger than contact angles viewed from the 90° direction. The maximum contact angle of 120° is reached at 1-μm pitch setting. These results indicate that there is a strong tendency of water droplets to flow to the direction that provides the greatest distance between grooves. Hence, the maximum contact angle is always normal to this direction, that is, for line, diagonal, and interval patterns, the maximum contact angles are normal to the 0°, 22.5°, and 0° directions, respectively.

These results show that the water contact angle on aluminum surface can be modified rapidly by micro/nanostructures fabricated using the NPF-CRI technique. In addition, the contact angle measurement results suggest that it is possible to effectively control the shape of water droplets on the surface of materials by introducing asymmetrical patterns.

12.5 Conclusion

The NPF-CRI technique has been utilized to rapidly modify the engineering surface of a metal in terms of surface optical and wettability properties. To show the capability of this technique, four kinds of patterns, namely, line, cross, diagonal, and interval patterns, have been successfully fabricated on the surface of mirror-finished aluminum. Comparison between the master mold and rolling results reveals a good transferability of pattern geometries with an average transfer ratio of 91%. In terms of the surface optical property, measurement results show that variations in reflectance values are successfully observed on the fabricated patterns. In addition, fabricated patterns with 1-mm pitch setting are shown to have the capability to filter the reflectance of light of particular wavelengths. In the case of surface wettability, we have shown that the contact angles of patterned surfaces can be successfully modified. In addition, the possibility of controlling the shape of water droplets on a patterned surface using asymmetrical structure shapes has also been shown. On the basis of these results, the capability of the NPF-CRI technique to modify the surface optical and wettability properties by creating nano/microstructures is validated. This combined technology is well suited to fabricate such micro- and nanostructures on other metals and also to exploit the specially induced functionalities.

Acknowledgments

This research work is financially supported by Naito Taisyun Science and Technology Foundation. The authors express their humble thanks to Indo-Japan (DST-JSPS) collaborative project.

References

1. Yoshino, M., Matsumura, T., Umehara, N., Akagami, Y., Aravindan, S., and Ohno, T. 2006. Engineering surface and development of a new DNA micro array chip. *Wear* 260(3): 274–286.
2. Bruzzone, A.A.G., Costa, H.L., Lonardo, P.M., and Lucca, D.A. 2008. Advances in engineered surfaces for functional performance. *CIRP Annals—Manufacturing Technology* 57(2): 750–769.

3. Nubile, P. 1999. Analytical design of antireflection coatings for silicon photovoltaic devices. *Thin Solid Films* 342(1): 257–261.
4. Kanamori, Y., Kikuta, H., and Hane, K. 2000. Broadband antireflection gratings for glass substrates fabricated by fast atom beam etching. *Japanese Journal of Applied Physics, Part 2: Letters* 39(7): 735–737.
5. Zaidi, S.H., Chu, A.-S., and Brueck, S.R.J. 1996. Optical properties of nanoscale, one-dimensional silicon grating structures. *Journal of Applied Physics* 80(12): 6997–7008.
6. Gombert, A., Glaubitt, W., Rose, K., Dreibholz, J., Blasi, B., Heinzel, A., Sporn, D., Doll, W., and Wittwer, V. 1999. Subwavelength-structured antireflective surfaces on glass. *Thin Solid Films* 351(1–2): 73–78.
7. Saarikoski, I., Suvanto, M., and Pakkanen, T.A. 2009. Modification of polycarbonate surface properties by nano-, micro-, and hierarchical micro-nanostructuring. *Applied Surface Science* 255(22): 9000–9005.
8. Wenzel, R.N. 1936. Resistance of solid surfaces to wetting by water. *Industrial & Engineering Chemistry* 28(8): 988–994.
9. Cassie, A.B.D. and Baxter, S. 1944. Wettability of porous surfaces. *Transactions of the Faraday Society* 40: 546–551.
10. Oner, D. and McCarthy, T.J. 2000. Ultrahydrophobic surfaces. Effects of topography length scales on wettability. *Langmuir* 16(20): 7777–7782.
11. McHale, G., Shirtcliffe, N.J., Aqil, S., Perry, C.C., and Newton, M.I. 2004. Topography driven spreading. *Physical Review Letters* 93(3): 036102.
12. Shirtcliffe, N.J., Aqil, S., Evans, C., McHale, G., Newton, M.I., Perry, C.C., and Roach, P. 2004. The use of high aspect ratio photoresist (SU-8) for super-hydrophobic pattern prototyping. *Journal of Micromechanics and Microengineering* 14(10): 1384–1389.
13. Pozzato, A., Zilio, S.D., Fois, G., Vendramin, D., Mistura, G., Belotti, M., Chen, Y., and Natali, M. 2006. Superhydrophobic surfaces fabricated by nanoimprint lithography. *Microelectronic Engineering* 83(4–9): 884–888.
14. Chaowei, G., Lin, F., Jin, Z., Guojie, W., Yanlin, S., Lei, J., and Daoben, Z. 2004. Large-area fabrication of a nanostructure-induced hydrophobic surface from a hydrophilic polymer. *ChemPhysChem* 5(5): 750–753.
15. Zhu, L., Feng, Y., Ye, X., and Zhou, Z. 2006. Tuning wettability and getting superhydrophobic surface by controlling surface roughness with well-designed microstructures. *Sensors and Actuators, A: Physical* 130–131: 595–600.
16. Yoshimitsu, Z., Nakajima, A., Watanabe, T., and Hashimoto, K. 2002. Effects of surface structure on the hydrophobicity and sliding behavior of water droplets. *Langmuir* 18(15): 5818–5822.
17. Coulson, S.R., Woodward, I., Badyal, J.P.S., Brewer, S.A., and Willis, C. 2000. Super-repellent composite fluoropolymer surfaces. *The Journal of Physical Chemistry B* 104(37): 8836–8840.
18. Shibuichi, S., Onda, T., Satoh, N., and Tsujii, K. 1996. Super water-repellent surfaces resulting from fractal structure. *The Journal of Physical Chemistry* 100(50): 19512–19517.
19. Xiu, Y., Zhu, L., Hess, D.W., and Wong, C.P. 2006. Biomimetic creation of hierarchical surface structures by combining colloidal self-assembly and Au sputter deposition. *Langmuir* 22(23): 9676–9681.
20. Kurnia, W. and Yoshino, M. 2009. Nano/micro structure fabrication of metal surfaces using the combination of nano plastic forming, coating and roller imprinting processes. *Journal of Micromechanics and Microengineering* 9(12): 125028.
21. Yoshino, M. and Aravindan, S. 2004. Nanosurface fabrication of hard brittle materials by structured tool imprinting. *Journal of Manufacturing Science and Engineering, Transactions of the ASME* 126(4): 760–765.
22. Aravindan, S., Rao, P.V., and Yoshino, M. 2010. Focused ion beam machining. *Introduction to Micromachining*, ed. V.K. Jain. India: Narosa Publishing House.
23. Kurnia, W. and Yoshino, M. 2009. Nano plastic forming-coating-roller imprinting (NPF-CRI) process for rapid fabrication technique of nano and micro structures. *Proceedings of the 5th International Conference on Leading Edge Manufacturing in 21st Century (LEM21)*. Japan, pp. 285–288.

24. Chen, S.-T. and Luo, T.-S. 2011. Development of high precision wear resistant micro holes structure. *Journal of Material Processing Technology* 211: 285–293.
25. Sondergaard, T., Bozhevolnyi, S.I., Beermann, J., Novikov, S.M., Devaux, E., and Ebbesen, T.W. 2010. Resonant plasmon nanofocusing by closed tapered gaps. *Nano Letters* 10(1): 291–295.
26. Kuttge, M., De Abajo, F.J.G., and Polman, A. 2009. How grooves reflect and confine surface plasmon polaritons. *Optics Express* 17(12): 10385–10392.
27. Roach, P., Shirtcliffe, N.J., and Newton, M.I. 2008. Progress in superhydrophobic surface development. *Soft Matter* 4(2): 224–240.
28. He, B., Patankar, N.A., and Lee, J. 2003. Multiple equilibrium droplet shapes and design criterion for rough hydrophobic surfaces. *Langmuir* 19(12): 4999–5003.

13

Microextrusion

U.S. Dixit and R. Das
Indian Institute of Technology Guwahati

CONTENTS

13.1 Introduction

The demand for microcomponents has encouraged the scaling down of traditional macromanufacturing to micro-length-scale manufacturing. Fundamental issues concerning micrometal forming have been studied intensively over the past 10 years. The major issues examined were the mechanism of material deformation, tool–interface conditions, material properties characterization, process modeling and analysis, prediction of forming limit, and process design optimization. Geiger et al. (2001) defined microforming as the production of parts or structures with at least two dimensions in the submillimeter range. Microforming can be defined more precisely as the process of manufacturing a part or a feature by plastic deformation, at least one orthogonal view of which can be enclosed in a square of side 1 mm. These small components are used in microsystem technologies (MST) and microelectromechanical systems (MEMS). Typical examples of such components are pins for IC carriers, fasteners, microscrews, lead frames, sockets, and various kinds of connecting elements (Engel and Eckstein, 2002).

Although conventional bulk and sheet metal forming processes can be used to fabricate microcomponents, the small size of the components poses certain problems. The material, as well as friction behavior, becomes different at microscale. Nevertheless, a properly designed microforming process can produce microcomponents at a high production rate, with good mechanical properties, and a low cost. Extrusion is a conventional metal forming process, in which a block of metal is pushed through a die orifice to form a product of reduced cross section. In microextrusion, the cross section of the product can be enclosed in a square of side 1 mm. This chapter discusses the state of the art of microextrusion and pertinent issues.

The chapter is organized as follows. In Section 13.2, the effect of size on flow stress and friction is described. Understanding the size effect is important to develop an efficient microextrusion process. Section 13.3 describes various types of microextrusion processes developed by researchers. Section 13.4 discusses issues related to fabrication of dies and rams for microextrusion. Section 13.5 reviews the literature on experimental and theoretical studies on microextrusion. A multihole microextrusion process is described in Section 13.6. The limitations of this process and other challenging issues are discussed in Section 13.7. Section 13.8 is the concluding section of this chapter.

13.2 Size Effect

Investigations on microforming processes have shown significant differences in the forming behavior of small parts compared to conventional macro-scale forming. This poses a challenge in using metal-forming technology at the microscale. Some attempts have been made to model microforming processes and study the size effect on material flow and friction (Geiger et al., 2007). Size effects are the deviations from the expected results that occur when the dimension of a workpiece or sample is reduced.

Vollertsen (2008) grouped size effects into three categories—density, shape, and microstructure size effects. Density size effects occur when the absolute number or integral value of features per unit volume is kept constant, which is independent of the size of the object. These features could be small pores, dislocation lines, or interface areas. When the density of the grains is kept constant, the number of grains gets reduced because of the scaling down. Because there are fewer grains, the orientation of individual grains becomes more important and different strengths are observed with different samples. The most popular example of this size effect is the size dependence of the strength of brittle materials. As the probability of existence of defects (such as cracks) decreases with decreasing size, the strength increases. This effect is expected to occur in samples of size range 1–10 mm.

Shape size effects are related to the surface area and volume. When the shape is kept constant, owing to the reduction in the size of an object, the ratio between the total surface area and volume increases, because the volume of a part is proportional to the cube of its size, while the surface area is proportional to the square of its size. Depending on the size, either the volume-related or the surface-related effect will be dominant. A reduction in flow stress is observed with decreasing sample size as the measured flow stress is the sum of the strength of grains at the free surface of the part and the strength of grains in its inner region. This effect has been explained by surface layer model theory (Engel and Eckstein, 2002). According to this theory, the grains located at the free surface are less restricted than those located inside the material. Therefore, there is less hardening and lower resistance to

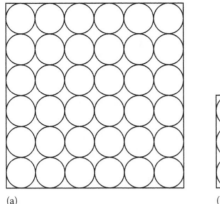
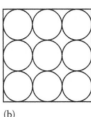

(a)   (b)

FIGURE 13.1
Schematic microstructural diagram showing grains in (a) a large sample and (b) a small sample.

deformation of the surface grains, making their deformation easier than that of the grains inside the material. This is due to a pileup of dislocations during deformation at the grain boundaries but not at the free surfaces. For the same grain size, with decreasing size of the specimen, the share of surface grains increases, which results in decreasing flow stress of the material. This can be understood from the schematic diagram in Figure 13.1. In the square in Figure 13.1a, 20 of 36 grains touch the boundary. If the area of the square is now reduced to one-fourth the original keeping the grain size same, eight of nine grains touch the boundary, as shown in Figure 13.1b. Thus, the share of surface grain increases with decreasing size and the flow stress of a smaller specimen is lower than that of a larger sample. This effect is prominent in the size range of 100 μm to 1 mm.

Yu et al. (2006) have developed a simple mathematical model for predicting the flow stress as a function of grain size and shape of the specimen. This model is briefly explained here. According to this model, the flow stress of a billet can be considered as the weighted average of the flow stresses of the inner and outer grains. Thus, the flow stress is given by

$$\sigma_f = \alpha\sigma_{fi} + (1 - \alpha)\sigma_{fs},\tag{13.1}$$

where σ_{fi} and σ_{fs} are the inner and outer grain flow stresses, respectively, and α is the inner grain area fraction in the billet cross section. We assume that σ_{fi} is equal to traditional flow stress and use the Hall–Petch equation

$$\sigma_{fi} = \sigma_0 + Kd^{-1/2}\tag{13.2}$$

where d is the grain diameter. We also assume that σ_{fs} is equal to σ_0, the flow strength of a single grain. With these assumptions, Equation 13.1 can be written as

$$\sigma_f = \sigma_0 + \alpha Kd^{-1/2}\tag{13.3}$$

If C is the perimeter and A is the area of cross section, α is given approximately by

$$\alpha = \left(1 - \frac{Cd}{A}\right)\tag{13.4}$$

Substituting Equation 13.4 into Equation 13.3,

$$\sigma_f = \sigma_0 + \left(1 - \frac{Cd}{A}\right)\frac{K}{d^{1/2}}$$

(13.5)

Equation 13.5 expresses the flow stress as a function of grain size d, cross-sectional area A, and perimeter C.

Equation 13.5 shows that the billet shape has a significant influence on the flow stress. For the same cross-sectional area, the total number of grains will be the same across the cross section. The number of surface grains increases with an increase in the circumference. Therefore, it can be observed that the flow stress decreases with an increase in the ratio of the circumference to the area of cross section of the billet. Thus, for the same cross-sectional area and grain size, flow stress for a billet of rectangular cross section is lesser than that for a billet of circular cross section. This has been verified experimentally.

Chen and Tsai (2006) conducted experiments on two cylindrical specimens with dimensions of 5 mm diameter and 7.5 mm height, and 2 mm diameter, and 3 mm height. They observed that the hardness decreases with decrease in size. Also, the hardness is nearly proportional to the flow stress. Thus, microindentation can be a viable method of assessing the flow stress of smaller parts.

In microstructure size effect, the microstructural features are not scaled down in the same manner as the macroscopic size of the object. One example is the intrinsic material length. The concept of intrinsic material length was introduced to include size effect in constitutive laws. Intrinsic material length is the length scale associated with materials. For each metal, there exists a particular intrinsic material length scale. For example, the intrinsic material length for polycrystalline copper is 1.54 μm. Therefore, in polycrystalline copper, strain gradients of the order $1/1.54$ (μm)$^{-1}$ are significant. The theory that takes into account the strain gradient effects is known as *strain gradient plasticity*. This effect is usually observed in the size range of 10–100 μm or during severe plastic deformation (SPD) processes.

The friction behavior between the work interfaces of a die is greatly affected by miniaturization. Engel and Eckstein (2002) investigated the effect of miniaturization using the ring compression test and the double-cup extrusion setup. They found that the value of the friction factor increases as the size of the billet decreases. Calculated values of the friction factors at different die locations were in good agreement with the simulated ones. An increase in friction was observed with decreasing specimen size. This effect was studied in detail by Engel et al. (1998). In an extrusion process, the friction factor increased by 20 times for reduced size when using extrusion oil as lubricant (Gieger et al., 2001). This behavior has been explained by the open and closed lubricant pockets model. According to this model, pockets that are not connected to the edges of the specimen can retain the lubricant during the process. Such pockets are called *closed pockets*. On the other hand, pockets connected to the edges are called *open pockets* and are unable to retain the lubricant when pressure is applied. In a smaller component, the proportion of closed to open pockets is low. Therefore, the lubricant is not retained effectively. Because of this, the pressure on the peaks increases and they get flattened. This increases the friction force. When a dry lubricant is used, friction does not vary significantly with size, thus confirming the model of closed and open lubricant pockets.

13.3 Different Extrusion Processes Developed by Researchers

Forming processes are based on plastic deformation that does not involve addition or removal of material. These processes are suited for the mass production of metallic parts because of their well-known advantages of higher production rates, minimized or zero metal loss, excellent mechanical properties of the final products, close tolerances, and ability to make near net-shaped or net-shaped parts. Forming of small metallic parts or microparts comes with challenges when the size gets reduced to tens/hundreds of microns or the precision requirement for miniature parts increases. Microforming is a better option for mass production at reduced cost. Conventional metal-forming processes such as forging, stamping, coining, deep drawing, and extrusion may be used for the forming of miniaturized parts. However, the process capabilities are constrained owing to material, interfacial, and tooling considerations in microforming.

Extrusion is a well-established macroscale manufacturing process that has been applied to produce microparts also. It is a fast process with minimal wastage of raw material and is suitable for mass production. Microextrusion is classified as either *forward* or *backward extrusion process*. Some special processes have also been developed. In the following subsections, we briefly discuss a few microextrusion processes.

13.3.1 Microbillet Formation by Extruding the Sheet

Hirota (2007) proposed a methodology to form microbillets by extruding a sheet in the direction of the thickness. This process is called *microforging*. He carried out experiments with an extrusion ratio of 2 by using a pure aluminum sheet of 2 mm thickness. A billet of 1 mm diameter was successfully formed from the aluminum sheet of 2 mm thickness. The billet remains attached to the sheet surface. This billet can be cut or can remain as an integral part of the sheet, depending on the application.

A schematic diagram of the process is shown in Figure 13.2. The die diameter d (equal to the diameter of the formed billet) is 1 mm, while the punch diameter D is 2 mm. The punch penetrates into the specimen with the application of ram force. The punch penetration h is adjusted properly using a spacing ring. A blank holder prevents vertical deformation of the specimen. A counterpunch is used to eject the specimen after completion of the process. For lubrication, graphite in tallow (hard animal fat) is used. In some cases, piping defects were observed at the bottom of the billet; these can be prevented by keeping the punch penetration less than the critical value predicted in the simulation.

13.3.2 Manufacturing of Micropins Using Segmented Dies

Krishnan et al. (2007) used a novel microextrusion apparatus for the extrusion of brass pins, with the extruded diameter ranging from 1.33 to 0.57 mm. A forward extrusion process was used, in which the movements of the ram and billet were in the same direction. The force–displacement response for the extrusion of the brass pin has been obtained. The authors fabricated dies using conventional drilling and polishing processes. Die cavities of submillimeter size were made by micro-electro-discharge machining. The dies were segmented to facilitate the removal of pins after extrusion. Each segmented die was then mounted on a yoke that slides along linear bearings. The forming assembly was placed inside the loading substage equipped with a load cell and an LVDT to measure the

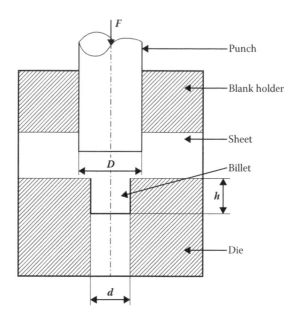

FIGURE 13.2
Schematic of microbillet formation by sheet extrusion.

extrusion force and the corresponding ram displacement. All dies were polished to ensure that they had similar surface roughness values. The surface roughness of each die was measured using a Micro XAM surface-mapping microscope, and all the dies were found to have surface roughness (Ra) in the range of 0.8–1 μm. Samples used in the experiments were heat-treated at various temperatures for 1 h to obtain the desired grain sizes ranging from 32 to 211 μm. The microextrusion apparatus enabled the authors to record the force–displacement response of the extrusion process. Finite element method simulations and analytical models were then used to make comparisons with the experimental data to obtain an estimate of the friction conditions during extrusion.

Parasiz et al. (2007) studied the grain size effect on bending and microhardness of the extruded micropins. They found that pins extruded from a larger grain-sized material have a greater tendency to bend as compared to pins extruded from a smaller grain-sized material. This is because a smaller grain-sized material deforms more uniformly than a larger grain-sized material. They also performed a microindentation test on the extruded pins to obtain an estimate of strain hardening. They found that pins fabricated with coarse-grained material had a higher value of hardness than those fabricated with fine-grained material. This is in contradiction to Equation 13.5.

13.3.3 Manufacturing of Microgear Shaft

Saotome and Iwazaki (2001) fabricated a microextrusion apparatus. A backward extrusion process was used in which the ram and extruded part moved in opposite directions. There was an arrangement for heating the die and billet, and the temperature was controlled by a PID controller. Microdies were produced from photochemically machinable glass. Piezoelectric actuators were employed as punch drivers. The microextrusion system was small enough to be held in the palm. The container diameter was kept at 2 mm. The authors used $La_{55}Al_{25}Ni_{20}$, an amorphous alloy, to produce a microgear shaft with gear dimension of 10 μm module and 100 μm pitch diameter. They also extruded Al–78Zn, a superplastic alloy, and produced gears of 20 μm module and 200 μm pitch diameter.

13.3.4 Fabrication of Microcondenser

Cho et al. (2007) developed a direct extrusion process to make microcondenser tubes for the application of ecofriendly refrigerants with higher working pressure. The raw materials used were Al100 and A3003 alloys. A 1250-ton horizontal extrusion press was used. It is to be noted that although the cross section of the condenser tube is of several millimeters, there are several rectangular channels in it. A typical web cross section was around 200 μm × 600 μm. Thus, this is an example where the size of the product is bigger but the features produced in it are of the submillimeter range. The authors observed that A3003 inhibits coarse grain growth and is more suitable for ecofriendly refrigeration systems.

13.3.5 Microextrusion of Bulk Metallic Glass

Bulk metallic glass (BMG) is a class of metal alloys having an amorphous structure instead of a crystalline one. Bulk metallic glasses based on lanthanum, magnesium, zirconium, palladium, iron, cobalt, and nickel have been developed (Basu and Ranganathan, 2003). They have better strength, corrosion, and wear properties. Recently, Wu et al. (2010) carried out microextrusion of $Zr_{55}Cu_{30}Ni_5Al_{10}$ BMG in its supercooled liquid state. A backward extrusion process was used at a temperature of 725 K and a strain rate of about 1×10^{-2}/s. The authors produced a 3D cup-shaped object with wall thickness of 0.05 mm and outside diameter of 2.2 mm.

13.4 Fabrication of Dies and Rams for Microextrusion

The most crucial part in the design and fabrication of a microextrusion machine is the fabrication of the die and ram. The small size of the die and ram, along with the stringent accuracy requirement, needs suitable manufacturing processes. Conventional technology may not be compatible for the manufacturing of miniature/microproducts. It is necessary to reduce the scale of the equipment to reduce the energy consumption and the equipment costs. Some traditional forming machine designs can be scaled down for microforming needs, as long as the machines can operate with microtools of acceptable quality and efficiency.

Microdies, containers, punches, and actuators are the major parts designed and fabricated for the microextrusion process. Depending on the level of miniaturization, different processes are used to fabricate microdies. Saotome and Iwazaki (2001) used photochemically machinable glass to fabricate microdies for the production of microgears. The surface roughness of the die was about 2 μm owing to the etching of the fine crystals of Li_2O–SiO_2. Cao et al. (2004) developed a forward extrusion setup to extrude micropins. The die geometry has a base diameter of 0.76 mm, an extruded diameter of 0.57 mm, and a bearing surface length of 201 μm. Die cavities of submillimeter size were made by micro-electro-discharge machining (μ-EDM). The extrusion die was made as a segmented block in order to facilitate the removal of the micropins after extrusion. Krishnan et al. (2007) used conventional drilling and polishing to fabricate extrusion dies. The segmented die was mounted onto a forming assembly consisting of a ram mounted on a yoke that slides along linear bearings and guides the ram into the segmented die. A load cell and an LVDT were used to measure the extrusion force and the corresponding ram displacement. The authors fabricated four dies with base diameters of 0.76, 1.50, 1.75, and 2.00 mm. The corresponding extruded diameters were 0.57, 1.00, 1.17, and 1.33 mm, respectively, and the die half-angle was 30° in each case. The larger dies were made by drilling, and the smaller dies were made by the micro-EDM process. All the dies were polished to ensure similar roughness values.

Mahayotsanun et al. (2009) developed a high-speed microextrusion machine to obtain high-production rates. They described the challenges and development procedure of the high-speed microextrusion machine, which was designed to provide consistent and reliable results. The actuator and forming assembly that were initially designed provided up to 19 kN force, 25 mm/s speed, and a stroke length of 30 mm. Plastic deformation was observed in the die because of the misalignment of the punch and the die. The failure of the ejector pin was because of strong adhesion between the formed sample and the die surface, which exceeded the buckling load of the ejector pin. These difficulties were eliminated with the newly developed microextrusion machine. The new machine has more precise parts with tolerances within 1 mm. The punch has a stepped diameter to reduce buckling.

A major difficulty is that when the forming process is scaled down to microdimensions, the microstructure of the workpiece and the surface topologies of the workpiece and the tooling remain unchanged. Conventional methods of tool manufacturing can be applied to manufacture microforming tools with some limitations. Wire-cut EDM with wires of 10 μm diameter allows the production of very narrow extrusion dies. The punches/rams can be made by grinding. Structures with a size of 10 μm can be produced by laser ablation. Electron beam lithography makes it possible to create structures as small as 200 nm in width (Engel and Eckstein, 2002). Precision manufacturing, tolerance, assembly process, and handling of the microparts are hot topics for investigation.

13.5 Parametric Study and Modeling of Microextrusion

When a macroscale forming process is scaled down to microdimension, certain factors such as microstructure of the workpiece, surface topology of the workpiece, as well as the tool play more significant roles. A number of authors have carried out experimental and theoretical studies for understanding the influence of various process parameters. Engel and Eckstein (2002) studied the effect of miniaturization on bulk metal forming through double-cup extrusion experiments. They observed that decrease in sample dimension caused a significant increase in friction factor values. Higher friction was observed with workpieces of smaller grain size. Inhomogeneous deformation was observed with workpieces of larger grain size. Grain size plays an important role in the quality of the final product.

Krishnan et al. (2007) carried out experiments with a novel microextrusion apparatus to obtain the force–displacement response of brass pins with extruded diameter ranging from 1.33 to 0.57 mm for a ram stroke of 3.5 mm. They compared the force profile and pin lengths with the results obtained from FEM simulations and analytical models of Avitzur (1965) and Altan et al. (1983) Avitzur's model is based on the upper bound method using the von Mises criterion. According to this model, the extrusion load per unit area, p, is given as

$$p = p_i + p_s + p_{sf} + p_f$$

(13.6)

where p_i is the contribution due to internal deformation, p_s is the contribution due to velocity discontinuity, p_{sf} is the contribution due to die friction, and p_f is the contribution due to friction at the billet–container interface. Furthermore,

$$p_i = \sigma_0 f(a) \ln (R) \tag{13.7}$$

$$p_s = \frac{2\sigma_0}{\sqrt{3}} \left(\frac{\alpha}{\sin^2 \alpha} - \cot \alpha \right) \tag{13.8}$$

$$p_{sf} = \frac{\sigma_0}{\sqrt{3}} m_3 \cot \alpha \ln (R) \tag{13.9}$$

$$p_f = \frac{2\sigma_0}{\sqrt{3}} m_4 \left[\frac{L}{r_0} - \left(1 - \frac{1}{\sqrt{R}} \right) \cot \alpha \right] \tag{13.10}$$

where σ_0 is the flow stress, $f(a)$ is the geometric factor, R is the extrusion ratio (the ratio of initial cross-sectional area of the billet to final cross-sectional area after extrusion), α is the semi-cone-angle of the die, L is the billet length, r_0 is the billet radius, and m_3 and m_4 are the friction factors along the die and container surfaces, respectively. The geometric factor is equal to 1.00625 for $\alpha = 30°$, and it can be approximated to 1 for most practical applications. The slab method model of Altan et al. (1983) also provides similar expressions. Both models differ in terms of their calculation of frictional components along the container and the die.

Krishnan et al. (2007) used the commercial package ABAQUS/Explicit to compute the external load. They used CAX4R elements, which are four-noded bilinear axisymmetric quadrilateral elements with a reduced integration scheme to avoid locking of the elements. The adaptive meshing option is used to avoid mesh distortion. The die and container surfaces are modeled as rigid analytical solids and their contact with the workpiece is modeled using the "hard contact" model.

For larger pins, the ram force obtained from FEM simulation closely matches the experimental results. For small pins, FEM simulation overpredicts the ram force as compared to the experimental results. From the force comparison study, it was observed that the friction coefficient and the friction factor tend to decrease with smaller size of the extruded pins. This trend is opposite to what is reported by Engel and Eckstein (2002). The pins extruded from the smallest die tend to curve. This may be due to the large ratio of grain size to the sample dimensions of the pin. The authors used different low-friction coatings such as diamond-like carbon with silicon (DLC–Si), chromium nitride (CrN), and titanium nitride (TiN) on the dies to investigate the feasibility of such coatings in microextrusion. The die coated with DLC–Si produced longer pin size as compared to other nitride-coated dies. This longer pin length indicates a lower level of friction between the die and the workpiece.

Cao et al. (2004) investigated the effect of grain size and surface roughness on the extrusion process for micropins. Extrusion of the sample with different grain sizes shows that the process depends on the surface structure of the die. The authors used the reproducing kernel element method (RKEM) to simulate the extrusion of micropins for a 211 μm grain-sized sample. FEM simulation was also carried out to compare the results obtained from RKEM. It was observed that RKEM has the ability to capture deformation characteristics more accurately.

Eichenhuller and Engel (2009) carried out full backward extrusion to extrude a single form feature to study the individual effect and the interaction of parameters such as temperature, size of the form feature, lubrication, grain size, and velocity. The effects of the interactions between temperature and width of the fin and between temperature and lubrication are significant to the maximum attainable aspect ratio (ratio between the feature height and width or diameter of the feature). The largest variation in the maximum possible aspect ratio was observed with variation in temperature. Lubrication has more profound effect on the extrusion of products with higher surface area. Thus, it is more significant in fin extrusion than in pin extrusion. The effect of punch velocity was found to be insignificant compared to other parameters and their interaction. In short, the authors have highlighted the effect of size, parameter settings, and combinations of parameter settings in microextrusion processes.

Rosochowska et al. (2010) carried out finite element analysis of the microextrusion of conical pins whose larger diameters were 340 μm. They compared the performance of ultra-fine-grained material with that of coarse-grained material. Although ultra-fine-grained material produces a better quality product, the extrusion load is about 25% higher. Figure 13.3 shows the pin geometries used for simulation. The pin is conical and has a head diameter of 0.34 mm. Figure 13.4 shows the displacement versus extrusion force curves for different pins. A coarse-grained pin having 15° semi-cone-angle provides the lowest extrusion load, and an ultra-fine-grained pin having a 5° semi-cone-angle provides the highest extrusion load among the cases studied.

13.6 Microextrusion Using Multihole Extrusion

In microextrusion, billet preparation is also a difficult task. If the billet size is many times more than the extruded product size, extrusion load will be high. Making a billet of only a

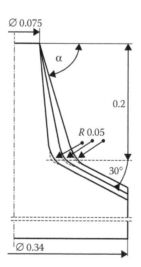

FIGURE 13.3
Pin geometries (dimensions in millimeters). (From Rosochowska, M., et al., *International Journal of Materials Forming*, 3, 423–426, 2010.)

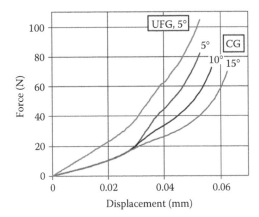

FIGURE 13.4
Force in the backward extrusion of a coarse-grained material with pin angle 5°, 10°, and 15° and an ultra-fine-grained material Al1070 with pin angle 5°. (From Rosochowska, M., et al., *International Journal of Materials Forming*, 3, 423–426, 2010.)

slightly larger cross-sectional area than the extruded product will provide a low extrusion load. However, fabrication of small-sized billets is not easy. One way to sort out this problem is to carry out multihole extrusion. Extrusion through a die having more than one opening is known as *multihole extrusion*. In this process, several products are manufactured simultaneously. Although the size of each product may be very small, the overall extrusion ratio is low. Hence, the load requirement is low. Moreover, in one setup and with one billet, one can fabricate many components. Thus, the overall time for production is reduced.

At the Indian Institute of Technology Guwahati, attempts have been made to produce small-sized pins by a multihole extrusion process. Up to 15 pins of 1.5-mm diameter have been produced from a billet of 20-mm diameter (Sinha et al., 2009b). The authors have proposed a simplified mathematical model to estimate the ram force in a multihole extrusion process. In the model, the multihole extrusion process is broken down into parallel single-hole extrusion processes. For each single-hole extrusion process, the effective inner diameter of the die is calculated by a simplified approach. In the multihole square die, the holes are arranged at different pitch circles. It is assumed that a material particle will flow through the holes situated on the pitch circle diameter nearest to it in order to encounter the least resistance. Furthermore, flow through each hole on a pitch circle is assumed to be identical. Thus, the total billet cross-sectional area is divided into a number of cross-sectional areas corresponding to different holes. The equivalent inner diameter for a hole in a die is the diameter of the cylinder whose cross-sectional area corresponds to the cross-sectional area of the billet passing through the hole.

The procedure is illustrated for a 13-hole die. The arrangement of the holes on the flat bottom surface of the die is shown in Figure 13.5. The inner diameter of the die is R_3. It has one hole in the center, four holes on a pitch circle radius of R_1, and eight holes on a pitch circle radius of R_2. The radial distance of a general particle from the center of the die is r. According to the assumptions, the flow of the material through various holes is as follows:

- The particles with $0 < r \leq R_1/2$ flow through the hole at the center of the die. In the figure, the zone of these particles is depicted by the inner circle of radius $R_1/2$.
- The particles with $(R_1/2) < r \leq (R_1 + R_2)/2$ flow through the holes having their centers on the pitch circle of radius R_1. These particles lie in the annular region

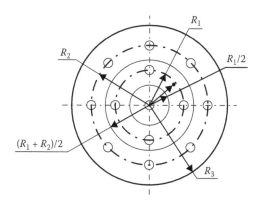

FIGURE 13.5
Assumed zones of material flow for holes located on different pitch circles. The boundaries between the zones are shown by thin, solid-line circles. (From Sinha, M.K., et al., *Materials & Design*, 30, 2386–2392, 2009b.)

between the inner circle of radius $R_1/2$ and middle circle of radius $(R_1+R_2)/2$ in the figure.

- The particles with $(R_1 + R_2)/2 < r \le R_3$ flow through the holes that have their centers on the pitch circle of radius R_2. These particles lie in the region outside the circle of radius $(R_1 + R_2)/2$ in the figure.

With these assumptions, the cross-sectional area A_1 of the billet flowing through the hole in the center of the die is

$$A_1 = \frac{\pi R_1^2}{4}$$

(13.11)

The equivalent die inner diameter d_1 for the material flowing through the hole in the center of the die is

$$d_1 = \sqrt{\frac{4A_1}{\pi}} = R_1$$

(13.12)

Similarly, the cross-sectional area A_2 of the billet flowing through holes with centers on the pitch circle of radius R_1 is

$$A_2 = \pi \left(\frac{R_1 + R_2}{2} \right)^2 - \frac{\pi R_1^2}{4} = \frac{\pi}{4} \left(R_2^2 + 2R_1 R_2 \right)$$

(13.13)

As there are four holes on the pitch circle of radius R_1, the area of the billet associated with each hole is $A_2/4$. Hence, the equivalent die inner diameter d_2 for the material flowing through holes with centers on the pitch circle radius of R_1 is

$$d_2 = \sqrt{\frac{A_2}{\pi}} = \frac{1}{2}\sqrt{R_2 \left(R_2 + 2R_1 \right)}$$

(13.14)

Finally, the cross-sectional area A_3 of the billet flowing through holes on the pitch circle of radius R_2 is

$$A_3 = \pi \left(R_3^2 - \frac{(R_1 + R_2)^2}{4} \right)$$

(13.15)

As there are eight holes on the pitch circle of radius R_2, the area of the billet associated with each hole is $A_3/8$. Hence, the equivalent inner diameter of the die for the material flowing through holes with centers on the pitch circle of radius R_2 is

$$d_3 = \sqrt{\frac{A_3}{2\pi}} = \sqrt{\frac{1}{2}\left(R_3^2 - \frac{(R_1 + R_2)^2}{4} \right)}$$

(13.16)

The exit velocity of the extruded product through holes of diameter d_f can be calculated from the continuity equation. Thus, if the ram velocity is V_1, the velocity v_1 of the product coming out of the central hole is

$$v_1 = \frac{d_1^2 V_1}{d_f^2}$$

(13.17)

Similarly, the velocity v_2 of the product coming out of the holes on the pitch circle of radius R_1 is

$$v_2 = \frac{d_2^2 V_1}{d_f^2}$$

(13.18)

and the velocity v_3 of the product coming out of the holes on the pitch circle of radius R_2 is

$$v_3 = \frac{d_3^2 V_1}{d_f^2}$$

(13.19)

At any instant, the lengths of the extruded products coming out through various holes will be proportional to the velocities of the products.

The power required for extruding the material through a hole is obtained by an upper bound model for single-hole extrusion (Sinha et al., 2009a) through the hole in the equivalent cylindrical die. The calculation of the equivalent inner diameter of the die for a hole has already been explained. The details of the upper bound model are available in Appendix A. The total power is the sum of the powers required for forcing the material through various holes. The ram force is obtained by dividing the total power by the ram velocity.

Several experiments were carried out with multihole dies having 9, 13, and 15 holes. Different billet lengths (20, 25, and 30 mm) were used in multihole extrusion to study the effect of billet length on ram force. A typical result of multihole extrusion for a billet of 20 mm length and 1.5 mm diameter is given in Table 13.1. It is observed that extrusion load decreases with increase in the number of holes in the die.

Several experiments conducted on single and multihole extrusion have shown that the extrusion load is considerably reduced in multihole extrusion compared to single-hole extrusion. Initial experiments were carried out on lead and wax. Cold extrusion of aluminum was

TABLE 13.1

Extrusion Load in the Multihole Extrusion of Lead Metal

Billet Length (mm)	Extrusion load (kN)					
	9-Hole Die		13-Hole Die		15-Hole Die	
	Theoretical	Experimental	Theoretical	Experimental	Theoretical	Experimental
20	122.67	101.31	111.43	95.30	106.08	92.00

Source: Sinha, M.K., Modelling and experimental investigation of multi-hole extrusion process. M.Tech. Thesis, Indian Institute of Technology Guwahati, 2008.

also carried out recently. It was observed that the product length from each hole is different in a multihole extrusion process and also that there is bending of the product. The amount of bending varies from hole to hole. To eliminate bending, a constrained extrusion process is introduced. In this process, instead of allowing the extruded products to come out freely from the die, the products are guided through blind holes, as shown in Figure 13.6. Thus, bending is eliminated and the length of each product remains the same. Figure 13.7 shows photographs of extruded products. Compared to free extrusion, the load is higher in constrained extrusion. Nevertheless, several products get fabricated simultaneously.

Attempts are on to extrude products less than 500 μm in diameter. Holes of size 378 μm have been fabricated using the EDM process with a copper electrode. Using a constrained, multihole extrusion process, it is planned to produce 378-μm-diameter pins from a 5-mm-diameter billet.

13.7 Limitations and Challenges of Microextrusion Processes

With the scaling down of dimensions and increasing geometric complexity of the objects, currently available technologies and systems may not be able to meet the development

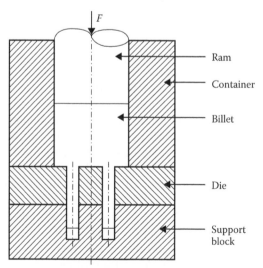

FIGURE 13.6
A schematic of constrained multihole extrusion process.

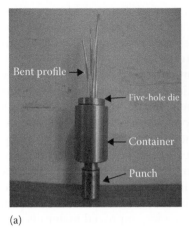 

(a)   (b)

FIGURE 13.7
Extruded products in (a) free extrusion and (b) constrained extrusion.

needs. New measuring devices, principles and instrumentation, tolerance rules, and procedures have to be developed. Material databases with detailed information on various materials and their properties/interface properties including microstructures and size effect would be very useful for product innovation and process design. More studies are necessary on micro/nanowear and damages/failures of the micromanufacturing tools (Qin, 2006; Qin et al., 2010). The forming limits for different types of materials at the microlevel must be prescribed. More specific considerations must be incorporated into the design of machines that are scaled down for microforming to meet engineering applications and requirements.

The following points need special attention from researchers and developers of microextrusion processes and machines:

- Manufacturing and assembly of small components is difficult.
- Clearance or backlash between machine parts, which may be tolerable in a conventional extrusion machine, plays a significant role in a microextrusion machine.
- Handling raw material and extruded parts is difficult because of less grip on the surface area.
- Microextrusion requires precise manufacturing processes to produce smaller dies and punch/ram.
- There is a need to develop adequate measuring technology suitable for measuring small components.

13.8 Conclusion

Research in recent years has tried to gain fundamental knowledge to focus on the challenges and difficulties with microforming processes. Many researchers have tried to produce components of submillimeter range with microextrusion processes. Both simulation and experimental investigations have been carried out considering different process parameters. Production of microcomponents such as micropins and microgears has been attempted. Process parameters and their combined effect on punch load, surface quality, and strength of the parts produced have been studied.

Fabrication of microextrusion machine components, their assembly, and tolerances are the challenges to date. Some specific properties of traditional materials often set limits to their direct use in microproducts. Material behavior is often affected by scale effects as a result of reduced dimensions. Special attention must be given to size effect and its importance to the scaling down of the process. Although microextrusion is able to produce components of microsize, further detailed studies are required on size effect, work material behavior at microlevels, and the limitation to the use of concepts of macroextrusion. The approach to development of microproducts is still not systematized. The peculiar material behavior at micro- and nanoscales must be studied to succeed in designing extrusion machine tools and processes. New design principles, improved materials, and technology are necessary for mass production at low cost.

Appendix A: Estimation of Ram Force by Upper Bound Method

It is assumed that the flow of a metal through a square die having a single hole may be considered as flow through a conical die, as shown in Figure 13.A1. The boundary of the dead-metal zone acts like an inner surface of the conical die. The die semi-angle is obtained by minimizing the total power. For the calculation of the power through the conical die, the upper bound model of Reddy et al. (1996) has been adopted, with a modification to include the effect of bending of the material of the billet at entry and exit.

Taking line 1–2 (Figure 13.A1) and the axis of symmetry as the two streamlines, and assuming that all the streamlines are straight, the appropriate assumed velocity field is

$$V_z = \frac{V_1 R_1^2}{R^2}$$

(13.A1)

$$V_r = -\frac{V_1 R_1^2 r}{R^3}\tan\alpha$$

(13.A2)

where V_1 is the ram velocity, R_1 is the radius of the billet, R_2 is the radius of extruded product, α is the die semi-angle, and R is the radius of the cone at location z, with the r–z

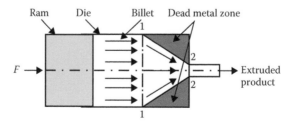

FIGURE 13.A1
Schematic diagram of the flow of a metal through a square die. (From Sinha, M.K., et al., *Materials & Design*, 30, 330–334, 2009.)

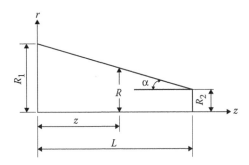

FIGURE 13.A2
Enlarged view of the upper half of the conical portion. (From Sinha, M.K., et al., *Materials & Design*, 30, 330–334, 2009.)

coordinate system being shown in Figure 13.A2. The power of deformation in the assumed conical portion is given by

$$\dot{W}_d = \dot{W}_i + \dot{W}_f + \dot{W}_s \tag{13.A3}$$

where $\dot{W}_i$ is the internal power of deformation in the conical zone having continuous velocity, $\dot{W}_f$ is the friction power, and $\dot{W}_s$ is the power loss due to velocity discontinuities at sections 1–1 and 2–2 shown in Figure 13.A1.

The internal power of deformation in the continuous velocity zone is given by

$$\dot{W}_i = \int_0^L \int_0^R \sigma_y \dot{\varepsilon}_{eq} \, 2\pi r \, dr \, dz \tag{13.A4}$$

where σ_y is the flow stress of the material, $\dot{\varepsilon}_{eq}$ is the equivalent strain rate, and L is the length of the die. Further,

$$\dot{\varepsilon}_{eq} = \sqrt{\frac{2}{3}\left(\dot{\varepsilon}_{rr}^2 + \dot{\varepsilon}_{\theta\theta}^2 + \dot{\varepsilon}_{zz}^2 + \dot{\varepsilon}_{rz}^2\right)} \tag{13.A5}$$

where

$$\dot{\varepsilon}_{rr} = \frac{\partial V_r}{\partial r}; \quad \dot{\varepsilon}_{\theta\theta} = \frac{V_r}{r}; \quad \dot{\varepsilon}_{zz} = \frac{\partial V_z}{\partial z}; \quad \dot{\varepsilon}_{rz} = \frac{1}{2}\left(\frac{\partial V_r}{\partial z} + \frac{\partial V_z}{\partial r}\right) \tag{13.A6}$$

For a strain-hardening material, flow stress may be assumed to be related to strain in the following manner:

$$\sigma_y = K\varepsilon_{eq}^n \tag{13.A7}$$

where K is the strength coefficient of the material, n is the strain-hardening exponent, and ε_{eq} is the equivalent strain defined by

$$\varepsilon_{eq} = (\varepsilon_{eq})_i + \int_0^t \dot{\varepsilon}_{eq} \, dt \tag{13.A8}$$

where $(\varepsilon_{eq})_i$ is the initial equivalent strain due to bending of material at the die inlet and t is the time. The initial equivalent strain is calculated by the following expression of Iwahashi et al. (1996):

$$\left(\varepsilon_{eq}\right)_i = \frac{2}{\sqrt{3}} \tan\left[\frac{1}{2}\tan^{-1}\left(\frac{r}{R}\tan\alpha\right)\right]$$

(13.A9)

Knowing that along a streamline

$$dt = \frac{dz}{V_z} = \frac{dr}{V_r}$$

(13.A10)

Equation 13.A8 can be written as

$$\varepsilon_{eq} = \frac{2}{\sqrt{3}}\left[\tan\left\{\frac{1}{2}\tan^{-1}\left(\frac{r}{R}\tan\alpha\right)\right\} + \left\{\sqrt{3 + 2.25\frac{r^4}{R^4}\tan^2\alpha}\right\}\ln\frac{R_1}{R}\right]$$

(13.A11)

Internal power dissipation in the continuous velocity zone is found by numerically integrating Equation 13.A4.

Frictional power dissipation at the surface of the dead metal zone is given by

$$\dot{W}_f = \int_{R_1}^{R_2}\frac{\sigma_y}{\sqrt{3}}\frac{V}{\cos\alpha\sin\alpha}2\pi r\,dr$$

(13.A12)

Power dissipation due to velocity discontinuities at sections 1–1 and 2–2 is given by

$$\dot{W}_s = 2\int_0^{R_1}\frac{\sigma_y}{\sqrt{3}}\frac{2\pi r^2 V_1}{R_1}\tan\alpha\,dr$$

(13.A13)

Power dissipation due to friction at the billet–container interface is given by

$$\dot{W}_{Fc} = \frac{m\left(\sigma_y\right)_0 2\pi R_1 L_c V_1}{\sqrt{3}}$$

(13.A14)

where $(\sigma_y)_0$ is the yield stress of the material, m is the friction factor at the billet–die interface, and L_c is the billet length outside the assumed conical deformation zone.

The total power is given by

$$\dot{W}_t = \dot{W}_d + \dot{W}_{Fc}$$

(13.A15)

Ram force can be obtained by dividing the total power by the ram velocity. The value of the die semi-angle α is obtained by a one-dimensional optimization procedure for minimizing the total power. The average ram pressure is calculated as

$$P_{avg} = \frac{\dot{W}_t}{\pi R_1^2 V_1}$$

(13.A16)

References

Altan, T., Oh, S., and Gegel, L.H. 1983. *Metal Forming: Fundamentals and Applications*. Metals Park, Ohio: American Society of Metals.

Avitzur, B. 1965. Analysis of metal extrusion. *ASME Journal of Engineering Industry* 87(1): 57–70.

Basu, J. and Ranganathan, S. 2003. Bulk metallic glasses: A new class of engineering materials. *Sadhana* 28: 783–798.

Cao, J., Krishnan, N., Wang, Z., Liu, W., and Swanson, A. 2004. Microforming: Experimental investigation of the extrusion process for micropins and its numerical simulation using RKEM. *Journal of Manufacturing Science and Engineering* 126: 642–652.

Chen, F.K. and Tsai, J.W. 2006. A study of size effects in micro-forming with micro-hardness tests. *Journal of Materials Processing Technology* 177: 146–149.

Cho, H., Hwang, D., Lee, B., and Jo, H. 2007. Fabrication of micro condenser tube through direct extrusion. *Journal of Materials Processing Technology* 187–188: 645–648.

Eichenhuller, B. and U. Engel. 2009. Investigation of the effect and parameter interactions in metal microforming processes. *4M/ICOMM 2009—The Global Conference on Micro Manufacture*. pp. 245–248, DOI:10.1243/17547164C0012009048.

Engel, U. and Eckstein, R. 2002. Microforming—From basic research to its realization. *Journal of Materials Processing Technology* 125–126: 35–44.

Engel, U., Messner, A., and Tiesler, N. 1998. Cold forging of microparts—Effect of miniaturization on friction. *Proceedings of the 1st ESAFORM Conference on Materials Forming*, ed. J.L., Chenot, J.L. Chenot, J.F. Agassant, P. Montmitonnet, B. Vergnes, and N. Billon Sophia Anitpolis, France, pp. 77–80.

Geiger, M., Geibdorfer, S., and Engel, U. 2007. Mesoscopic model: Advanced simulation of microforming process. *Production Engineering Research and Development* 1: 79–84.

Geiger, M., Kleiner, M., Eckstein, R., Tiesler, N., and Engel, U. 2001. Microforming. *CIRP Annals—Manufacturing Technology* 50(2): 445–462.

Hirota, K. 2007. Fabrication of micro-billet by sheet extrusion. *Journal of Materials Processing Technology* 191: 283–287.

Iwahashi, Y., Wang, J., Hortia, Z., Nemoto, M., and Langdon, T.G. 1996. Principle of equal-channel angular pressing for the processing of ultra-fine grained materials. *Scripta Materials* 35: 143–146.

Krishnan, N., Cao, J., and Dohda, K. 2007. Study of the size effect on friction conditions in microextrusion—Part I: Microextrusion experiments and analysis. *Transaction of the ASME, Journal of Manufacturing Science and Engineering* 129: 669–676.

Mahayotsanun, N., Lee, H.C., Cheng, T.J., Chuang, Y., Tu, K.Y., Ehmann, K., and Cao, J. 2009. Development of the high-speed micro-extrusion machine to investigate strain rate and size effects, 4M/ICOMM 2009, Karlsruhe, Germany.

Parasiz, S.A., Kinsey, B., Krishnan, N., Cao, J., and Li, M. 2007. Investigation of deformation size effects during Microextrusion. *Transaction of the ASME, Journal of Manufacturing Science and Engineering* 129: 690–697.

Qin, Y. 2006. Micro-forming and miniature manufacturing systems—Development needs and perspectives. *Journal of Materials Processing Technology* 177: 8–18.

Qin, Y., Brockett, A., Ma, Y., Razali, A., Zhao, J., Harrison, C.S., Pan, W., Dai, X. and Loziak, D. 2010. Micro-manufacturing: Research, technology outcomes and development issues. *International Journal of Advanced Manufacturing Technology* 47: 821–837.

Reddy, N.V., Dixit, P.M., and Lal, G.K. 1996. Central bursting and optimal die profile for axisymmetric extrusion. *Transactions of the ASME, Journal of Manufacturing Science and Engineering* 118: 579–584.

Rosochowska, M., Rosochowski, A., and Olenik, L. 2010. FE simulations of micro-extrusion of a conical pin. *International Journal of Materials Forming* 3: 423–426.

Saotome, Y. and Iwazaki, H. 2001. Superplastic backward microextrusion of microparts for micro-electro-mechanical systems. *Journal of Materials Processing Technology* 119: 307–311.

Sinha, M.K. 2008. Modelling and experimental investigation of multi-hole extrusion process, MTech Thesis, Indian Institute of Technology, Guwahati.

Sinha, M.K., Deb, S., and Dixit, U.S. 2009a. Design of a multi-hole extrusion process. *Materials & Design* 30: 330–334.

Sinha, M.K., Deb, S., Das, R., and Dixit, U.S. 2009b. Theoretical and experimental investigations on multi-hole extrusion process. *Materials & Design* 30: 2386–2392.

Vollertsen, F. 2008. Categories of size effects. *Production Engineering Research and Development* 2: 377–383.

Wu, X., Li, J.J., Zheng, L., and Li, Y. 2010. Micro back extrusion of a bulk metallic glass. *Scripta Materialia* 63: 469–472.

Yu, S., Hu-ping, Y., and Xue-yu, R. 2006. Discussion and prediction on decreasing flow stress scale effect. *Transactions of Nonferrous Metals Society of China* 16: 132–136.

14

Microbending with Laser

U.S. Dixit, S.N. Joshi, and V. Hemanth Kumar
Indian Institute of Technology Guwahati

CONTENTS

14.1 Introduction

Starting with its invention in 1960, light amplification by stimulated emission of radiation (LASER) has been applied in a number of manufacturing processes. Laser beam machining is now a well-established industrial process. Laser welding and heat treatment also find applications in industry. A relatively recent application of laser is in metal forming. This application can be classified into two parts—laser-assisted forming and laser forming (LF). In laser-assisted forming, the parts to be formed are heated by a beam of laser, while the actual forming is done by conventional tools. LF is a process that forms the sheet metal by means of stresses induced by external heat instead of external force (Geiger et al., 1993).

LF is used to deform sheets made of materials with high thermal expansion coefficients such as stainless steels and light alloys of aluminum, magnesium, and titanium. During the process, deformation is induced in a controlled manner by tracking the laser beam across one surface of the sheet metal (Shen et al., 2006). From the distribution of the thermal stress generated by laser heating, a precise and complex shape of the workpiece can be formed by a controllable laser beam (Hsieh and Lin, 2005). This is a novel way of bending a metal sheet without the use of a die or additional loading.

LF has the following advantages over traditional metal-forming technologies:

- It requires no tools or external forces in the process.
- Work material degradation is less because of the production of a small heat-affected zone (HAZ) by a focused heat source of the laser.
- The process can be conveniently and accurately controlled by adjusting laser parameters including laser power and laser beam diameter and processing parameters such as scanning speed of the laser beam (Gillner et al., 2005).
- It can be used to produce complex three-dimensionally formed components by using combinations of straight and curved scan lines (Shen and Vollertsen, 2009).
- It offers excellent reproducibility, less manufacturing time and cost, as well as relatively low thermal influence on the mechanical properties of a metal (Labeas, 2007).

LF is gradually finding its place in the manufacturing industry, which previously relied on expensive stamping dies and presses. Industries benefitting from LF are aerospace, automotive, shipbuilding, and microelectronics (Shen et al., 2009). In this chapter, we focus on microbending processes that utilize the laser beam either as an assistive component or as a replacement for the mechanical punch.

14.2 Laser-Assisted Microforming

In laser-assisted forming, a laser light is used to increase the temperature of the material during forming. This increases the formability in the required area of the part and reduces the flow stress and anisotropy of the material. As material softening is done locally, a major portion of the sheet remains unaffected. To enable the transmittance of laser light into the workpiece, sapphire tools are used. The temperature of local regions of the workpiece can be increased quickly by direct heating. Use of transparent tools such as sapphire guides the laser radiation directly on to the workpiece through the closed tool. The workpiece can be heated in the areas where the material needs to flow during the forming process, while other zones of the material remain cold (Gillner et al., 2005). Experimental investigations have shown that the use of sapphire tools in laser-assisted microforming processes is a suitable method for the production of microparts. Sapphire has the required transparency for laser irradiation, at a wavelength of 809 nm, and excellent mechanical properties, high hardness, and compressive strength. Thus, it is a suitable tool for laser-assisted metal forming. The melting temperature of sapphire is 2300 K, which allows warm or hot forming of steel without damaging the tool (Samm et al., 2009). Figure 14.1 shows a schematic diagram of the laser-assisted microbending method. A preshaped metal sheet is placed over a die with a V-shaped slot in it. A conical sapphire tool is used to achieve the required bend angle.

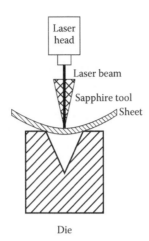

FIGURE 14.1
Schematic diagram of laser-beam-assisted microbending.

Initially, thermal energy is applied on the metal sheet using the laser beam. Then the transparent sapphire tool is plunged inside the die along with the metal sheet to bend it to the required angle. It is to be noted that the laser continuously heats the work sheet during the process of mechanical bending using a tool.

Samm et al. (2009) have proposed a similar kind of laser-assisted microforming method to create a microfeature on two concentric cylinders with a depth of 100 μm and diameters of 400 and 200 μm. A frequency-quadrupled Nd:YVO$_4$ laser beam was used to supply thermal energy to the microstructures during the process. The laser beam was focused to a diameter of 10 μm using a high-precision galvanoscanner. The authors studied the effect of different laser sources on the precision of the formed feature.

Another way of using laser in microbending is in the so-called hybrid processing (Magee and De Vin, 2002). In hybrid forming, the initial shape can be formed by a conventional tool, which can be corrected by a laser beam. The forming can also be followed by welding and machining by a laser beam.

Laser has also been used for localized heating in incremental forming (Duflou et al., 2007). In incremental forming, a mechanical tool keeps moving over the workpiece and localized heating is done by a moving laser beam. The surrounding zone is kept at room temperature only, providing the necessary support during the process.

In laser-assisted microbending, a mechanical tool is also required. The use of a mechanical tool can increase the production rate compared to that achieved in microbending using the heating effect of the laser beam alone. However, the fabrication of a small mechanical tool is a challenge. Owing to this, most researchers have attempted to carry out microbending using a laser beam as a replacement for the mechanical tool as described in the following sections.

14.3 Laser Microbending

At present, the manufacturing industry faces a challenge in providing a fast, accurate, and economical process for producing very small parts with microlevel bending that are used in micro electro mechanical systems (MEMS) and microelectronics, sectors such as medicine and automotives, and in the optical, chemical, and sensor industries. Electronics production is especially characterized by a progressive miniaturization because of a general trend toward higher integration and package density (Chen et al., 1998). With increasing demand on the geometrical complexity and a variety of parts produced with microlevel deformation, there is an increasing interest in selecting a suitable manufacturing process for mass production. To meet these challenges, laser technology has provided a promising option to the microtechnology industry through its high flexibility and ability to focus on a spot of a few microns. The availability of new processes and beam sources qualifies the laser as a universal tool in this area (Gillner et al., 2005).

Laser-assisted forming has found application in many forming processes including incremental forming. The process can be explored further to enhance its performance in microbending. However, LF without the use of any mechanical tool appears to be a more suitable method of microbending.

In microelectronics applications, accurate bending of thin plates is very important for manufacturing accurately shaped parts and assembling them precisely. For example, the gap between the disk surface and the head of a hard disk drive is adjusted by bending the supporting beam. It is, however, difficult to give a very small bending angle using this mechanical method with external force because the spring-back angle is much greater than the resulting permanent angle. LF seems to be a suitable method for microbending because the sheet is bent by the thermal stress induced by laser irradiation, and thus the effect of spring back is very small (Otsu et al., 2001). Thus, the required microlevel deformations in the work parts can be achieved by using the controlled radiation of laser on the designed locations. In this chapter, we describe the fundamentals of and recent advances in laser microbending.

During LF, the irradiated material is formed because of the local plastic strains induced by laser heating of the material and not because of the mechanical forces and moments applied using common sheet bending techniques. Figure 14.2 shows a schematic diagram of a straight-line irradiation process that produces a bend angle from a flat sheet metal piece. The localized nature of laser irradiation yields high temperature gradients between the irradiated surface and the neighboring material that force the material to expand nonuniformly, resulting in irregular thermal expansion between the top and bottom surfaces. As a result, the specimen initially bends negatively, as viewed from the laser beam.

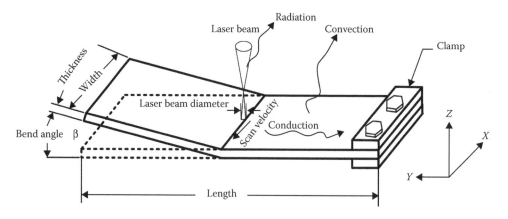

FIGURE 14.2
Schematic diagram of the laser microbending process.

The nonuniform expansion of the material leads to nonuniform thermal stresses, which results in plastic deformation at locations where thermal stresses exceed the material's yield point. During cooling, the upper material layers shrink more than the lower ones, resulting in permanent specimen bending toward the laser beam. From this description, it is evident that the LF process is composed of many simultaneous physical mechanisms and is affected by several process and material parameters. The most important parameters are the laser scanning path, the heating conditions including the laser power and type, the scanning velocity, the material thermal parameters, namely, emissivity and conductivity, as well as the material coefficient of thermal expansion (Labeas, 2007). Lasers such as carbon dioxide (CO_2) and neodymium:yttrium aluminum garnet (Nd:YAG) are generally used as a beam source with an irradiation intensity of about 100 W/mm^2.

LF is a complex transient process that involves thermodynamics, heat transfer, elastoplastic mechanics, and metallurgy. It involves various process parameters such as laser energy parameters, shape and size of workpiece, and work material properties. These important parameters are described in the following subsections.

14.3.1 Laser Beam Energy

Laser beam energy is the energy absorbed by a metal sheet during surface heating by a laser beam. The energy absorbed into the material is mainly determined by the laser power, the scanning speed, and the absorptivity of the material. In general, the bend angle increases with an increase in the laser power and absorptivity because of the higher absorbed energy.

The heat flux caused on the surface of the sheet metal by a laser beam obeys normal distribution and is expressed as a function of the beam radius as follows (Holzer et al., 1994):

$$q(r) = \frac{2\eta P}{\pi R^2} e^{-2r^2/R^2}$$

(14.1)

where q is the thermal heat flux density of the laser beam (in W/m^2); η is the absorptivity of the sheet metal surface; P is the laser beam power (in W); R is the effective radius of the laser beam, which is defined as the radius at which the power density is reduced from the peak value by a factor of the square of the natural exponent $1/e^2$; and r is the distance from the center of the heat source (laser beam) in meters.

14.3.2 Scanning Velocity (Feed Rate)

Scanning velocity is the feed rate of the CNC machine on which the plate is placed. It is used to control the energy absorption in LF (Vollertsen, 1998). In general, as the feed rate increases, the required bend angle decreases since a lower amount of heat energy per unit time is transferred to the sheet metal, thereby resulting in a smaller bend angle. If the energy input by laser heating is constant, the bend angle increases as the path feed rate increases (Hu et al., 2001). In the case of low scanning speed, the temperature field is more homogeneous because of the longer time taken for heat conduction into the lower layer of the sheet. Therefore, the difference between the plastic strain of the irradiated layer and the nonirradiated layer is smaller. For high path feed rates, a very steep temperature gradient occurs in the plate, which generates high gradients of the plastic compression.

14.3.3 Laser Beam Radius

In most cases, the heat flux of the laser beam follows a Gaussian distribution with a peak at the center of the beam. The radius of the beam is equal to the distance from the center of the beam, where the heat flux is $1/e^2$ of its peak value. It is usually interpreted as the radius of a focused laser beam at the surface of a workpiece. Increasing the beam radius lowers the energy density of the laser spot, which leads to a reduced bend angle.

14.3.4 Material Properties

The material properties of a metal sheet play an important role in LF because the process is based on the thermal expansion caused by laser beam heating. The important thermal properties are thermal conductivity, specific heat capacity, density, and coefficient of thermal expansion. There exists a nearly linear relationship between the bend angle and the ratio of the coefficient of thermal expansion to the volumetric heat capacity (expressed by the product of the specific heat and the density) (Olowinsky and Bosse, 2003). The magnitude of thermal expansion is determined by the temperature increase and the coefficient of thermal expansion for all LF mechanisms.

14.3.5 Plate Geometry

The geometry parameters comprising sheet thickness, length, and width influence the angular change during laser bending. The most important parameter among these is the sheet thickness (Vollertsen, 1998). The bend angle is approximately inversely proportional to the square of the sheet thickness (Geiger et al., 1993).

14.4 Mechanisms of Laser Forming

LF is a complex transient process that involves thermodynamics, heat transfer, elastoplastic mechanics, and metallurgy to control the deformation of the metal workpiece. Three key mechanisms of LF demonstrate the forming behavior in the process. These mechanisms are determined by the temperature field, and each is associated with specific combinations of component geometries and laser process conditions. These are as follows (Geiger and Vollertsen, 1993):

1. Temperature gradient mechanism
2. Buckling mechanism
3. Upsetting mechanism

14.4.1 Temperature Gradient Mechanism

Temperature gradient mechanism (TGM) is the most widely reported laser bending mechanism. It is illustrated in Figure 14.3. In TGM, the diameter of the heated area is equal to the sheet thickness. TGM occurs in two stages, namely, heating of the workpiece using a laser beam and natural convection cooling. During the heating phase, a laser beam quickly scans the workpiece, producing a steep temperature gradient in the workpiece across

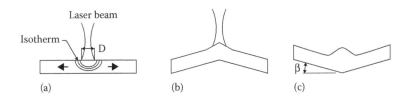

FIGURE 14.3
Temperature gradient mechanism in the laser bending process during (a) initial heating, (b) intermediate stage, and (c) cooling.

its thickness. As a result, a differential thermal expansion occurs in the direction of the thickness. The material expands in the heated zone, and the whole material bends away from the beam, as shown in Figure 14.3b. This is called *counterbending*. During cooling, the material contracts again in the upper layers and further plastically deforms. It leads to local shortening of the upper layers, and bending occurs in the opposite direction, as shown in Figure 14.3c. A positive bending angle is obtained.

To achieve higher efficiency in the process, the thermal expansion has to be converted into a more plastic strain than an elastic one. The amount of elastic strain may be minimized by using high temperatures. Thus, repeated scanning by the laser beams will provide a higher bending angle. This can be achieved by using an appropriate combination of laser power, beam diameter, and scanning velocity. It is to be noted that in TGM, the workpiece bends toward the laser beam.

14.4.2 Buckling Mechanism

Buckling mechanism (BM) is another mechanism used in LF to manufacture three-dimensional complex components. It is generally used to deform thin metal sheets. In BM, unlike in TGM, deformation is achieved by employing low feed rate and larger beam diameters (approximately 10 times that of the sheet thickness) to avoid a steep temperature gradient across the sheet thickness. Figure 14.4 shows the developing stages of a bend angle from a sheet metal under BM. Heating leads to thermal compressive stresses in the sheet, which results in a large amount of thermoelastic strain, which in turn causes local thermoelastic–plastic buckling of the material.

Figure 14.4a illustrates the compressive thermal stresses generated by a laser beam. It can be seen from the figure that there is no steep temperature gradient in the sheet metal. Figure 14.4b shows buckling due to thermal stresses. Plastic buckling occurs preferentially at the top of the metal sheet because the flow stress is low in this region owing to the temperature rise, and elastic deformation occurs at the region neighboring that of the plastic buckling because of the lower heating. Forces caused by the elastic deformation are

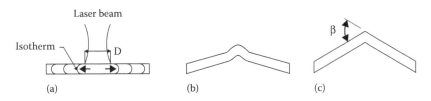

FIGURE 14.4
Laser forming stages by buckling mechanism: (a) generation of thermal compressive stresses by a laser beam; (b) development of buckling by thermal stresses; and (c) development of a bend angle.

counteracted by the constraints from the surrounding material. Buckling increases along the scanning line. Figure 14.4c shows the full development of a bend angle. The buckle is generated across the whole sheet when the laser beam leaves the sheet surface.

The direction of the bending angle is not defined by the process itself as it is for TGM. The part can bend in either the positive or the negative direction depending on a number of factors such as laser parameters, prebending orientation of the sheet, preexisting residual stresses, the direction in which any other elastic stresses are applied, internal stresses, and external or gravitational forces. The combined effect of these factors decides the direction of buckling. However, it is possible, albeit with some difficulty, to bend a sheet of metal in a defined way using BM, which allows the mechanism to be used as a flexible forming process. Therefore, it is suggested that the BM may be employed for bending thin sheets along a straight line toward or away from the laser beam (Holzer et al., 1994; Hu et al., 2002). It is also proposed that it may be used for a tube bend (Li and Yao, 2001; Hao and Li, 2003). BM produces bend angles generally in the range 1°–15°.

14.4.3 Upsetting Mechanism

Upsetting mechanism (UM) is a shortening (or thickening) mechanism, as shown in Figure 14.5. In UM, the process parameters are similar to those of BM but the dimension of the heated area is much smaller compared to the sheet thickness. The thermal gradient across the workpiece thickness is low because of the comparatively slow scanning by the laser beam over the work surface. During UM, the growth of the buckle will not take place as the geometry of the workpiece is stiffer; instead, a shortening of material in the heated area and a simultaneous thickening along the laser-irradiated path occur. In the UM process, the sheet metal almost constantly expands through the thickness direction during heating and contracts in the width direction of the sheet metal during cooling. Thus, the thickness of the plate increases. This heating process is repeated across the whole width of the material to change the thickness of the plate. This mechanism can be used to form a plane sheet into a specially formed part with a proper heating strategy. Aligning, adjustment, and rapid prototyping are also possible.

14.5 Machines for Laser Microbending

A conceptual laser microbending machine is shown in Figure 14.6. A specimen is clamped to a CNC table to facilitate the movement of the specimen beneath a laser beam, in the X-, Y-, and Z-directions. The specimen is formed to the required shape/angle using a repeated

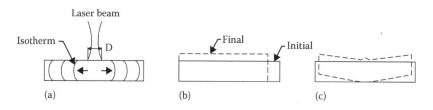

FIGURE 14.5
Laser forming by upsetting mechanism: (a) initial stage, (b) intermediate stage, and (c) final stage.

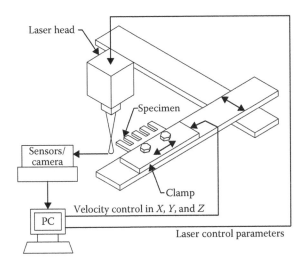

FIGURE 14.6
Schematic representation of a laser beam microbending machine.

traverse under the laser, along the required length, to obtain the desired deflection of the specimen. The tip deflection is monitored using contact/noncontact-type sensors, and the signal is fed into a host PC where deflections are compared with predetermined targets to generate an error signal. The appropriate signals for error correction are generated and used as input to control the laser parameters. This process is continued till the deflection formed by laser irradiation is equal to the desired value.

Thus, the machines for laser microbending process can be easily developed. They can be similar to other CNC machine tools except that the conventional cutting tool is replaced by a laser head. The laser microbending machine has to be properly shielded to avoid exposure to radiation. There has to be a provision for cooling, as much heat is generated in the process. CO_2 laser is convenient for industrial purposes, as it has relatively high efficiency (10%–15%), as against the single-digit efficiency of the Nd:YAG laser. A recent trend is to use a fiber-optic laser.

14.6 Mathematical Modeling of Laser Bending Processes

Some closed-form analytical expressions have been obtained for estimating the bend angle in microbending (Kraus, 1997; Shen and Vollertsen, 2009). A detailed analysis can be carried out using the finite element method (FEM). For an accurate modeling, the mechanical and thermal analyses have to be carried out together. However, sequential analysis (thermal analysis followed by mechanical analysis) can also provide fairly accurate results. The finite element analysis procedure is described briefly in this section.

14.6.1 Governing Equations for Thermal Analysis

Heat transfer needs to be considered in the LF process. As shown in Figure 14.2, the heat transfer modes in the problem are classified into conduction, convection, and radiation.

A part of the laser heat is transferred inside the plate by conduction and a part of it is transferred to the surrounding by convection and radiation effects.

The governing differential equation for transient heat conduction is

$$\rho c_p \frac{\partial T}{\partial \tau} = \frac{\partial}{\partial x}\left(k_x \frac{\partial T}{\partial x}\right) + \frac{\partial}{\partial y}\left(k_y \frac{\partial T}{\partial y}\right) + \frac{\partial}{\partial z}\left(k_z \frac{\partial T}{\partial z}\right) + \dot{q}$$

(14.2)

where ρ is the density of the conducting medium; c_p is the specific heat of the medium; k_x, k_y, k_z are the thermal conductivities of the medium in the x-, y-, and z-directions, respectively; τ is the time; and $\dot{q}$ is the rate of heat generated per unit volume. For solving the differential equation, boundary and initial conditions are required. The initial condition is that the temperature of the plate at each and every point at time $\tau = 0$. The general form of the boundary conditions can be written as

$$-\left(k_x \frac{\partial T}{\partial x}n_x + k_y \frac{\partial T}{\partial y}n_y + k_z \frac{\partial T}{\partial z}n_z\right) = h_{\text{conv}}\left(T - T_\infty\right) + \sigma_{\text{st}}\varepsilon_r\left(T^4 - T_\infty^4\right) - \alpha q_r$$

(14.3)

where n_x, n_y, n_z are the direction cosines of the normal to the surface, T_∞ is the ambient temperature, h_{conv} is the convective heat transfer coefficient, σ_{st} is Stephan's constant (equal to 5.67×10^{-8} W/m^{2}K^4), ε_r is the emissivity of the surface, α is the absorptivity, and q_r is the radiant laser heat flux function. Note that the radiant heat flux function keeps changing because of the traverse of the laser beam.

14.6.2 Governing Equations for Mechanical Analysis

The thermal expansion caused by heating a laser beam is enough to produce large plastic strains. Hence, elastoplastic constitution equations are to be used. The following are the equations that need to be solved by a numerical method such as the FEM:

1. Equilibrium equations:

   The force balance in the three directions provides

$$\frac{\partial \sigma_{xx}}{\partial x} + \frac{\partial \sigma_{xy}}{\partial y} + \frac{\partial \sigma_{xz}}{\partial z} = 0$$

(14.4a)

$$\frac{\partial \sigma_{yx}}{\partial x} + \frac{\partial \sigma_{yy}}{\partial y} + \frac{\partial \sigma_{yz}}{\partial z} = 0$$

(14.4b)

$$\frac{\partial \sigma_{zx}}{\partial x} + \frac{\partial \sigma_{zy}}{\partial y} + \frac{\partial \sigma_{zz}}{\partial z} = 0$$

(14.4c)

where σ_{ij}, i, and j that take on values x, y, and z, are components of the stress tensor. In index notation, the equilibrium equations are expressed as

$$\frac{\partial \sigma_{ij}}{\partial x_j} = 0$$

(14.5)

where Einstein's summation convention has been employed. According to this convention, as j is a repeated index, it takes on values x, y, and z to generate three terms. The three terms are added. Thus, Equation 14.5 is actually

$$\frac{\partial \sigma_{i1}}{\partial x_1} + \frac{\partial \sigma_{i2}}{\partial y_2} + \frac{\partial \sigma_{i3}}{\partial z_3} = 0$$

(14.6)

Replacing i by x, y, and z in a sequence, three scalar equations (Equation 14.4) are obtained.

2. Incremental elastic–plastic stress–strain relations after yielding:

$$d\sigma_{ij} = C_{ijkl}^{Ep} d\varepsilon_{kl}$$

(14.7)

where

$$C_{ijkl}^{Ep} = 2G\left[\frac{v}{1-2v}\delta_{ij}\delta_{kl} + \delta_{ik}\delta_{jl} - \frac{9G}{2}\frac{\sigma_{ij}'\sigma_{kl}'}{(H'+3G)\sigma_y^2} \right]$$

(14.8)

In Equation 14.8, δ_{ij} is the Kronecker's delta, which equals 1 if both the indices are equal and 0 otherwise; G is the shear modulus; v is the Poisson's ratio; and H' is the rate of change of flow stress σ_y with respect to the equivalent plastic strain. The equivalent plastic strain can be obtained by using the relation

$$d\varepsilon_{eq}^P = \sqrt{\frac{2}{3}d\varepsilon_{ij}^P d\varepsilon_{ij}^P}$$

(14.9)

The volume constancy provides

$$d\varepsilon_{ii}^P = d\varepsilon_{xx}^P + d\varepsilon_{yy}^P + d\varepsilon_{zz}^P = 0$$

(14.10)

3. Incremental elastic–plastic stress–strain relations before yielding and after unloading:

$$d\sigma_{ij} = C_{ijkl}^E d\varepsilon_{kl}$$

(14.11)

where

$$C_{ijkl}^E = \frac{2v}{1-2v}G\delta_{kl}\delta_{ij} + G\delta_{ik}\delta_{jl}$$

(14.12)

4. Incremental strain–displacement relations:

$$d\varepsilon_{ij} = \frac{1}{2}\left(\frac{\partial(du_i)}{\partial x_j} + \frac{\partial(du_j)}{\partial x_i} \right)$$

(14.13)

where du is the incremental displacement vector in time dt.

14.6.3 Finite Element Formulation and Implementation

There are two types of analyses that need to be carried out—thermal analysis and mechanical analysis. The domain is discretized into small finite elements. The temperature field variable within the element can be interpolated by

$$T = [N]\{T_e\} \tag{14.14}$$

where $[N]$ is the row vector of shape functions and T_e is the column vector of nodal temperatures. By the Galerkin FEM, Equation 14.2 with internal heat generation $\dot{q} = 0$ gets converted to

$$[c]\{\dot{T}\} + [k]\{T\} = \{q\} \tag{14.15}$$

where

$$[c] = \int_V \rho c_p [N]^T [N] dV^e \tag{14.16}$$

$$[k] = \int_{V^e} \left[k_x [N,x]^T [N,x] + k_y [N,y]^T [N,y] + k_z [N,z]^T [N,z] \right] dxdydz \tag{14.17}$$

where $[N, x]$ is the row vector of partial derivatives of the shape function and others and $\{q\}$ is the heat flux vector. The above elemental finite element equations are assembled to yield the following global system of equations:

$$[C]\{\dot{T}\} + [K]\{T\} = \{Q\} \tag{14.18}$$

Equation 14.18 is a system of ordinary differential equation that is solved by the finite difference technique.

After carrying out thermal analysis, structure analysis is carried out. The formulation can be carried out by the Galerkin method or by following the virtual work principle. According to the virtual work principle,

$$\int_V \sigma_{ij} \delta \varepsilon_{ij} dV = \int_V q_i \delta u_i dV + \int_V p_i \delta u_i dS \tag{14.19}$$

The above equation can written as

$$\int_V \{\delta \varepsilon\}^T \{\sigma\} dV = \int_V \{\delta u\}^T \{q\} dV + \int_V \{\delta u\}^T \{p\} dS \tag{14.20}$$

Here, $\{\delta \varepsilon\}$ is the virtual strain vector, $\{\delta u\}$ is the virtual displacement, where $\{\delta u\} = \lfloor \delta u\ \delta v\ \delta w \rfloor^T$, $\{q\}$ is the body force in volume V, and $\{p\}$ is the surface traction on surface S. The displacements $\{u\}$ can be interpolated over an element as

$$\{u\} = [N]\{d\} \tag{14.21}$$

where $\{\delta u\} = \lfloor u \; v \; w \rfloor^T$ and $\{d\}$ is the nodal displacement degree of freedom of an element. Strains are determined from displacements as

$$\{\varepsilon\} = [B]\{d\} \tag{14.22}$$

where $[B]$ contains the partial derivatives of shape functions. In mechanical analysis, one side of a metal plate is fully constrained, that is, the boundary conditions are zero displacement at one side of the metal plate.

The eight-node solid element exhibits shear locking, that is, because of spurious shear strain, it is excessively stiff in the beam bending mode. To account for the large deformation including large deflection, large rotation, and large strain in the 3D eight-node brick element, the following interpolation formulae are used in terms of the natural coordinates ξ_1, ξ_2, ξ_3:

$$u = \frac{1}{8}[u_1(1-\xi_1)(1-\xi_2)(1-\xi_3)+u_2(1+\xi_1)(1-\xi_2)(1-\xi_3)+u_3(1+\xi_1)(1+\xi_2)(1-\xi_3)$$

$$+u_4(1-\xi_1)(1+\xi_2)(1-\xi_3)+u_5(1-\xi_1)(1-\xi_2)(1+\xi_3)+u_6(1+\xi_1)(1-\xi_2)(1+\xi_3)$$

$$+u_7(1+\xi_1)(1+\xi_2)(1+\xi_3)+u_8(1-\xi_1)(1+\xi_2)(1+\xi_3)+a_1(1-\xi_1^2)+a_2(1-\xi_2^2)+a_3(1-\xi_3^2)]$$

$$v = \frac{1}{8}[v_1(1-\xi_1)(1-\xi_2)(1-\xi_3)+(\text{terms similar to } u) \tag{14.23}$$

$$w = \frac{1}{8}[w_1(1-\xi_1)(1-\xi_2)(1-\xi_3)+(\text{terms similar to } u)$$

Using the summation operator, the above equations can be written as,

$$u = \sum_{i=1}^{8} N_i u_i + a_1(1-\xi_1^2)+a_2(1-\xi_2^2)+a_3(1-\xi_3^2)$$

$$v = \sum_{i=1}^{8} N_i v_i + a_4(1-\xi_1^2)+a_5(1-\xi_2^2)+a_6(1-\xi_3^2) \tag{14.24}$$

$$w = \sum_{i=1}^{8} N_i w_i + a_7(1-\xi_1^2)+a_8(1-\xi_2^2)+a_9(1-\xi_3^2)$$

In the above equations u_i, v_i, and w_i are nodal degrees of freedom, that is, nodal displacements, and a_i are generalized degrees of freedom. They may also be called *nodeless* degrees of freedom. Displacement modes associated with the a_i are called incompatible or nonconforming because at locations other than nodes, they allow overlaps or gaps between adjacent elements.

Substituting the approximation of Equation 14.24 for an element and the constitutive relation in Equation 14.20, we obtain a group of nonlinear equations that can be written in matrix form as

$$[K]\{\Delta u\}=\{F\} \tag{14.25}$$

where

$$[k] = \int_{V^e} [B]^T [D^{ep}][B] dxdydz$$

(14.26)

is the global tangential stiffness matrix, $[D^{ep}]$ is the elastoplastic stress–strain matrix form of Equation 14.8 also called a *constitutive matrix*, $\{\Delta u\}$ is the incremental displacement vector at the element nodes, and $\{F\}$ is the global force vector, including nodal force and force caused by thermal strain. The nonlinear Equation 14.25 can be solved by Newton–Raphson method.

A flowchart of the sequential analysis procedure is shown in Figure 14.7. First, the thermal analysis is completed, and then the mechanical analysis is carried out. Kumar (2008) carried out the FEM modeling of laser bending using the ANSYS package. The workpiece is a square plate made of 1.0584 (D36) shipbuilding steel. The plate is divided into 40 elements along each side and three elements along the direction of thickness. The total number of elements is 4800, that is, 40 elements along the x-direction (in width) × 40 elements along the y-direction (in length) × 3 elements along the z-direction (in thickness). A uniformly mapped mesh is generated, as shown in Figure 14.8. X–Y–Z represents the global Cartesian coordinate system, and local cylindrical coordinate systems are defined

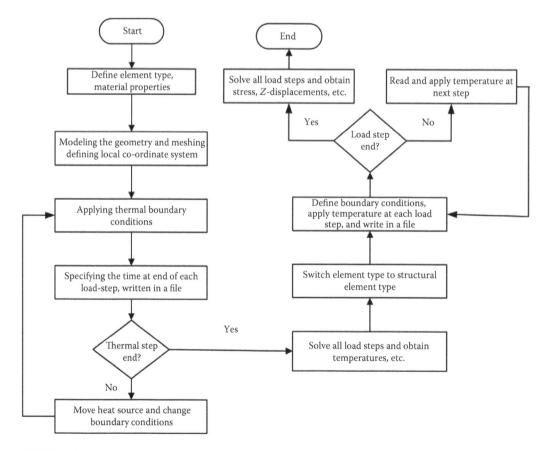

FIGURE 14.7
Flow chart of finite element analysis.

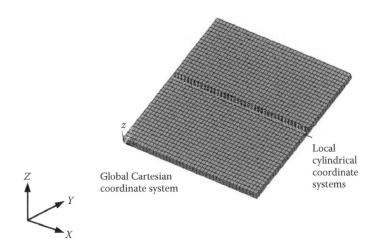

FIGURE 14.8
A typical finite element mesh.

at each node through the centerline along the x-direction. Heat flux is applied at these local cylindrical coordinate systems. The heat flux is shifted from one local coordinate system to the next local coordinate system at each load step to assume moving heat flux. The local coordinate systems are taken as cylindrical since here the heat flux is applied as a Gaussian distribution that is to be specified in the cylindrical coordinate system in ANSYS.

The element used for thermal analysis in ANSYS is SOLID70, which is an eight-node thermal mass, solid, brick element. SOLID70 has a 3D thermal conduction and convection capability. The element has eight nodes with a single degree of freedom, temperature at each node. The element is applicable to a 3D, steady-state or transient thermal analysis with or without material nonlinearities. The workpiece (D36 shipbuilding steel) is surrounded by air at room temperature T_0 which is 20°C. Tables 14.1 and 14.2 show the variation of the

TABLE 14.1

Thermal Properties and Density of D36 Shipbuilding Steel

Temperature (°C)	Thermal conductivity (W/m°C)	Specific heat (J/kg°C)	Density (kg/m³)
20	35.1	427	7860
200	36.8	502	7795
300	36.8	548	7759
400	36.2	602	7725
500	34.6	665	7690
600	32.0	741	7654
700	28.5	1481	7620
760	25.8	686	7609
900	26.6	653	7604
1000	27.2	653	7604
1100	27.8	653	7604
1200	28.4	653	7604
1300	29.0	653	7604
1400	29.6	653	7604

Source: Kumar, G.R.S., Estimation of beam traverse speed in the laser forming by using FEM with online learning. M.Tech. Thesis, IIT Guwahati, 2008.

TABLE 14.2

Mechanical Properties of D36 Shipbuilding Steel

Temperature (°C)	Young's modulus (GPa)	Poisson's ratio	Yield stress (MPa)	Tangent modulus (GPa)
20	199.67	0.3	371.9	10.78
100	199.6	0.309	369.2	10.77
200	199.3	0.32	362.1	10.71
300	198.04	0.331	347	10.58
400	192.9	0.34	317.5	10.1
500	174	0.353	267.9	8.87
600	125.3	0.364	202.8	6.52
700	61.7	0.375	139.8	3.72
800	26.3	0.386	94.89	1.8
900	14.9	0.397	69.2	0.94
1000	12.02	0.4	56.3	0.622
1100	11.32	0.419	50.35	0.514
1200	11.1	0.43	47.65	0.479
1300	11.1	0.441	46.4	0.467
1400	11.1	0.452	45.9	0.463

Source: Kumar, G.R.S., Estimation of beam traverse speed in the laser forming by using FEM with online learning. M.Tech. Thesis, IIT Guwahati, 2008.

properties of the material with respect to temperature. The heat flux generated by the laser beam is applied on the top surface. On the top surface, the convective boundary condition is also applied with a convective heat transfer coefficient $h = 10$ W/m^2 °C. The initial temperature of the specimen is taken as T_o.

Conduction of heat between the workpiece and the support is disregarded. The material may undergo phase transformation at elevated temperatures during LF. Here, the maximum temperature attained by the workpiece is kept well below the melting point of steels, which is around 1400°C. It is assumed that the surface of the workpiece will not melt, and the effects of phase change and latent heat are not considered. The analysis was performed up to the time point when the metal plate was cooled to room temperature. It is also assumed that the cooling of irradiated material occurs through free convection to air.

The element used for structural analysis in ANSYS is SOLID45, which is an eight-noded 3D solid brick element, with extra shape functions, having three degrees of freedom at each node, namely, translations in the nodal x-, y-, and z-directions. SOLID45 is used for the 3D modeling of solid structures. The element has plasticity, creep, swelling, stress stiffening, large deflection, and large strain capabilities. The material stress–strain behavior is approximated by means of a bilinear curve with a slope that depends on temperature. The bilinear curve is obtained by specifying the yield stress and tangential modulus varying with temperature. In ANSYS, the coefficient of thermal expansion was taken as constant, at 12×10^{-6}/°C, and the reference temperature as 20°C.

14.6.4 Results

Qualitative and quantitative validation with the results in the literature was carried out by Kumar (2008). A comparison of the FEM results of ANSYS with the experimental results (Kyrasanidi et al., 2000) is shown in Table 14.3. Here, a workpiece of dimensions 300 mm (length) × 150 mm (width) × 6 mm (thickness) is considered. The power of the

TABLE 14.3

Results of Experimental and FEM Simulations

S No.	Traverse speed (m/min)	Z-displacement of free end (m)	Bend angle by FEM	Bend angle by experiments
1	0.75	0.2323×10^{-3}	0.089°	0.1°
2	0.6	0.399×10^{-3}	0.1524°	0.13°
3	0.45	0.769×10^{-3}	0.294°	0.27°
4	0.3	0.1255×10^{-2}	0.479°	0.49°
5	0.25	0.162×10^{-2}	0.62°	0.70°
6	0.2	0.185×10^{-2}	0.71°	0.81°

Source: Kumar, G.R.S., Estimation of beam traverse speed in the laser forming by using FEM with online learning. M.Tech. Thesis, IIT Guwahati, 2008.

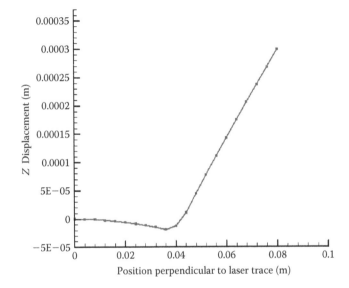

FIGURE 14.9
Z-displacement along the line perpendicular to the laser trace through the midpoint. (From Kumar, G.R.S., Estimation of beam traverse speed in the laser forming by using FEM with online learning. M.Tech. Thesis, IIT Guwahati, 2008.)

laser beam is considered to be 1500 W, and the beam diameter is 16 mm. Table 14.3 gives the values of experimental and FEM results, showing good matching.

Figure 14.9 shows a typical result from the thesis of Kumar (2008). A workpiece having dimensions of 80 mm × 80 mm × 2 mm is used for simulation. The parametric values are assumed as laser power 1000 W, absorption coefficient 0.8, laser beam diameter 4 mm, and scanning speed 50 mm/s. The Z-displacement of the midpoint along the free end of the plate was obtained as 300 μm, from which the final bend angle is calculated as 0.4325°. Note that although the workpiece is of a larger size, the maximum deflection is only 300 μm. Hence, it can be called a *microbending process*.

Figure 14.10 shows the temperature histories of the midpoints of the top and bottom surfaces for different velocities. It is observed that as the scanning velocity reduces, the peak temperature increases. The temperature rises at a very fast rate but reduces at a lower rate.

FEM can also compute plastic strains in a workpiece. Figure 14.11 shows the plastic strain histories in the *y*-direction of the midpoints on the top and bottom surfaces for

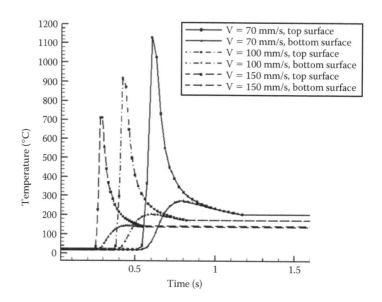

FIGURE 14.10
Temperature histories of the midpoints the on top and bottom surfaces with different velocities. (From Kumar, G.R.S., Estimation of beam traverse speed in the laser forming by using FEM with online learning. M.Tech. Thesis, IIT Guwahati, 2008.)

different lengths of the specimen. The plastic strains have been plotted for two plates of sizes 60 mm × 40 mm × 2 mm and 40 mm × 40 mm × 2 mm. It is observed that as the length of the workpiece increases, the maximum temperature attained by the workpiece decreases, hence the plastic strains in the *y*-direction on the top and bottom surfaces are reduced, resulting in a decrease in the final bend angle of the specimen.

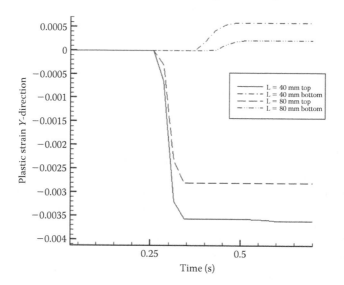

FIGURE 14.11
Plastic strain histories in the *y*-direction of the midpoints on the top and bottom surfaces for different lengths of the specimen. (From Kumar, G.R.S., Estimation of beam traverse speed in the laser forming by using FEM with online learning. M.Tech. Thesis, IIT Guwahati, 2008.)

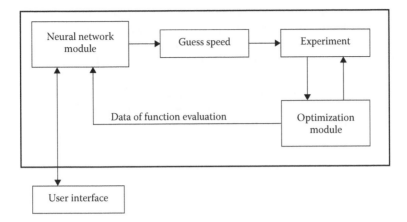

FIGURE 14.12
Methodology for the estimation of traverse speed using hit and trial experiments with online learning.

14.7 Soft-Computing-Based Methods

The bend angle can also be predicted using soft computing methods, such as neural network and fuzzy logic. However, these methods require much experimental data. In fact, finite element analysis involves long computational times. Kumar (2008) proposed a method of online learning. In this method, initially, the parameters for the desired bend angle are obtained by solving an optimization problem. Function evaluation in the optimization is carried out by solving the FEM code. During the optimization process, the function evaluation by FEM is preserved and set to a neural network module. Once sufficient data gets stored in this module, prediction can be carried out by a trained neural network. A trained neural network takes less time for prediction.

This method can be employed for online learning at the shop floor. A block diagram for this case is shown in Figure 14.12. Suppose we want to select a proper laser scanning speed for getting a desired bend angle. Here, the real experiments are carried out according to the optimization algorithm. The information gained by the experiments is stored in a neural network module that gains the capability to model the process after getting sufficient number of training and testing data. A sufficiently trained neural network will be able to provide a proper estimate of the scanning or traverse speed of the laser. The estimated value may provide the desired results, or a couple of experiments may be needed to arrive at the proper value of the traverse speed. Instead of a neural network module, a fuzzy-based module can be used. Soft-computing-based modeling is expected to gain importance in the near future.

14.8 Conclusion

In this chapter, a study of the emerging techniques of microbending with the help of laser is presented. The techniques are classified into laser-assisted microbending and laser microbending. In the former, laser plays an assistive role, with the actual bending carried out by a mechanical tool. In the latter, no mechanical tool is needed. The fabrication of a small-sized tool needs a precision manufacturing process, and the life of the tool is limited. Thus, laser microbending without the use of a mechanical tool will be more suitable than laser-assisted microbending. CNC machines can be developed for carrying out the microbending operation. There is a need and scope for rigorous mathematical and/or soft-computing-based modeling of the process. One can also explore the microstructural changes occurring in the workpiece because of microbending.

References

Chen, G., Xu, X., Poon, C.C., and Tam, A.C. 1998. Laser-assisted microscale deformation of stainless steels and ceramics. *Society of Photo-Optical Instrumentation Engineers* 37(10): 2837–2842.

Duflou, J.R., Callebaut, B., Verbert, J., and De Baerdemaeker, H. 2007. Laser assisted incremental forming: Formability and accuracy improvement. *CIRP Annals—Manufacturing Technology* 56(1): 273–276.

Geiger, M. and Vollertsen, F. 1993. The mechanisms of laser forming. *CIRP Annals* 42(1): 301–304.

Geiger, M., Vollertsen, F., and Deinzer, G. 1993. Flexible straightening of car body shells by laser forming. *International Congress and Exposition, Detroit.* Vol. 930, *Michigan,* MI, USA: Publ by SAE. p. 279.

Gillner, A., Holtkamp, J., Hartmann, C., Olowinsky, A., Gedicke, J., Klages, K., Bosse, L., and Bayer, A. 2005. Laser applications in microtechnology. *Journal of Materials Processing Technology* 167: 494–498.

Hao, N. and Li, L. 2003. An analytical model for laser tube bending. *Applied Surface Science* 208–209: 432–436.

Holzer, H., Arnet, M., and Geiger, M. 1994. Physical and numerical modeling of the buckling mechanism, *Proceedings of Laser Assisted Net Shape Engineering* 1: 379–386.

Hsieh, H.S. and Lin, J. 2005. Study of the buckling mechanism in laser tube forming with axial pre-loads. *International Journal of Machine Tools & Manufacture* 45: 1368–1374.

Hu, Z., Kovacevic, R., and Labudovic, M. 2002. Experimental and numerical modeling of buckling instability of laser sheet forming. *International Journal of Machine Tools and Manufacture* 42: 1427–1439.

Hu, Z., Labudovic, M., Wang, H., and Kovacevic, R. 2001. Computer simulation and experimental investigation of sheet metal bending using laser beam scanning. *International Journal of Machine Tools and Manufacture* 41: 589–607.

Kraus, J. 1997. Basic processes in laser bending of extrusions using the upsetting mechanism, laser assisted net shape engineering 2. *Proceedings of the LANE* 2: 431–438.

Kumar, G.R.S. 2008. Estimation of beam traverse speed in the laser forming by using FEM with online learning, M.Tech. Thesis, IIT Guwahati.

Kyrasanidi, A.K., Kermanidis, T.B., and Pantelakis, S.G. 2000. An analytical model for the prediction of distortions caused by the laser forming process. *Journal of Materials Processing Technology* 104: 94–102.

Labeas, G.N. 2007. Development of a local three-dimensional numerical simulation model for the laser forming process of aluminium components. *Journal of Materials Processing Technology* 207: 248–257.

Li, W. and Yao, Y.L. 2001. Laser bending of tubes: Mechanism, analysis, and prediction. *Journal of Manufacturing Science and Engineering* 123(4): 674–681.

Magee, J. and De Vin, L.J. 2002. Process planning for laser-assisted forming. *Journal of Materials Processing Technology* 120: 322–326.

Olowinsky, A.M. and Bosse, L. 2003. Laser beam micro forming as a new adjustment technology using dedicated actuator structures. *Proceedings of SPIE* 5116: 285–294.

Otsu, M., Wada, T., and Osakada, K. 2001. Micro-bending of thin spring by laser forming and spark forming. *CIRP Annals—Manufacturing Technology* 50(1): 141–144.

Samm, K., Terzi, M., Ostendorf, A., and Wulfsberg, J. 2009. Laser-assisted micro-forming process with miniaturised structures in sapphire dies. *Applied Surface Science* 255: 9830–9834.

Shen, H., Shi, Y., Yao, Z., and Hu, J. 2006. An analytical model for estimating deformation in laser forming. *Computational Materials Science* 37: 593–598.

Shen, H. and Vollertsen, F. 2009. Modelling of laser forming—An review. *Computational Material Science* 46: 834–840.

Vollertsen, F. 1998. *Forming and Rapid Prototyping. Handbook of the EuroLaser Academy*, Schuocker, D, Vienna: Chapman & Hall.

Section VI

Miscellaneous

15

Dimensional Metrology for Micro/Mesoscale Manufacturing

Shawn P. Moylan

National Institute of Standards and Technology

CONTENTS

15.1 Introduction

Lord Kelvin said, "To measure is to know," and "If you can not measure it, you can not improve it." In fact, it is not possible to really know what is manufactured until it is measured. This has been a particular challenge in micro/mesoscale manufacturing because development of production processes has considerably outpaced development of measurement processes. Decisions that need to be made regarding a part or product being manufactured are based on measurements. These cannot be informed decisions if one does not have confidence in, or does not fully understand, the uncertainty of the measurement process.

This chapter provides an overview of the most common methods used for geometric dimensional metrology in three dimensions (3D) at the micro/mesoscale. The references provided in this chapter provide examples of the work in the area, and not an exhaustive list. A similar study focusing on academic pursuits in the area can be found elsewhere, with no fewer than 175 references [1]. The processes discussed here are by no means an exhaustive list of all measurement processes used for micro/mesoscale metrology. However, they are the processes used most frequently; they all have multiple commercially available systems that measure the 3D topography of the surface of the part under test. The strengths, limitations, and challenges of each method are presented. The methods are grouped into three categories, touch probe, optical measurement, and scanning probe microscopy (SPM). To properly frame this discussion, we start by defining the scale being investigated, then introduce the concept of measurement uncertainty, and finally discuss some general considerations that need to be made for all measurement methods.

15.1.1 Defining the Scale

For the purposes of the current discussion, the micro/mesoscale is defined as ranging from 1 μm to 50 mm. Most parts manufactured at the micro/mesoscale have overall sizes ranging above 100 μm. However, designed features on these parts can be as small as tens of micrometers and the tolerances of such features can be close to 1 μm [1,2]. Figure 15.1 shows a variety of measurement processes that fit within this size range.

15.1.2 Uncertainty

Just as no part can be made perfectly, no measurement can be made perfectly. Uncertainty is both the concept and the quantity that expresses the amount of doubt in a particular measurement. The formal definition of uncertainty (of a measurement) in the *International Vocabulary of Basic and General Terms in Metrology* (*VIM*) is a "non-negative parameter characterizing the dispersion of the quantity values being attributed to the measurand, based

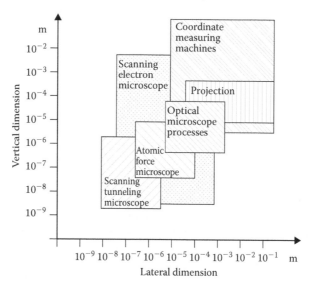

FIGURE 15.1

Size scale of typical measurement techniques employed at the micro/mesoscale. Acronyms are described in the text. (From Hansen, H., et al., *CIRP Annals*, 55(2), 721–743, 2006 and Weckenmann, A. and Ernst, R., *Technisches Messen*, 7–8, 334–342, 2000.)

on information used" [3]. A general procedure for calculating measurement uncertainty is described in the *Guide to the Expression of Uncertainty in Measurement* (GUM) [4].

In general, the uncertainty of a measurement is due to various sources. For example, a measurement of distance is influenced by the device used to measure it, the environment in which it is measured, the setup of the measurement components, and the number and positions of samples collected. Uncertainty in any of these inputs contributes to uncertainty of the overall measurement. How measurement uncertainty affects acceptance or rejection of a part based on geometric dimensioning and tolerancing (GD&T) is governed by standards [5].

Measurement uncertainty encompasses many traditional concepts in measurement including resolution, accuracy, and repeatability. Care should be taken when investigating device specification sheets that terminology is used properly. For example, most specification sheets provide a number associated with the "accuracy" of the device. However, the VIM defines accuracy as "closeness of the agreement between the result of a measurement and a true value of the measurand," and states that "accuracy" is a qualitative concept that cannot be quantified. Therefore, the term *accuracy* on data sheets often refers to the uncertainty associated with the device, but can also be more in line with resolution or repeatability. This chapter focuses on the factors that contribute to the uncertainty in the measurement device.

15.1.3 Environment

The environment in which a measurement is made can greatly influence the measurement, especially at the micro/mesoscale. Specific influences on specific measurement methods are discussed later in more detail. However, to minimize measurement uncertainty, some general considerations that apply to all measurements should be made. Vibration isolation of the part being measured, the device used for measurement, and the room in which the measurement is taking place should all be considered. Temperature can have a large effect on measurement results. All components of a measurement, including the measurement device and the part being measured, can expand or contract (depending on their coefficients of thermal expansion) as a result of a change in their temperature. The temperature in the environment directly affects the temperatures of the device and the part. For example, a part measured at one temperature will have a different size when transferred to an environment with a different temperature because of thermal expansion/contraction. Most standards and best practices recommend calibrating devices and performing measurements at 20°C. In addition to the actual temperature, the stability of the temperature and the uncertainty of temperature measurement are also of concern.

15.1.4 Unique Challenges

Many people expect that as a part scales down in size, its tolerances will also scale down. However, this is generally not the case. It is common for parts on the traditional scale to have dimensions on the order of 100 mm with tolerances in the range of 25 μm. However, a part 100 times smaller (1 mm dimension) will rarely have a tolerance 100 times smaller (0.25 μm) because of the difficulty not only in producing a part with that tolerance but also in verifying that the desired tolerance is achieved.

Consider the rule of ten. A general rule of thumb is that to measure a feature with a specified tolerance, the uncertainty associated with the measurement device should be at least 10 times smaller than the tolerance. It is common for micro/mesoscale parts as large as tens of millimeters to have tolerances close to 1 μm. Thus, in order to measure this part following the rule of ten, the measurement device must have an uncertainty less than 100 nm over a range of tens of millimeters, a ratio of 10^5. Although devices with such a scale do exist, they are often very specialized and expensive. Therefore, the rule of ten, considered by many metrologists to be overkill, is often relaxed at the micro/mesoscale. A better guideline endorsed by standards [5] is a ratio of 4:1.

15.1.5 Scanning Electron Microscopy (SEM)

Scanning electron microscopy (SEM) of micro/mesoscale parts is often found in literature. Electron microscopy is similar to light or optical microscopy except that to achieve better resolution and depth of focus, an electron source is used instead of light for illumination. In SEM, a focused beam of high-energy electrons is rastered over a sample surface in two dimensions. The interaction of the electron beam with the sample surface results in secondary electrons being emitted from the surface. These secondary electrons are collected by a detector and converted into an image of the sample surface [6].

SEM is an excellent tool for imaging and qualitatively assessing small parts because of the excellent resolution and high depth of focus achievable. However, this method is generally not well suited to 3D measurement. Because of the high depth of focus and the fact that the electron beam is only scanned in one plane, it is difficult to interpret the height of features imaged under SEM. In addition, low-uncertainty measurement would require the microscope to be calibrated at every magnification and every setting. Microscopy settings (such as magnification, spot size, and astigmation) are continually changed to improve image quality, and the magnetic microscope "lenses" tend to drift. Thus, an SEM continually needs recalibration to perform low-uncertainty dimensional measurements. SEM may be well suited to certain 1D measurements (e.g., line width) and 2D measurements, but care should be taken that in addition to proper calibration, the feature being measured should be truly normal to the detector. 3D measurement by SEM would require multiple images or beams (stereo pairs), and such systems continue to be developed.

15.2 Touch Probe Measurement

Touch probe measurement is a logical place to start the examination of micro/mesoscale dimensional metrology methods because the method is extremely powerful and very well understood at the traditional scale. Coordinate measuring machines (CMMs) provide low uncertainty postprocess 3D measurements of traditional-scale parts. In addition, many new traditional-scale machine tools come equipped with spindle-mounted touch probes both for workpiece setting and for part dimensional measurement. A considerable amount of effort has gone into characterizing the coordinate measuring process, as evidenced by an entire book written about the method [7] and by the 208 references in a review paper on the subject [8]. The micro/mesoscale process is not dramatically different from the traditional-scale process. The range of the measurement is limited only by the places the probe can reach, so the output of point coordinates is a true 3D measurement. There is considerable effort going into enabling CMMs to operate on the micro/mesoscale [9–17].

However, significant obstacles still remain in micro/mesoscale measurement by a touch probe. The most significant of these is the size of the probe itself. A touch probe cannot

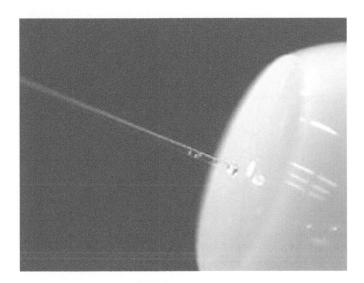

FIGURE 15.2
Fiber probe entering a 125-μm hole.

measure an internal feature (such as bore, slot, and hole) if the probe cannot fit inside the feature. While this concept is obvious to many, it is still a significant hurdle because probes smaller than 25 μm in diameter are rare, but micro/mesoscale features of that size are becoming common (Figure 15.2). Adding to the problem is the fact that even as smaller probes become more available, uncertainty in the calibration of the probe tip profile becomes more important and more troublesome [12]. Also to be considered is an averaging effect due to the relative size of the probe tip to the size of the feature being measured.

At the much smaller size scales of micro/mesoscale manufacturing, electrostatic, Van der Waals, and meniscus forces play a much larger role, affecting touch probe measurement. These forces cause the probe to stick to the surface after initial contact. More importantly, small, less stiff probes can be pulled toward the surface or in undesirable directions, resulting in increased measurement uncertainty (Figure 15.3) and possible damage to the small, often fragile probes. A good deal of research has led to some unique solutions in this area, and modern CMMs designed specifically for the micro/mesoscale often use vibrating probes to break the contact at the surface [9,15–17]. As a vibrating probe approaches a surface, a change in vibration amplitude or frequency can signal contact. However, vibrating the probe greatly increases the effective tip diameter because the surface detection point is governed not only by the physical size of the probe but also by the amplitude of the vibration (Figure 15.4).

The size scales of mesoscale machine tools and micro/mesoscale parts lead to additional complications. Contact detection is more important at the micro/mesoscale because contact areas are so small that local stresses can be rather large. In fact, contact by a 100-μm sphere can leave a permanent indentation in an aluminum surface at forces as low as 500 μN. Thus, keeping contact forces low and quickly detecting the probe contact is highly important, but at the same time, more difficult. Several studies focus entirely on this problem [10–12].

The uncertainty of CMMs is well understood and is a function of several components [18]. Contributing to the combined uncertainty of a measurement are the uncertainty in the machine positioning, uncertainty in the probing system, the sampling of coordinates

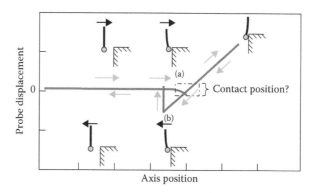

FIGURE 15.3
As a microscale touch probe approaches a surface, forces can draw the probe tip toward the surface (a) and can cause the probe tip to adhere to the surface as the probe retracts (b). These phenomena can result in an increase in uncertainty of measuring the surface's position.

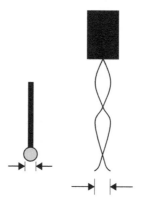

FIGURE 15.4
Apparent probe diameter differs greatly with traditional probes and vibrating probes. With traditional probes, the apparent diameter is slightly smaller than sphere diameter owing to time delay in triggering. With vibrating probes, the apparent diameter is more a function of the vibration amplitude.

(the number and distribution of coordinates measured on the part), and uncertainty in the fitting of geometric shapes to the sampled coordinates. National and international standards govern the determination of machine and probing system uncertainty components [19,20]. Some recent work has focused on qualification and verification tests and standards for micro/mesoscale-specific systems [21]. Modern CMMs provide excellent positioning accuracy, and their positioning errors are also well mapped, allowing for compensation. Uncertainty in the probing system is a function of the probe tip profile and the sensitivity of the triggering mechanism. Under proper conditions, uncertainty in dimensional measurements using CMMs can be as small as tens of nanometers [15].

CMM measurement requires a high degree of environmental control and is rather slow. Because contact is required for measurement, a thin layer of water on the surface can affect measurement. Therefore, a high level of humidity control is required along with temperature control and vibration isolation. Individual part coordinates are measured in series, and measuring hundreds of coordinates to reduce sampling and fitting uncertainties can take hours. The number of points still pales in comparison to the number of points used in optical processes that can collect millions of points in a matter of seconds.

15.3 Optical Measurements

Review papers [22,23] discussing optical methods for dimensional metrology mention 14 different methods. The methods used most commonly for 3D metrology of micro/meso-scale parts and features are white-light interferometry, confocal microscopy, fringe projection, and optical microscopy. As would be expected, each has its own relative strengths and limitations. However, certain aspects of optical measurements apply to all methods. Many of these concepts are covered in great detail in the *Handbook of Optical Metrology* [24].

One overall advantage is that optical processes are generally noncontact. This is an advantage at the micro/mesoscale because, as seen in the discussion on touch probes, the small sizes involved often make the parts more fragile or susceptible to damage. However, these being noncontact means, the part is imaged remotely, usually with the optical axis normal to the surface under measurement. This means that optical processes are usually limited to 2.5D.

2.5D means that optical processes can provide information in the third, vertical dimension but with some limitations. The particulars of how the different processes provide vertical measurement are discussed further in the individual subsections that follow. However, the limitations of all the processes are similar. The main limitation deals with the line of sight. If there is no direct line of sight from the optical sensor to the feature to be measured, no measurement can take place. This limitation primarily affects vertical sidewalls and overhangs (e.g., T-slots). Because the vertical sidewalls are usually parallel to the optical axis, there is no direct line of sight from a point on the sidewall to the optical sensor. As such, optical processes may be able to measure the depth of a hole or slot but usually cannot provide information on the taper of a slot or the cylindricity of a hole. A true 3D measurement would provide all this information.

Another issue common to all optical processes is lighting. If the feature or point to be measured cannot be properly illuminated, and the light cannot return to the optical sensor, no measurement can take place. This makes the measurement of high-aspect-ratio holes and slots difficult. Features measured by optical processes usually need to be reflective and should have minimal slope to reflect adequate light back to the sensor. It is often difficult to optically measure the surfaces of transparent and translucent parts. A more troublesome aspect of lighting is that the quality of the sample illumination often directly affects the measurement. If a certain point is illuminated too much, a portion of the image can appear washed out due to the optical sensor becoming saturated. Alternatively, if a point or feature to be measured is illuminated too dimly, the features in the measured image may be less sharp. Both circumstances affect the apparent resolution of the measurement. Unfortunately, providing quality lighting to a measurement scene is often more of an art than a science.

All optical processes discussed in this chapter use intensity measurement in the visible light spectrum as a primary component of the measurement. Intensity deals with the amount of light reaching an optical sensor. For example, edge detection, phase decomposition, and thresholding, all involve the interpretation of intensity in some manner. Thus, uncertainty in intensity measurement will directly affect uncertainty in the optical measurement.

Uncertainty in intensity measurement is rather complicated and comes from several, often correlated, components. The two primary sources of uncertainty in intensity measurement are quantization and noise. Most optical measurements now involve digital electronics that utilize a focal plane array (FPA). The most common FPA device utilized in optical metrology cameras is a charge-coupled device (CCD), and most commercially available optical measurement systems utilize a midlevel camera (as opposed to a high-level camera with superior specifications and characteristics) to keep system prices down. An FPA is actually an array of individual optical sensors termed *pixels*. Each pixel captures light emitted or reflected from the target being observed. Because these devices are digital, the intensity of the light reaching each pixel is broken into digital levels. An eight-bit system divides intensity into 256 digital levels from black (0) to saturation (255), although some cameras use more levels. This is known as *quantization*. However, the division is often nonlinear (Figure 15.5). If linearity is assumed, significant uncertainty will exist at low and high intensities. If nonlinearity is compensated for, uncertainty may be smaller, but it will still exist in the uncertainty of the compensation. The approach with least uncertainty usually involves only working in the linear portion of the intensity plot. However, this limits the range the process can work in and is often impractical.

The two most common noise components are known as *thermal noise* and *shot noise*. Thermal noise in an FPA device is a result of heat from the imaging components causing a change in the free electrons that are mistakenly recorded by the imaging device. Shot noise is a characteristic of all devices that convert a measurement to charge and is defined as uncertainty due to random fluctuations of the charge carriers. Camera data sheets will often attempt to quantify these errors, but unfortunately, it is nearly impossible to measure the two components independently. Many high-end cameras combat thermal noise using a variety of methods to cool the sensor array.

Other sources of uncertainty in optical measurement may come from the construction of the FPA itself. Typically, the size of the FPA chip and the number of pixels in the chip are known, and the pixel size is obtained by dividing the size of the chip by the number of pixels. However, in actuality, there are tiny wires between the pixels that carry the charge into the chip's electronics. Thus, only a certain percentage of the chip (the fill factor) is able to gather light (or measure intensity), and even if that percentage is known, there remains some uncertainty in the size and distribution of the pixels.

Most optical processes are surface profilers that have been adapted to provide dimensional measurement. Because of this, measurements of the vertical dimension typically take advantage of various optical properties and provide measurements with lower uncertainty than measurements of lateral dimensions that almost always involve counting of sensor pixels. While lateral resolution is generally theoretically limited by the wavelength of light used, in most cases, this limit is much smaller than the size of the optical sensor pixel projected onto the feature. As such, uncertainty in lateral dimensional measurements is often a function of the size of the pixel in the FPA, which is not an optical property.

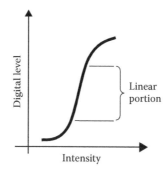

FIGURE 15.5
Intensity response is nonlinear at low and high digital levels, which can lead to large uncertainty.

15.3.1 Scanning White-Light Interferometry (SWLI)

Scanning white-light interferometers are readily available, and the process has been widely applied in characterizing micro/mesoscale parts and features [25–29]. Phase-shifting interferometry, common in the field of optics, provides exceptionally low vertical measurement uncertainty (<1 nm possible), but is limited to only surfaces with small slopes. Scanning white-light interferometry (SWLI) can handle very large slopes, even steps, but does so at the expense of higher uncertainty. SWLI extends the vertical range by vertically translating the interferometric objective. Typically, a piezoelectric actuator translates a Mirau interferometric objective (Michelson and Linnik objectives are used less commonly) along the optical axis, and interferograms are collected as the objective is translated (Figure 15.6). Fast Fourier transformation and phase evaluation, searching pixel by pixel for the maximum intensity contrast in the interferogram and comparing this with the z-position of that interferogram, allow determination of the surface profile (Figure 15.7) [29]. White light is used because it has a shorter coherence than monochromatic laser light, avoiding fringe ambiguity and allowing better measurement.

Benefits of SWLI are that vertical resolution is independent of optical magnification and that uncertainty in the vertical dimension can be 20 nm or better. However, lateral resolution is associated with the objective's numerical aperture (NA). A smaller NA typically permits a larger field of view and also results in a larger (poorer) resolution. Furthermore, at low magnification, the FPA in the camera collecting the interferograms typically limits the resolution of lateral measurement. Even at these low magnifications, the lateral field of view is typically only a couple of millimeters, much smaller than the larger micro/mesoscale parts. Thus, to completely measure parts larger than the field of view (or if a high-resolution image of smaller features is desired), several neighboring images must be taken individually and stitched together. The machine must move the part between consecutive

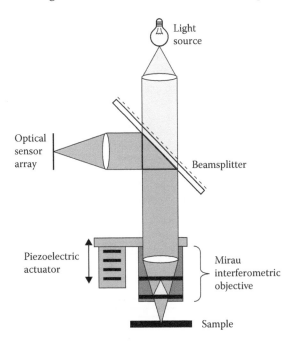

FIGURE 15.6
Schematic of a typical scanning white-light interferometry setup.

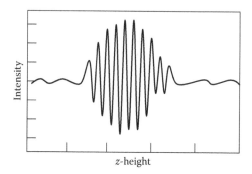

z-height

FIGURE 15.7

As the interferometric objective is scanned along the vertical axis, the contrast between fringes is measured and compared to the z-height, allowing the determination of vertical dimensions.

images, introducing the potential for machine tool errors to affect the measurements. This stitching adds another layer of uncertainty to the measurements. Lateral SWLI reduces the severity of this limitation [30].

15.3.2 Confocal Microscopy

In confocal microscopy, the field of view is limited to a very small area. This limitation results in improved lateral resolution over that small area when compared to conventional optical microscopy [31]. The process has found considerable applications in the biological field (live-cell imaging) and has been applied to measurement of micro/mesoscale parts and features [32–34]. The process operates by projecting light through a pinhole onto the workpiece and also collecting light reflected from the workpiece through the pinhole (Figure 15.8). This has the effect of returning light to the detector only when the measurement surface is exactly in focus. As in SWLI, the objective lens is translated along the optical axis. Because light only reaches the detector when the surface is in focus, the height of the surface is found by comparing the peak intensity with the z-height of the objective (Figure 15.9). This is done pixel by pixel to create a topographical map of the surface.

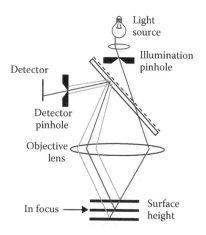

FIGURE 15.8

Schematic of a typical confocal microscope. Note that if the specimen height is not in focus, the light reflected from the surface does not reach the detector.

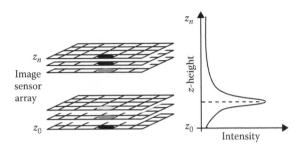

FIGURE 15.9
Schematic demonstrating the determination of the vertical dimension in confocal microscopy. This process is performed for every pixel in the sensor array.

Because the process only views one pinhole worth of a surface at a time, one might think the process would be rather slow. This is not the case, however, because creative solutions allow multiple pinholes to quickly scan over a surface. Older models of confocal microscopes would raster the beam over the surface, similar to the SEM. This process would require several seconds to fill the entire imaging array at only one z-height. However, modern confocal microscopes use a Nipkow disk. A Nipkow disk has a large number of pinholes arranged in a spiral along the surface of a disk (Figure 15.10). Thousands of pinholes ~50 μm in diameter can be illuminated simultaneously, and the disk is spun to distribute the pinholes over the entire field of view. A spinning disk confocal microscope can acquire an entire image in one plane in less than a millisecond. As such, the limiting factor for the process lies more in the vertical scanning and data processing. Acquiring a full 3D image typically takes on the order of several seconds.

Similar to most other optical measurements, vertical and lateral resolutions are different, but lateral resolution of confocal microscopy is better than that of any other optical process. In general, lateral resolution is a function of the pinhole geometry and NA. Pinholes must be arranged such that stray light from an out-of-focus surface does not return to the detector through neighboring pinholes. At 40x magnification with an NA of 1.2, a lateral resolution of 250 nm is attainable [35]. Unlike in SWLI, uncertainty in the vertical dimension also changes with NA. Uncertainty in the vertical dimension is typically around 50 nm when vertical translation is accomplished by a piezoelectric actuator, but if longer travel is desired to increase the vertical measuring range, a stepper motor can be used for the translation, resulting in poorer uncertainty. The measurement area, when using a Nipkow disk, is similar to that with SWLI, again requiring stitching for larger parts.

15.3.3 Fringe Projection

Fringe projection or structured light processes, although used for inspection of micro/ mesoscale features [28,36], are much less common in the field of micro/mesoscale metrology. Fringe projection is based on photogrammetry and provides 3D measurement through a solution to the stereo correspondence problem [37–39]. Typical photogrammatic solutions to stereo correspondence involve imaging a part with two cameras from two different perspectives and need some type of reference in both images. Fringe projection, however, replaces one camera with a projector that projects a known pattern. This pattern becomes distorted by the shape and contours of the surface on which it is projected (Figure 15.11). The pattern provides the reference and allows corresponding points to be more easily determined automatically.

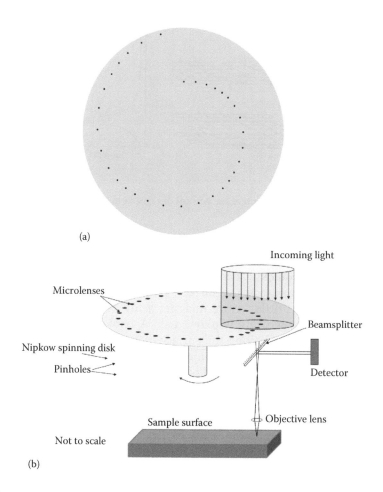

FIGURE 15.10
The Nipkow spinning disk greatly improves the speed of confocal microscopy by allowing multiple pinholes to be scanned over the entire field of view. (a) Spiral design of Nipkow disk; (b) Nipkow disk incorporation into a confocal microscope.

Far fewer commercial systems are available for micro/mesoscale fringe projection than for SWLI or confocal microscopy (though the process is becoming more popular at the traditional scale for noncontact 3D measurement), but literature shows that fringe projection can be equally effective. Windecker et al. [36] reviewed all three processes and measured surfaces of varying roughness, comparing results with output from a stylus measuring device. While SWLI and confocal microscopy proved superior at very low (<500 nm) roughness, fringe projection performed equally well at roughness scales above 1 μm.

The intrigue with fringe projection stems from the fact that the strengths and limitations of the process are often opposite to those of other optical processes. Fringe projection measurement is very flexible, allowing the setup to be adjusted to accommodate a variety of part shapes. Also, the measurement can be configured to project patterns of various sizes, accommodating a range of part sizes. Digital projection and imaging can easily scale in size to envelop the entire part or feature, that is, the entire part can be measured without stitching and without any movement of the measuring device.

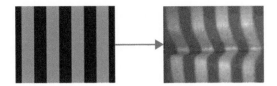

FIGURE 15.11
Demonstration of a typical known fringe pattern and how that pattern is distorted when projected onto a contoured part.

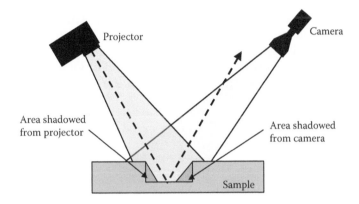

FIGURE 15.12
Shadowing and reflective surfaces are more problematic for fringe projection than other optical measurement processes.

However, shadowing is a larger problem with fringe projection, and highly reflective surfaces cannot be measured (Figure 15.12). Because there are multiple lines of sight in fringe projection, if any of these are obscured from a certain area of the part, that area cannot be measured. Depending on the setup, this may limit the aspect ratio of internal features that can be measured. Also, if the surface reflects all the light from the projector in a direction away from the camera, measurement cannot take place. To avoid these problems, the surface being measured may be coated with a material that scatters light better, and multiple projector angles and camera views may be used.

Determining uncertainty for fringe projection is a complicated matter [40] but the uncertainty is generally larger than in SWLI or confocal microscopy. In addition to the sensor uncertainties discussed earlier, the resolution of fringe projection is limited by the size of projector pixels. With the current state of the art, projector pixels are much larger than camera pixels, but the technology is constantly evolving.

15.3.4 Optical Microscopy

Optical microscopy is a very powerful tool for quick, low-uncertainty measurements in two dimensions. Features and patterns can be identified within the field of view, and a well-calibrated vision system can provide low-uncertainty measurements and quick comparison to templates. However, for 3D measurement, optical microscopy is used in a different manner.

To provide 3D measurement, the optical microscope performs more like a CMM except that it has an optical probe instead of a touch probe. As in SEM, it is difficult to obtain information about a part in the vertical dimension. However, in contrast to SEM, an optical microscope has a relatively small depth of focus, especially at higher magnification. Because of this, the feature being inspected is in focus at only one particular height. That position of focus can translate to a measurement in the third dimension with the use of a precise scale on the positioning device. With this type of measurement method, the position of focus is determined automatically (autofocus). The method of determining this position is usually proprietary to the particular system, but a through-focus technique similar to the one described for confocal microscopy is common, although with larger measurement uncertainty. This vertical measurement is combined with the position of the x- and y-axes scales to locate the coordinate of the in-focus feature at the center of the microscope's field of view. If the image system is calibrated, coordinates of features elsewhere in the field of view that are also in focus can be determined by combining the distance from the center of the image with the axes scales' positions.

15.4 Scanning Probe Microscopy

The strength of SPM lies in its superior measurement uncertainty. Under proper conditions, SPM can be used to resolve individual atoms. However, SPM processes are limited in their range to only hundreds of micrometers laterally and tens of micrometers vertically. As such, SPM may be better suited for nanomanufacturing metrology. However, there is important overlap with micro/mesoscale manufacturing that warrants a brief discussion here. A more thorough introduction to SPM and the variety of SPM processes can be found elsewhere [41].

All SPM processes involve bringing a probe tip into near-field proximity with the part's surface, scanning the probe over the surface, and monitoring a specific interaction between the probe and the surface. Operating in the near-field avoids common limitations on resolution (e.g., Reyleigh's criterion) seen in the far-field processes discussed in the previous section. Scanning is accomplished with piezoelectric actuators. The height of the probe is controlled to keep the interaction between the probe and the surface constant. Height adjustments as a function of scanning position provide a topographic map of the measured surface. There are a variety of interactions that can be monitored, leading to as many as 16 different types of SPM.

Scanning tunneling microscopy (STM), scanning force microscopy (SFM), and scanning near-field optical microscopy (SNOM) are the most commonly encountered SPM processes. SFM requires the least amount of sample preparation, can measure electrically nonconductive as well as conductive samples, and has the largest measurement range. Therefore, it is the most pertinent technique to micro/mesoscale metrology. In SFM, the forces between the probe and workpiece are monitored. It is assumed that in the near-field, this interaction is primarily between the atoms at the probe tip and the atoms at the workpiece surface, leading to the common name *atomic force microscopy* (AFM).

AFM probes are typically SiN tips, etched to a near-atomic dimension, mounted at the end of a cantilever (Figure 15.13). When the probe tip is brought into close proximity with the surface, the cantilever deflects by an amount proportional to the atomic force between

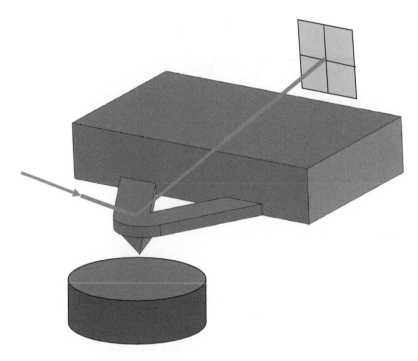

FIGURE 15.13
Beam deflection is a common method of monitoring cantilever deflection in AFM.

the probe and the surface. The amount of deflection is commonly monitored by beam deflection (Figure 15.13). A laser beam is deflected from a source, off the top of the cantilever, and to a position-sensitive photodiode. As the cantilever deflection changes, the position of the beam on the photodiode also changes. Less commonly, cantilever deflection can be monitored through embedded piezoresistive or piezoelectric material, interferometry, capacitance, or tunneling current.

There are also several different types of AFM, classified into dynamic and static methods. The most common static method is referred to as *contact AFM*. With this variant, the probe tip is brought to within the region where the forces between the tip and the surface are repulsive. As the probe is moved, the probe height is controlled to result in the same amount of cantilever deflection. Noncontact AFM is a dynamic process. The probe tip is oscillated near its eigenfrequency, and the probe height is controlled to keep this frequency constant. Noncontact AFM is the only SFM process to achieve atomic resolution. Between these two variants is another dynamic process termed *tapping AFM*. The probe tip is vibrated by approximately 20–100 nm such that the probe tip moves in and out of the repulsive force regime. The probe height is controlled to provide a constant amplitude of vibration. Tapping AFM has the advantage of being limited in resolution only by tip geometry, similar to contact AFM, but with significantly lower lateral forces than encountered in contact AFM.

The resolution of an AFM, when operated under proper environmental conditions, is primarily limited by tip geometry. The tip can be thought of as a cone or a pyramid. The height and angle of the pyramid as well as the sharpness of the tip are the important limiting factors (Figure 15.14). Because of this, uncertainty in the calibration of AFM probe tips is important to low-uncertainty measurement.

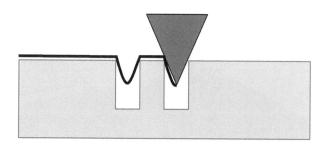

FIGURE 15.14
Tip geometry is a main factor in contact AFM and tapping AFM resolution.

15.5 Hybrid Processes

It is easy to imagine that in the absence of the development of a new measurement process, all future commercial micro/mesoscale metrology devices will be hybrid systems. As seen in the previous sections, every measurement process has trade-offs. Micro-CMMs trade speed for range and low uncertainty. Other processes trade measurement area (lateral range) for lateral uncertainty or resolution. No one process can meet the high demands of micro/mesoscale part and feature manufacturing. Hybrid processes draw from the strengths of multiple processes to overcome the limitations of the involved processes and meet the measurement demands of the part.

Presently, almost all CMMs marketed toward micro/mesoscale parts offer a multiprobe solution. These machine builders realize that manufacturers producing micro/mesoscale parts often produce a small volume of a wide variety of parts. These manufacturers require a measurement device capable of handling this variety. Parts that require quick measurement or are too fragile for contact measurement can be measured with an optical probe or a laser probe. Parts that require real 3D measurement or do not reflect light adequately can be measured by the machine's touch probe.

Alternatively, many micro/mesoscale parts have the high demand of a large number of measured points with extremely low uncertainty measurement over a long range. The concepts of precision engineering will be necessary to conduct these types of measurement [42]. There is considerable research toward utilizing the range of CMMs and combining it with the low-uncertainty measurements capable of quickly measuring a large number of points. AFM probes have been mounted on CMMs [43,44]. The capabilities of SWLI are being extended to utilize them more like CMMs [45]. When extending the range of these processes, fiducials can be used to reduce measurement uncertainty [46].

15.6 On-Machine Metrology

In situ or on-machine metrology has been called the *holy grail* of discrete part manufacturing and will have a large impact on high-value manufactured products. Typical micro/mesoscale parts are inherently more valuable than typical traditional-scale parts.

On-machine measurement can reduce the number of scrapped parts by immediately identifying and correcting for errors in the manufacturing process.

Many manufacturers think of metrology as a "non-value-added" process because postprocess metrology requires taking the part away from the production line. Despite the flaws in this reasoning, on-machine measurement appeases these manufacturers because the measurement process takes place within the production line, eliminating the time lost in moving the part to a separate measurement machine, possibly in a separate, dedicated room or facility. In fact, on-machine measurement could make a separate metrology facility unnecessary, allowing a manufacturer to use space originally allotted to metrology for further production.

An additional impact is the fact that on-machine metrology does not require reregistration of the workpiece. When the workpiece is first fixtured to the machine tool, key datum surfaces are established to register the part within the machine's coordinate system. Features are then machined relative to these surfaces. With postprocess measurement, these datum surfaces would need to be reestablished on the metrology device. With on-machine measurement, because the part is not removed from the machine's fixture, the time required to reregister the part is eliminated, but more importantly, the errors in the workpiece are better connected to the errors in the machine tool.

Better connection between machine tool and part errors has several impacts. Of primary importance is that after part errors have been detected, the user or the controller can correct these errors on the current part or compensate for the errors on subsequent parts. This leads to fewer scrapped parts. Also, part errors can help to more quickly diagnose a machine or a tool trending out of tolerance. A change in surface roughness between two parts might indicate that a tool change is necessary.

While these advantages will impact all scales of manufacturing, micro/mesoscale manufacturing further benefits from on-machine inspection. The virtue of not needing to reestablish datum surfaces was previously mentioned, but micro/mesoscale parts further benefit because components are much smaller and absolute tolerances are smaller, making part registration more difficult and time consuming. Also, fixturing of micro/mesoscale parts is significantly more difficult than traditional-scale parts, often requiring unique fixturing solutions for individual parts. With postprocess measurement techniques, a fixturing solution for measurement would be necessary in addition to the fixturing solution for machining, adding significant time and cost. With on-machine measurement, only one fixturing solution suffices.

An additional benefit to micro/mesoscale parts is the reduction in handling that results from on-machine measurement. Micro/mesoscale parts are small and often fragile, presenting unique handling issues not encountered on the traditional scale. Pick-and-place robots with gripping end effectors aid in automation at the traditional scale and could avoid problems created by human handling at the micro/mesoscale. However, issues with adhesion discussed in the touch probe section are more troublesome in handling. If the part does not move from its fixture, as is the case with on-machine measurement, the location of the features is already known, allowing immediate start of the measurement cycle. Of course, any time a part is handled, there is the chance of damaging the part or surface. The lack of handling between machining and measurement greatly reduces the likelihood that the part is harmed in this phase of production.

A major challenge of on-machine metrology is the decoupling of machine tool errors from measurement uncertainty. No machine is perfect, so when the machine moves to produce the part, error motions will result in part imperfections. If the same machine again moves to perform the measurements, the same error motions will likely repeat, obscuring the part imperfections from the measurement. This is a significant obstacle, and if not properly addressed, will defeat any advantage on-machine measurement can offer.

Another consideration should be the harsh environment in which the metrology device would reside. Postprocess measuring devices typically reside and operate in a controlled environment ideal for measurement. On-machine devices would obviously reside on a machine tool and must be able to properly operate whether the machine tool is in a conditioned environment or on the shop floor. The metrology device would need to be robust enough to not be damaged by the harsh machining environment that could see flying chips and cutting fluids. Proper measurement requires removal of any debris from the machining process, deburring, and thorough cleaning of the part surfaces. However, this cleaning must be accomplished without removing the part from the fixture—no trivial task. In addition, thermal expansion could more significantly affect on-machine measurements than postprocess measurements. The machine likely just moved a significant amount with the spindle on, producing a considerable amount of heat. This heat could cause expansion of important machine components or even the part itself.

Again, the unique attributes of micro/mesoscale machining present further important considerations. The recent high demand for smaller parts has led to the development of new micro/mesoscale machine tools, significantly smaller than their traditional-scale cousins and often with different designs. An on-machine measurement device would need to operate within the confined work volume of any micro/mesoscale machine tool, often only 25 mm × 25 mm × 25 mm. This does not provide much space for large optics, supporting electronics, or stray cables. Also, while an on-machine measurement solution that succeeds on only one specific machine design would be progress, a more robust solution that fits many machine designs is more desirable.

15.7 Summary

This chapter presented an overview of processes typically used for 3D metrology at the micro/mesoscale. None of the processes discussed was specifically developed to operate with micro/mesoscale manufactured parts (see Table 15.1 for typical uses). Because of this, trade-offs exist with each of the processes. Table 15.2 summarizes the capabilities

TABLE 15.1

Since Many Micro/Mesoscale Metrology Processes Are Adaptations of Existing Methods, They Are also Used in a Variety of Other Fields

Process	Common Fields of Application
CMM	Traditional-scale geometric part metrology microhole characterization
SPM	Biological applications, semiconductor wafer measurement, microstructural analysis, nanomanufacturing
WLI	MEMS devices, surface texture metrology, semiconductors (step heights, critical dimensions, topographical features)
CM	Biotech, molecular biology, living cell observation, printer head measurement, high aspect ratio samples, MEMS
FP	Molds and plastic molded parts, reverse engineering

CMM, coordinate measuring machines; SPM, scanning probe microscopy; WLI, white-light interferometry; CM, confocal microscopy; and FP, fringe projection.

TABLE 15.2

Summary of the Capabilities of the Discussed Micro/Mesoscale Metrology Processes

Process	Contact (C) or Non-contact (N)	Vertical Uncertainty (nm)	Lateral Uncertainty (nm)	Measurement Size	Commercial Systems	Strengths	Weaknesses
CMM	C	25	25	Probes down to 100 mm	Few in microscale	Low uncertainty, very well characterized	Probe size; slow; environmental control
SPM	N	1 or less	1 or less	200 mm × 200 mm; 15-mm z-range	Many	Superior uncertainty, variety of measurements	Small scan size; limited z-range; environment control
WLI	N	20	300[a] at high mag.	2 mm × 2 mm down to[a] 70 mm × 70 mm	Many	Vertical resolution	Diffraction limited; small working distance
CM	N	50	1000[a] 150	Up to 2 mm × 2 mm[a], 550 mm × 550 mm, 70 mm × 70 mm	Many	Lateral resolution; long z-scan available	Diffraction limited; working distance
FP	N	<100	5000	100 mm and up	Few	Large field of view; fast	Large uncertainty; less developed

All values are approximate and are based on specification sheets of various commercially available systems.

[a] Measurement size and uncertainty vary with numeric aperture on optical systems. Cited values are for commonly available optics.

and availability of the discussed processes. Because of the inherent high value and high demands of micro/mesoscale parts, hybrid processes and on-machine metrology are likely the future of micro/mesoscale manufacturing.

References

1. Hansen, H., Carneiro, K., Haitjema, H., and De Chiffre, L. 2006. Dimensional micro and nano metrology. *CIRP Annals* 55(2): 721–743.
2. Weckenmann, A. and Ernst, R. 2000. Future requirements on micro- and nanomeasurement technique Challenges and approaches. *Technisches Messen* 7–8: 334–342.
3. ISO/IEC Guide 99 2007. *International Vocabulary of Metrology—Basic and General Concepts and Associate Terms (VIM)*. Published by ISO in the name of BIPM, IEC, IFCC, IUPAC, IUPAP, and OIML, Geneva, Switzerland.
4. ISO/IEC Guide 98-3 2008. *Uncertainty in Measurement—Part 3 Guide to the Expression of Uncertainty in Measurement (GUM 1995)*. Published by ISO in the name of BIPM, IEC, IFCC, IUPAC, IUPAP, and OIML, Geneva, Switzerland.
5. ISO 14253-1 1998, *Geometrical Product Specifications (GPS)—Inspection by Measurement of Workpieces and Measuring Equipment—Part 1: Decision Rules for Proving Conformance or Non-conformance with Specifications*.

6. Goldstein, J., Newbury, D., Joy, D., Lyman, C., Echlin, P., Lifshin, E., Sawyer, L., and Michael, J.R. 2003. *Scanning Electron Microscopy and X-Ray Microanalysis*. New York: Springer.
7. Bosch, J., ed. 1995. *Coordinate Measuring Machines and Systems*. New York: Marcel Dekker, Inc.
8. Weckenmann, A., Estler, T., Peggs, G., and McMurtry, D. 2006. Probing systems in dimensional metrology. *CIRP Annals* 53(2): 657–684.
9. Weckenmann, A., Peggs, G., and Hoffmann, J. 2006. Probing systems for dimensional micro- and nano-metrology. *Measurement Science and Technology* 17: 504–509.
10. Cao, S., Brand, U., Kleine-Besten, T., Hoffman, W., Schwenke, H., Butefisch, S., and Büttgenbach, S. 2002. Recent developments in dimensional metrology for microsystem components. *Microsystem Technologies* 8: 3–6.
11. Kung, A., Meli, F., and Thalmann, R. 2007. Ultraprecision micro-CMM using a low force 3D touch probe. *Measurement Science and Technology* 18: 319–327.
12. Lewis, A.J. 2003. Fully traceable miniature CMM with submicrometer uncertainty. *Proceedings of SPIE* 5190: 265–276.
13. Brand, U. and Kirchhoff, J. 2005. A micro-CMM with metrology frame for low uncertainty measurements. *Measurement Science and Technology* 16: 2489–2497.
14. Peiner, E., Doering, L., and Nesterov, V. 2004. Tactile probes for high aspect ratio micrometrology. *MST News* 2: 38–40.
15. Muralakrishnan, B., Stone, J., and Stoup, J. 2006. Fiber deflection probe for small hole metrology. *Precision Engineering* 30(2): 154–164.
16. Takaya, Y., Imai, K., Ha, T., Miyoshi, T., and Kinoshita, N. 2004. Vibrational probing technique for the nano-CMM based on optical radiation pressure control. *CIRP Annals* 53(1): 421–424.
17. Bauza, M.B., Hocken, R.J., Smith, S.T., and Woody, S.C. 2005. Development of a virtual probe tip with an application to high aspect ratio microscale features. *Review of Scientific Instruments* 76: 095112.
18. Phillips, S. 1995. Performance evaluations. *Coordinate Measuring Machines and Systems*, ed. J. Bosch. New York: Marcel Dekker.
19. ANSI/ASME B89.4.10360.2 2008. *Acceptance Test and Reverification Test for Coordinate Measuring Machines (CMMs) Part 2: CMMs Used for Measuring Linear Dimensions (Technical Report)*.
20. ISO 10360-2 2009. *Geometrical Product Specifications (GPS)—Acceptance and Reverification Tests for Coordinate Measuring Machines (CMM)—Part 2: CMMs Used for Measuring Linear Dimensions*.
21. Neuschaefer-Rube, U., Neugebauer, M., and Ehrig, W. 2008. Tactile and optical microsensors: Test procedures and standards. *Measurement Science and Technology* 19: 1–5.
22. Hocken, R.J., Chakraborty, N., and Brown, C. 2005. Optical metrology of surfaces. *CIRP Annals* 54(2): 169–183.
23. Schwenke, H., Neuschaefer-Rube, U., Pfeifer, T., and Kunzmann, H. 2002. Optical methods for dimensional metrology in production engineering. *CIRP Annals* 51(2): 685–700.
24. Yoshizawa, T. 2009. *Handbook of Optical Metrology: Principles and Applications*. Boca Raton: CRC Press.
25. Deck, L. and de Groot, P. 1994. High-speed noncontact profiler based on scanning white-light interferometry. *Applied Optics* 33: 7334–7338.
26. Vallance, R.R., Morgan, C.J., Shreve, S.M., and March, E.R. 2004. Micro-tool characterization using scanning white light interferometry. *Journal of Micromechanics and Microengineering* 14: 1234–1243.
27. Windecker, R., Fleischer, M., and Tiziani, H.J. 1999. White-light interferometry with extended zoom range. *Journal of Modern Optic* 46(7): 1123–1135.
28. Windecker, R., Fleischer, M., Körner, K., and Tiziani, H.J. 2001. Testing micro devices with fringe projection and white-light interferometry. *Optics and Lasers in Engineering* 36: 141–154.
29. Schmit, J. and Olszak, A. 2002. High-precision shape measurement by white-light interferometry with real-time scanner error correction. *Applied Optics* 41(28): 5943–5950.
30. Olszak, A. 2000. Lateral scanning white-light interferometry. *Applied Optics* 39(22): 3906–3913.
31. Wilson, T. 1990. *Confocal Microscopy*. London, England: Academic Press.

32. Jordan, H.-J., Wegner, M., and Tiziani, H. 1998. Highly accurate non-contact characterization of engineering surfaces using confocal microscopy. *Measurement Science and Technology* 9: 1142–1151.

33. Pfeifer, T., Schmitt, R., and Aleriano, U. 2005. Advanced optical metrology aimed at part inspection and reverse engineering form Mems and Nems. *Proceedings of SPIE* 5836: 587–598.

34. Tiziani, H.J., Achi, R., Krämer, R.N., and Wiegers, L. 1996. Theoretical analysis of confocal microscopy with microlenses. *Applied Optics* 35(1): 120–125.

35. Toomre, D., Langhorst, M., and Davidson M. Introduction to spinning disk confocal microscopy. http://zeiss-campus.magnet.fsu.edu/articles/spinningdisk/introduction.html (accessed in July 2012).

36. Windecker, R., Franz, S., and Tiziani, H.J. 1999. Optical roughness measurements with fringe projection. *Applied Optics* 38(13): 2837–2842.

37. Batlle, J., Mouaddib, E., and Salvi, J. 1998. Recent progress in coded structured light as a technique to solve the correspondence problem: A survey. *Pattern Recognition* 31(7): 963–982.

38. Salvi, J., Pages, J., and Batlle, J. 2004. Pattern codification strategies in structured light systems. *Pattern Recognition* 37(4): 827–849.

39. Posdamer, J.L. and Altschuler, M.D. 1982. Surface measurement by space-encoded projected beam systems. *Computer Graphics and Image Processing* 18: 1–17.

40. Moylan, S. and Vogl, G. 2009. Uncertainty analysis of a simple fringe projection system. *ASPE Summer Topical Meeting* 32–35.

41. Meyer, E., Hug, H.J., and Bennewitz, R. 2004. *Scanning Probe Microscopy: The Lab on a Tip.* Germany: Springer-Verlag.

42. Teague, E.C. 1989. The National Institute of Standards and Technology molecular measuring machine project-metrology and precision engineering design. *Journal of Vacuum Science and Technology B* 7(6): 1898–1902.

43. Hansen, H., Kofod, N., De Chiffre, L., and Wanheim, T. 2002. Calibration and industrial application of instrument for surface mapping based on AFM. *CIRP Annals* 51(1): 471–474.

44. Hansen, H., Bariani, P., and De Chiffre, L. 2005. Modelling and measurement uncertainty estimation for integrated AFM-CMM instrument. *CIRP Annals* 54(1): 531–534.

45. Ramasamy, S. and Kasper, A. 2010. Expanding white light interferometer capabilities to be used as a micro-CMM. *Proceedings of the American Society of Precision Engineers Annual Meeting.* Atlanta, GA, to appear.

46. Xu, Z., Shiguang, L., Burns, D., Shilpiekandula, V., Taylor, H., Yoon, S.-F., Youcef-Toumi, K., Reiding, I., Fang, Z., Zao, J., and Boning, D. 2009. Three-dimensional stitching based on fiducial markers for microfluidic devices. *Optics Communications* 282(4): 493–499.

16

Micromolding—A Soft Lithography Technique

B. Radha and G.U. Kulkarni

Jawaharlal Nehru Centre for Advanced Scientific Research

CONTENTS

16.1 Introduction

Photolithography is a well-established technique for patterning electronic circuits, biological assay devices, and plasmonic circuits and is currently used in the semiconductor industry for making integrated circuits [1]. The method has served for decades in patterning a variety of materials over large areas but is not compatible for patterning on surfaces that are sensitive to light or etchants. It has severe limitations when it comes to high-resolution patterning. An alternative to photolithography that could afford rapid prototyping as well as patterning of a wider range of materials is highly desirable. Soft lithography techniques, developed in 1993 by the Whitesides group at the Harvard University, meet some of these requirements and holds a great promise.

16.2 Soft Lithography

Learning from our ancestors and from the routinely used process of seal stamping, soft lithographic techniques take a step toward micro- and nanopatterning. If we look back at times around 900 BC, the Olmecs in Mexico were using baked clay with relief designs to print repeat patterns, possibly on cloth and perhaps on their bodies too! Various other materials such as metals, stones, and wooden pieces were also used as stamps and were later replaced by rubber [2]. A seal that is used to create the impression of a signature is another macroscopic analog to this technique.

Soft lithography is the modern version of creating impressions of this kind. It is a group of techniques in which an elastomeric stamp or mold transfers a pattern to the substrate [3]. The term *soft* refers to the use of materials involved in stamping—elastomers as well as organic molecules. To date, besides conventional methods, patterned microstructures have been produced on nonplanar surfaces. Complex and intricate 3D patterns have been fabricated and also shown to work as microelectronic devices [4]. At present, sub-100 nm patterning by soft lithography is a reality [5]. This chapter describes the basic aspects of soft lithography, citing latest examples from the literature.

To start with, this technique requires a master—the material on which the features to be transferred to the stamp are produced by photolithography, electron beam lithography, and the like (Figure 16.1) [6]. The strength of soft lithography is in replicating the master.

The soft lithography tool kit consists of the stamp, inks, and substrates, of which the key element is the stamp. One of the most successful stamp materials is poly(dimethylsiloxane) (PDMS), a resilient silicone elastomer [7]. It is chemically inert, thermally stable, permeable to gases, and simple to handle and manipulate. It is also environmentally friendly. PDMS reversibly deforms microscopically to conform to even nonplanar surfaces. Importantly, it can replicate the features on the master with high fidelity [8]. PDMS is optically

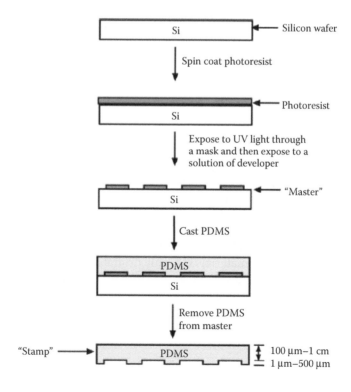

FIGURE 16.1
PDMS stamp fabrication. (Reprinted from *Biomaterials*, 20, Kane, R.S., et al., Patterning proteins and cells using soft lithography, 2363, Copyright 1999 Elsevier.)

transparent to as low as ~300 nm, is homogeneous and isotropic, and is ideal while combining soft lithography with optical methods. Precursors to PDMS are commercially available (e.g., Dow Corning) and are inexpensive. The precursors are the vinyl-terminated siloxane prepolymer (base material) and the siloxane cross-linking agent (curing agent). Curing or cross-linking involves an addition reaction, that is, hydrosilylation between the two components, thus forming a three-dimensionally cross-linked elastomer [7]. Curing temperature and the ratio of the two components decide the Young's modulus of the formed elastomer [9]. PDMS possesses a low surface energy because of the low intermolecular forces between the methyl groups. For PDMS stamp fabrication, the base material and curing agent are mixed in the desired ratio (the popular ratio is 1:10) and the mixture is stirred well and degassed by placing in vacuum; it is then poured onto the master pattern to be replicated (Figure 16.1). For cross-linking the PDMS, the setup is placed in an oven at the desired temperature and for the desired time, commonly 60°C for 6 h. (The reader is advised to further refer to Reference 10, where each step in this protocol is explained with photographs.)

Since the stamp is deformable (Figure 16.2), there are limits on the aspect ratios of the features that can be replicated [11]. If the height (H) is much greater than the width (L) of the feature, the stamp features collapse laterally while peeling off or during the inking process because of capillary forces. On the other hand, if the height is much less than the distance (D) between features, the roof of the stamp sags under its own weight, causing contact between the stamp and the substrate in regions where it is not desired. Sagging can be eliminated by fabricating a submicron-thick stamp on a rigid support [12]. It may be interesting to note that stamp deformations have been exploited widely in producing unusual patterns, as is detailed elsewhere (see Section 16.2.1.2).

Once the stamp is fabricated with the desired relief features, it can be used for many purposes, such as molding, printing, embossing, and imprinting. We discuss the molding method in detail, as an illustrative example for soft lithography, and provide a brief sketch of other related techniques.

16.2.1 Micromolding

Molding is casting the topographic pattern from one material to another by a liquid precursor that is solidified while being molded. This method is specifically called *replica*

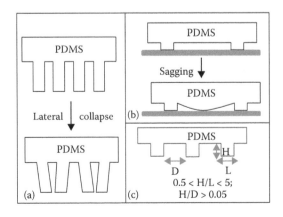

FIGURE 16.2
Schematic illustration of the failure events in PDMS.

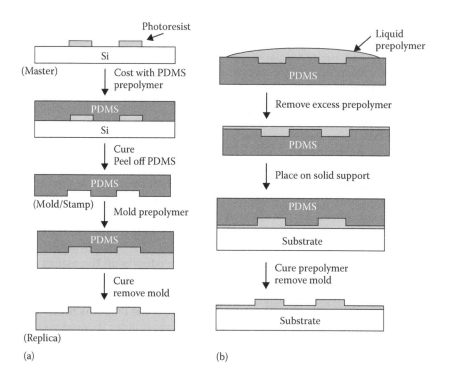

FIGURE 16.3
Schematic illustration of (a) REM and (b) μTM. (Adapted from *Materials Today*, 8, Gates, B.D., Nanofabrication with molds and stamps, 44, Copyright 2005 Elsevier.)

molding (REM) [13] (Figure 16.3). The name is justified as it shares similarity with the conventionally used wax molding. Another approach of molding is microtransfer molding (μTM) [14]. In this method, the polymer precursor is dropped onto the surface of the PDMS mold and excess material is removed. Then, the prepolymer is cured after placing the mold on another substrate. In this method, a residual thin layer of polymer is generally found connecting the isolated features, which could be removed by reactive ion etching (RIE). To prevent the formation of this residue, PDMS molds with appropriate surface chemistry can be used (e.g., fluorinated silanes) to dewet the polymer from the relief features of the PDMS mold [14].

Micromolding in capillaries (MIMIC) is yet another technique to mold isolated features [15]. When a patterned PDMS mold is placed on a substrate, the relief features conform to the substrate, thus forming a network of microchannels. When a liquid drop is placed at one end of the channel, the liquid fills the microchannels by capillarity, which deposits the material onto the substrate in the confined microchannels (Figure 16.4). Various materials such as polymers, prepolymers [16], colloids [17], proteins [18], and sol–gel precursors [19] have been patterned by this method. Using this technique, subwavelength optical devices, waveguides, and optical polarizers have been devised, which could be used in optical fiber networks and optical computers [20]. Microfluidic and nanofluidic devices for producing biochemical chips are other examples of MIMIC [21]. A few examples of patterned microstructures formed by this method are shown in Figure 16.5.

The criteria for choosing the "ink" and the substrate are well established. For micromolding, the viscosity of the liquid "ink" chosen should be low enough to allow fast filling of the microchannels, and the ink should at least partially wet the PDMS stamp.

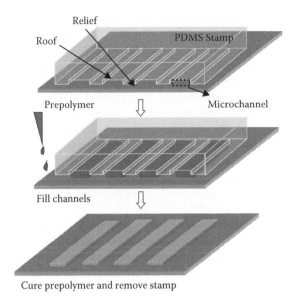

FIGURE 16.4
Schematic illustration of the MIMIC process.

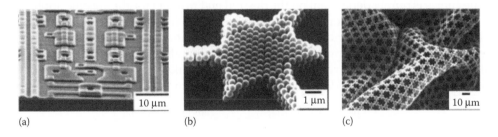

FIGURE 16.5
Scanning electron microscopy (SEM) images of microstructures of various materials fabricated using MIMIC. (a) Quasi-three-dimensional structures of polyurethrane formed on Si/SiO_2. (b) Patterned microstructures of polystyrene beads. (c) Freestanding microstructured membranes of polyurethane 3. (Adapted from Xia, Y. and Whitesides, G.M., *Angewandte Chemie International Edition*, 37, 550, 1998. Copyright 1998 Wiley-VCH.)

This can be further understood in terms of the following classical equation for the capillary filling [22]:

$$\frac{dl}{dt} = \frac{(\gamma r \cos \theta)}{4\eta l}$$

where, dl/dt is the rate of penetration of the liquid in the capillary, l is the distance penetrated into a cylindrical capillary, r is the capillary radius, t is the time taken, θ is the angle between the liquid and the surface, γ is the interfacial free energy of the liquid, and η is the viscosity of the liquid.

The solvent of the "ink" should be chosen to minimize interfacial energy. For example, PDMS being hydrophobic, water is not a good medium for MIMIC. If water has to be used as a solvent, either a surfactant or an organic solvent such as ethanol is required to achieve the capillary flow through the channel. Liquids with low interfacial free energies (e.g., methanol, ethanol) and low viscosity values fill the channels rapidly.

As regards the substrate, an important issue is the sticking of the material being molded (whether to the PDMS stamp or to the substrate), which is determined by the interfacial free energy (γ); the material sticks to a surface that has higher γ. As the interfacial free energy of the PDMS stamp is very low [3], (γ_{SV}(PDMS) = 21.6 dyn/cm) compared to the commonly used substrates for patterning (e.g., γ_{SV}(Si/SiO$_2$) = 72 dyn/cm), the molded material selectively adheres to the substrate rather than to the PDMS.

Apart from solution "inks," MIMIC has also been explored with gas-phase "inks" [23]. After placing the PDMS stamp in conformal contact with the substrate, the open channels are exposed to an organosilane-saturated environment. The silane molecules diffuse into the microchannels and condense preferentially along the interfaces on coming into contact with the substrate in the channels [23].

Conventional MIMIC consists of patterning a polymeric resist/etch mask onto a desired surface and etching away the exposed regions, followed by lift-off of the polymer. These days, direct-write techniques [24] are becoming popular in many beam-based techniques (electron beam, focused ion beam, laser lithography) and they reduce the number of steps in the lithography process flow (Figure 16.6). Very recently, direct methods have been considered for soft lithography as well [25]. In direct micromolding, the development, etching, and lift-off subprocesses are eliminated, thus making the process flow much simpler. The right choice of precursor "ink" is crucial for the success of direct micromolding, and the choice is purely dependent on the desired end product for the specific application. The solubility of the precursor in a chosen solvent, its viscosity and compatibility with the stamp, milder reaction conditions, and finally, the amount of impurities left behind in the end product are some important issues to be dealt with in direct MIMIC.

Pd alkanethiolate is one of the successful molecular inks for micromolding [26]. It is a metal-organic compound, and on thermolysis, neat decomposition takes place, leading

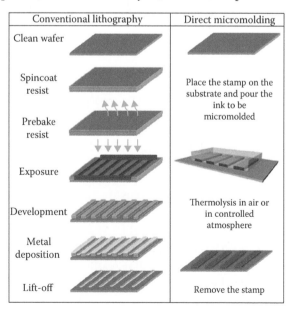

FIGURE 16.6
Conventional lithography versus direct micromolding.

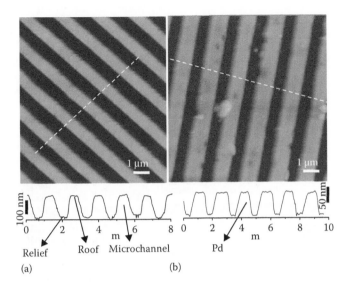

FIGURE 16.7
Atomic force microscopy (AFM) images of (a) PDMS mold, and (b) molded Pd microstripes. The corresponding z-profiles are also shown. The z-profile for PDMS stamp in (a) is shown inverted in order to represent the stamp during molding process.

to metallic Pd, with minimal carboncontamination [27]. As a result, the resistivity of the obtained Pd film is usually quite close to that of the bulk metal. Unlike other thiolates, Pd thiolates are soluble in organic solvents [28] and thus amenable to patterning by flowing in microchannels. Patterned Pd microstripes [26] (Figure 16.7) were formed by Kulkarni and coworkers on diverse substrates including flexible substrates using Pd thiolate as ink for molding [29]. Not only Pd metal but also Pd sulfides such as Pd_4S, $Pd_{16}S_7$, and PdS could be produced from the same precursor by thermolyzing it in a controlled atmosphere such as H_2 [30]. Thus, molded patterns of Pd_4S have been realized [30], which may be useful as corrosion-resistant and gas-impermeable conduits, owing to the known properties of Pd_4S.

In a similar manner, using a solution of Pt-carbonyl clusters, Greco et al. [31] have obtained conductive molecular wires of submicrometer width by micromolding (Figure 16.8a–c). On thermal annealing, decomposition of the precursor leads to the formation of metallic Pt, with conductivity increasing by two orders of magnitude. However, this value is much lower compared to bulk Pt as the thermolysis induces several defects in the microstripes [31].

Apart from metal–organic compounds, polymer-based precursor solutions have been utilized in certain cases [32]. Essentially, a metal nitrate is dissolved in an appropriate solvent and a polymer solution (e.g., polyacrylic acid) is added, which complexes with the metal, ensuring a homogenous distribution of the metal in the polymer. The polymer eases the flow in the microchannels while acting as a carrier for the metal complex. Post-thermolysis leads to the desired product; this methodology was used for molding ceramic materials such as Al_2O_3, ZnO, and $PbTiO_3$ [32]. Patterning of conductive Al-doped ZnO nanostructures has also been demonstrated [33] in a similar way (Figure 16.8d and e). The patterned films were highly porous with high resistivity (8.4 Ωcm), even though they possessed good transmittance.

Another attractive class of functional materials that can be directly micromolded is that of coordination polymers. Highly fluorescent coordination polymer patterns were obtained

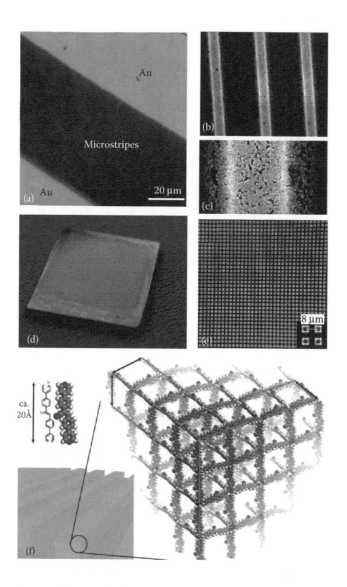

FIGURE 16.8
Various functional materials patterned by direct MIMIC. (a) Optical micrograph showing gold pads evaporated on Pt microstripes, (b) SEM image of the same, (c) magnified image of a single microstripe. (Adapted from Greco, P., et al., *Journal of the American Chemical Society*, 130, 1177, 2008. Copyright 2008 American Chemical Society.) (d) A photograph of patterned ZnO with an area of approximately 1 × 1 cm² and (e) optical microscope image of a dot-patterned ZnO film on glass. (Adapted from Göbel, O.F., et al., *ACS Applied Materials & Interfaces*, 2, 536, 2010. Copyright 2010 American Chemical Society.) (f) Schematic representation of the molecular super-structure of the coordination polymer patterns. (Adapted from You, Y., et al., *Angewandte Chemie International Edition*, 49, 3757. Copyright 2010 Wiley-VCH.)

on molding a nonfluorescent precursor solution and inducing controlled polymerization (Figure 16.8f) [34]. This indicates that one can carry out reactions inside the tiny micro-channel. Indeed, microchannels hosting a chemical reaction were employed to produce patterned superlattices of Au nanocrystals in another study [35]. Micromolding Au(PPh₃) Cl precursor at room temperature, followed by the injection of dodecanethiol at 120°C, led to the formation of Au(I) dodecanethiolate *in situ* inside the microchannel (this method

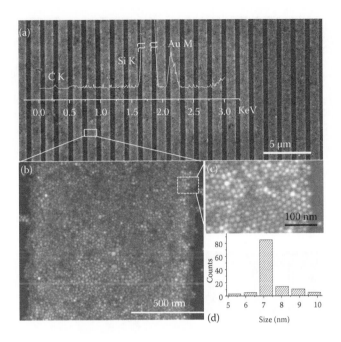

FIGURE 16.9

SEM image showing the patterned Au nanocrystal superlattice, with magnified views in (b) and (c); (d) is a histogram of the size of the Au nanocrystals from (c). The energy dispersive (EDS) data from the patterned region is also shown in (a) overlaid on the image. (Reprinted from Radha, B. and Kulkarni, G.U., *Nano Research* 3, 537, 2010. Copyright 2010 Springer.)

was followed as the precursor is insoluble in common solvents), which further decomposed when annealed at 250°C to give rise to nanoparticles (Figure 16.9) [35]. The nanoparticles ordered themselves hexagonally to form 2D superlattices inside the microchannel.

Another simple approach of direct MIMIC is to take the functional material in the form of ink for micromolding. Thus, fluorescent patterns composed of Au_{22} clusters using the cluster solution itself as ink were fabricated over large areas by micromolding [36]. A colloidal sol of Ag bonded with peptides was micromolded to give rise to fluorescent Ag patterns [37]. Functionalized surfaces for controlled cellular adhesion were prepared by molding polyethylene glycol [38].

Multilayer patterning is a technological requirement that can be easily met by direct MIMIC. Repeating the direct micromolding process of metal–organic precursors over prepatterned surfaces, stacks of micropatterned $BaTiO_3$ and Y-stabilized ZrO_2 films have been made [39]. Another example of the multilayer patterning is shown in Figure 16.10a and b, where Pd microstripes are patterned in the form of a crossbar (Figure 16.10a) by performing direct micromolding of Pd alkanethiolate twice. In the second step, the stamp was placed at 90° to the preformed Pd microstripe pattern. It is also possible to pattern complex microstructures in single step using 3D MIMIC, which was introduced by Whitesides and coworkers [40] for patterning of cells and proteins.

16.2.1.1 Devices

MIMIC is actively employed in the fabrication of devices. Pd microstripes patterned by direct MIMIC on flexible substrate have been used for H_2 sensing (Figure 16.11a) [29]. The mechanism of H_2 sensing by Pd was elucidated by using this as a model system

(a)   (b)

FIGURE 16.10
Multilayer patterning: (a) SEM image and (b) AFM image showing the doubly molded Pd patterns.

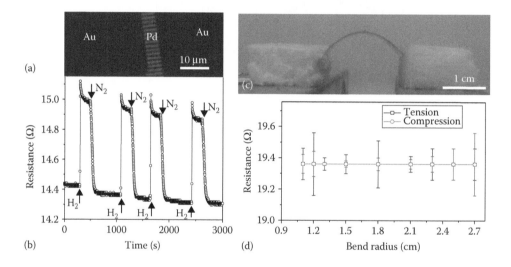

FIGURE 16.11
(a) An optical image of Pd microstripes between Au contact pads on a flexible polyimide substrate. The response to H_2 is shown in (b). (c) Photograph showing the bending of the Pd microstripe circuit. (d) Mechanical stability of a flexible Pd microstripe circuit.

(Figure 16.11b). Also, the stripes do not change their resistance on bending, which is a critical requirement for them to be useful as conduits in flexible electronics (Figure 16.11c and d).

Graphene, along with its derivatives, has become an important class of nanomaterials. Large-scale patterning of reduced graphene oxide (rGO) has been realized by direct micromolding on various substrates (Figure 16.12) [41]. This method was used to fabricate flexible field-effect transistors (FETs) that acted as biosensors [41]. Patterned polymer light-emitting diodes were fabricated by micromolding electroluminescent polymers such as poly(dioctylfluorene) [42]. Optical waveguides directly integrated with quantum cascade lasers have been fabricated using As_2S_3/propylamine ink for molding [43]. For an organic FET, Ag nanoparticle ink was used to define source/drain electrodes by direct micromolding [44]. On combining with screen printing, complete solution-based fabrication of organic FETs was realized by micromolding carbon or polymer electrodes [45].

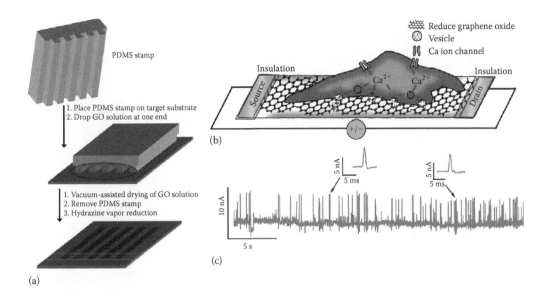

FIGURE 16.12
Schematic illustration of (a) patterning rGO films by MIMIC and reduction of the patterned features to graphene ribbons and (b) the interface between a PC12 cell and rGO FET. (c) Real-time response of rGO FET to the vesicular secretion of catecholamines from PC12 cells stimulated by high K^+ solution. (Adapted from He, Q., et al., *ACS Nano*, 4, 3201, 2010. Copyright 2010 American Chemical Society.)

Combining micromolding and microcontact printing, half-wave rectifier circuits have been made, with an yield of 90% [46]. It is noteworthy that these devices showed performance similar to those fabricated by photolithography.

16.2.1.2 Modified Micromolding Methods

Miniaturization has been the semiconductor industry's watchword since the mid-1950s. In this context, it is important to push the limit of resolution achieved by MIMIC in order to find its niche in next-generation lithography techniques. For improving the resolution of MIMIC, nanomolding in capillaries was recently introduced using composite PDMS stamps with sub-100 nm channels [47]. It is indeed attractive and cost effective only if the methodology of MIMIC could be somehow modified to produce nanosized features using the microchannels. Creating nanochannels for molding is not a solution, as the liquid flow in nanochannels poses many challenges. There are a few examples of MIMIC in which such high resolution has been achieved.

In one such study [48], dewetting of the polymer being molded from the PDMS stamp was employed to produce nanolines. While micromolding the polystyrene (PS), the entire setup was heated to above the glass-transition temperature of the PS, which resulted in its dewetting from the internal PDMS surface and creeping below the relief features (Figure 16.13). As the creeping does not reach completion, it produces V-shaped grooves (down to ~30 nm) with tiny gaps at the apex. These grooves could be post filled with a metal by physical deposition or chemical deposition, leading to essentially 30 nm wide metal lines [48]. The groove width itself is controllable by simply placing a weight over the PDMS stamp while molding.

In a related but new method named *nanoentrapment molding* (NEM) [26], the fragility of the PDMS stamp was exploited to produce nanochannels from microchannels *in situ*. The

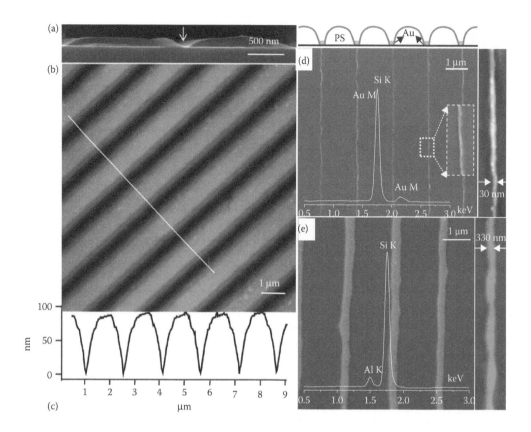

FIGURE 16.13
Dewetting-assisted patterning of PS. (a) Cross-section SEM image showing the groove formation in PS, (b) AFM image of the grooves, and (c) the corresponding z-profile. (d) SEM image of the Au metal line pattern formed by physical vapor deposition (PVD) of the metal onto the PS pattern followed by lift-off of the polymer in toluene with the overlay of the corresponding EDS spectrum. The associated schematic diagram on the top shows the deposited metal in the groove. The inset shows a higher magnification of a single Au line. (e) SEM image of the pattern of Al metal lines formed by PVD of the Al metal onto the polymer pattern with wider grooves. The corresponding EDS spectrum is shown overlaid. Wider grooves were created by pressuring the stamp while molding PS. AFM image of a single Al line is shown on the right. (Adapted from Radha, B. and Kulkarni, G.U. ACS *Applied Materials & Interfaces*, 1, 257, 2009. Copyright 2009 American Chemical Society.)

deformation of a PDMS stamp induced by on-top pressure in combination with a high-temperature ramp created fine capillaries that entrap a thiolate precursor. The trapped precursor could be thermolyzed to give rise to the metallic patterns at the nanometer scale in a single step. It was proposed that the confinement of the precursor takes place owing to microchannel roof collapse (Figure 16.14), the latter being a result of the combined effect of external pressure and differential swelling of PDMS because of skin-deep diffusion of the solvent under rising temperature [26]. By carefully controlling the on-top pressure, patterns with different line widths can be produced.

The concentration of the molding ink is another handy parameter. By controlling the concentration of the precursor ink being molded such that there is depletion of the ink before it could fill the capillary, edge-confined structures of functional materials have been obtained [25].

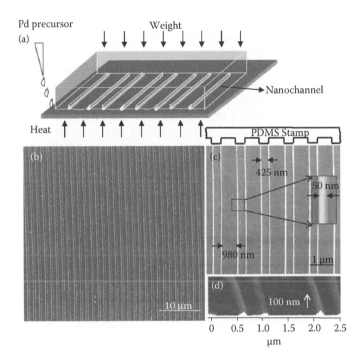

FIGURE 16.14
(a) Schematic representation showing the formation of nanochannels because of the near-complete microchannel roof collapse. (b) Large-area SEM micrograph of patterned Pd nanowires formed on Si substrate by NEM. (c) Magnified view with the inset showing an individual nanowire of width 50 nm. (d) AFM micrograph of the Pd nanowires. (Reprinted from Radha, B. and Kulkarni, G.U., *Small*, 5, 2271, 2009. Copyright 2009 Wiley-VCH.)

16.2.2 Micromolding Allied Technique—SAMIM

Solvent-assisted micromolding in capillaries (SAMIM) [49] produces patterns on the surface of a material using a solvent that can dissolve the material. The PDMS mold is wetted with the solvent and brought into contact with the surface of the material to be patterned (generally polymers). The solvent dissolves or swells a thin layer of the material, which gets molded by the PDMS. When the solvent evaporates, the patterned surface gets defined as the replica of the PDMS mold surface. The solvent chosen should not affect the PDMS mold. It should evaporate rapidly; thus, solvents having high vapor pressure are suitable for this purpose. For using solvents such as water, which do not wet the PDMS surface, a pretreatment to make it hydrophilic is generally employed. This technique generates features that are connected by a residual layer, unlike MIMIC. The patterns generated by SAMIM are affected by various parameters such as mold rising angle (angle between substrate and mold sidewall), concentration of the material being molded, and the amount of material being molded [50]. The capillary rise of the material on softening with the solvent can be controlled with the time of contact of the mold. Thus, miniaturized edge-confined structures have been prepared by bringing the swollen PDMS stamp into contact with PS for half a minute [51].

16.2.3 Soft Printing

Soft printing may refer to contact and transfer printing. Contact printing as a technique of soft lithography has gained relatively more attention than others. The principle underlying contact printing is similar to that used in day-to-day seal stamping. A patterned PDMS stamp is inked with specific molecular ink and transferred onto the desired substrate. As the relief features of the stamp contact the substrate, the ink gets transferred, defining a pattern similar to that of the stamp. This is the process of microcontact printing. This method is simple and straightforward, and multiple printing can produce even intricate patterns [52]. Microcontact printing was initially demonstrated by patterning a self-assembled monolayer (SAM) of thiol molecules onto Au surfaces based on chemisorption onto Au surface [53]. Now, contact printing technique is used to pattern molecules, colloids, nanoparticles, organic reactants, and biomolecules. [6]. The technique can be used to pattern discontinuous structures, but this is restricted to the surface. The stamp and surface can be planar or nonplanar, and roll-to-roll type of printing is possible with this technique (Figure 16.15). However, the limitation of this technique is that the recessed features of the stamp can deform at the nanoscale and cause distortions in the pattern [3]. To minimize the deformation, a composite stamp that retains the structure

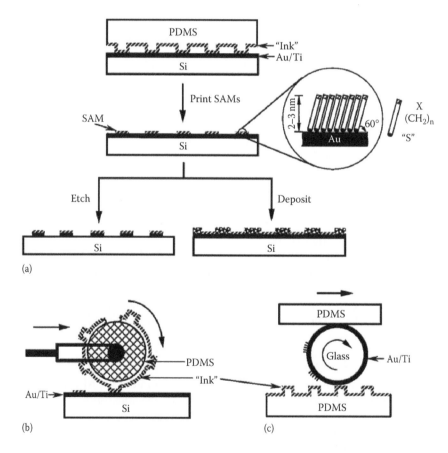

FIGURE 16.15
Schematic procedures for microcontact printing (μCP) of hexadecanethiol on Au surface: (a) printing on a planar surface with a planar stamp, (b) printing on a planar surface over large areas with a rolling stamp, and (c) printing on a nonplanar surface with a planar stamp. (Reprinted from Xia, Y. and Whitesides, G.M., *Angewandte Chemie International Edition*, 37, 550, 1998. Copyright 1998 Wiley-VCH.)

of nanoscale topographic patterns containing a base layer of hard-PDMS (h-PDMS) is employed [54]. Transfer of the material from the PDMS stamp to the substrate can also be assisted by applying a bias voltage, and this method is called *electrical microcontact printing* [55].

The PDMS stamp can be used to transfer not only inks but also metal patterns by nanotransfer printing (nTP) [56]. A patterned PDMS stamp is coated with the desired metal and brought in contact with the pretreated substrate. The pretreatment involves coating a SAM on the substrate, so as to mediate the transfer of the metal from PDMS to the substrate by covalent bonding. The metal gets transferred only from the relief features of the stamp, thus producing a pattern similar to that of the stamp. Interesting shapes such as Au nanocones have been fabricated using nTP [57]. The technique works fairly well with Au, but for metals such as Cu, the diffusion of PDMS oligomers makes the transfer-printed patterns nonconductive. Hence, conductive patterns can be printed by leaching the PDMS stamps before transfer printing [58]. One can substitute the covalent bonding step with heat treatment, which alters the surface energy of PDMS. This weakens the adhesion of material to the PDMS surface, thus enabling the transfer of the material to the substrate. Single-walled carbon nanotube networks have been transfer printed using this method [59]. Kinetic control of the adhesion is also exploited to transfer print a variety of materials [60]. For example, p–n junctions and photodiodes have been transfer printed onto highly curved surfaces by kinetically controlled switching between adhesion and release of materials [60].

16.3　Outlook

The chapter has dealt with various types of printing and molding techniques employing soft stamps. Before soft lithography becomes a general strategy for microfabrication, there are a few technical problems to be addressed. The volume of the stamp material, PDMS, shrinks by more than 1% after curing and can be readily swollen by various nonpolar organic solvents such as hexane, xylene, and toluene [54]. Such properties pose a serious threat to the high registration of the patterns, if the process flow involves these solvents. Besides, the softness of the PDMS limits the aspect ratio of relief structures. The aspect ratio must be between 0.2 and 2 in order to obtain defect-free stamps or molds. Because of such issues, soft lithography fails when multiple layers must be stacked "precisely" on top of one another [20].

This calls for a rigid stamp rather than an elastic one, leading to a technique called *nanoimprint lithography* (NIL) developed by Stephen Y. Chou, who employed a rigid master (generally made of Si or quartz) for embossing a film of polymer that is heated to a temperature close to its melting point to facilitate the embossing process [61]. In another technique called *step-and-flash imprint lithography (SFIL)* employing preferably a UV-transparent stamp, the stamp is pressed against a thin film of liquid polymer and is exposed to ultraviolet light, which solidifies the polymer to create the desired replica [62]. This technique is particularly useful when repeat periodic patterns have to be fabricated on a given substrate. Of course, as the stamp is rigid, these techniques employ relatively high pressures to ensure proper contact between the stamp and the substrate.

Nomenclature

AFM	Atomic force microscopy
EDS	Energy dispersive x-ray spectroscopy
FET	Field-effect transistor
mCP	Microcontact printing
MIMIC	Micromolding in capillaries
mTM	Microtransfer molding
NEM	Nanoentrapment molding
NIL	Nanoimprint lithography
nTP	Nanotransfer printing
PDMS	Poly(dimethylsiloxane)
h-PDMS	Hard-poly(dimethylsiloxane)
Pd	Palladium
PS	Polystyrene
PVD	Physical vapor deposition
REM	Replica molding
rGO	reduced graphene oxide
RIE	Reactive ion etching
SAM	Self-assembled monolayer
SAMIM	Solvent-assisted micromolding in capillaries
SEM	Scanning electron microscopy
SFIL	Step-and-flash imprint lithography

References

1. Madou, M.J. 2002. *Fundamentals of Microfabrication: The Science of Miniaturization*. New York, USA: CRC Press.
2. Carter, T.F. 1955. *The Invention of Printing in China and Its Spread Westward*. New York, USA: Ronald Press Co.
3. Xia, Y. and Whitesides, G.M. 1998. Soft lithography. *Angewandte Chemie International Edition* 37: 550.
4. Gates, B.D., Xu, Q., Love, J.C., Wolfe, D.B., and Whitesides, G.M. 2004. Unconventional nano-fabrication. *Annual Review of Materials Research* 34: 339.
5. Zhao, X.M., Xia, Y., and Whitesides, G.M. 1997. Soft lithographic methods for nano-fabrication. *Journal of Materials Chemistry* 7: 1069.
6. Kane, R.S., Takayama, S., Ostuni, E., Ingber, D.E., and Whitesides, G.M. 1999. Patterning proteins and cells using soft lithography. *Biomaterials* 20: 2363.
7. Gates, B.D. 2005. Nanofabrication with molds & stamps. *Materials Today* 8: 44.
8. Xia, Y. and Whitesides, G.M. 1998. Soft lithography. *Annual Review of Materials Science* 28: 153.
9. Armani, D., Liu, C., and Aluru, N. 1999. Re-configurable fluid circuits by PDMS elastomer micromachining. *Micro Electro Mechanical Systems* 12: 222.
10. Qin, D., Xia, Y., and Whitesides, G.M. 2010. Soft lithography for micro- and nanoscale patterning. *Nature Protocols* 5: 491.
11. Delamarche, E., Schmid, H., Michel, B., and Biebuvck, H. 1997. Stability of molded polydimethylsiloxane microstructures. *Advanced Materials* 9: 741.

12. Bietsch, A. and Michel, B. 2000. Conformal contact and pattern stability of stamps used for soft lithography. *Journal of Applied Physics* 88: 4310.
13. Xia, Y., McClelland, J.J., Gupta, R., Qin, D., Zhao, X.M., Sohn, L.L., Celotta, R.J., and Whitesides, G.M. 1997. Replica molding using polymeric materials: A practical step toward nanomanufacturing. *Advanced Materials* 9: 147.
14. Gates, B.D., Xu, Q., Love, J.C., Wolfe, D.B., and Whitesides, G.M. 2004. Unconventional nanofabrication. *Annual Review of Materials Research* 34: 339.
15. Kim, E., Xia, Y., and Whitesides, G.M. 1995. Polymer microstructures formed by moulding in capillaries. *Nature* 376: 581.
16. Kim, E., Xia, Y., and Whitesides, G.M. 1996. Micromolding in capillaries: Applications in materials science. *Journal of the American Chemical Society* 118: 5722.
17. Huang, W., Li, J., Luo, C., Zhang, J., Luan, S., and Han, Y. 2006. Different colloids self-assembly in micromolding. *Colloids Surfaces A* 273: 43.
18. Shim, H.W., Lee, J.H., Kim, B.Y., Son, Y.A., and Lee, C.S. 2009. Facile preparation of biopatternable surface for selective immobilization from bacteria to mammalian cells. *Journal for Nanoscience and Nanotechnology* 9: 1204.
19. Lochhead, M.J. and Yager, P. 1997. Patterned sol-gel structures by micro molding in capillaries. *Materials Research Society Symposium Proceedings* 444: 105.
20. Whitesides, G.M. and Love, J.C. 2001. The art of building small. *Scientific American* 285: 38.
21. Whitesides, G.M., Ostuni, E., Takayama, S., Jiang, X., and Ingber, D.E. 2001. Soft lithography in biology and biochemistry. *Annual Review of Biomedical Engineering* 3: 335.
22. Myers, D. 1991. *Surfaces, Interfaces and Colloids.* New York: Wiley-VCH.
23. George, A., Blank, D.H.A., and ten Elshof, J.E. 2009. Nanopatterning from the gas phase: High resolution soft lithographic patterning of organosilane thin films. *Langmuir* 25: 13298.
24. Kim, N.-S. and Han, K.N. 2010. Future direction of direct writing. *Journal of Applied Physics* 108: 102801.
25. Cavallini, M., Albonetti, C., and Biscarini, F. 2009. Nanopatterning soluble multifunctional materials by unconventional wet lithography. *Advanced Materials* 21: 1043.
26. Radha, B. and Kulkarni, G.U. 2009. A modified micromolding method for sub-100-nm direct patterning of Pd nanowires. *Small* 5: 2271.
27. Bhuvana, T. and Kulkarni, G.U. 2008. Highly conducting patterned Pd nanowires by direct-write electron beam lithography. *ACS Nano* 2: 457.
28. John, N.S., Thomas, P.J., and Kulkarni, G.U. 2003. Self-assembled hybrid bilayers of palladium alkanethiolates. *Journal of Physical Chemistry B* 107: 11376.
29. Radha, B., Sagade, A.A., Gupta, R., and Kulkarni, G.U. 2011. Direct micromolding of Pd-stripes for electronic applications. *Journal of Nanoscience and Nanotechnology* 11: 152.
30. Radha, B. and Kulkarni, G.U. 2010. Patterned synthesis of Pd$_4$S: Chemically robust electrodes and conducting etch masks. *Advanced Functional Materials* 20: 879.
31. Greco, P., Cavallini, M., Stoliar, P., Quiroga, S.D., Dutta, S., Zacchini, S., Iapalucci, M.C., Morandi, V., Milita, S., Merli, P.G., and Biscarini, F. 2008. Conductive sub-micrometric wires of platinum-carbonyl clusters fabricated by soft-lithography. *Journal of the American Chemical Society* 130: 1177.
32. Göbel, O.F., Nedelcu, M., and Steiner, U. 2007. Soft lithography of ceramic patterns. *Advanced Functional Materials* 17: 1131.
33. Göbel, O.F., Blank, D.H.A., and ten Elshof, J.E. 2010. Thin films of conductive ZnO patterned by micromolding resulting in nearly isolated features. *ACS Applied Materials & Interfaces* 2: 536.
34. You, Y., Yang, H., Chung, J.W., Kim, J.H., Jung, Y., and Park, S.Y. 2010. Micromolding of a highly fluorescent reticular coordination polymer: Solvent-mediated reconfigurable polymerization in a soft lithographic mold. *Angewandte Chemie International Edition* 49: 3757.
35. Radha, B. and Kulkarni, G.U. 2010. Micro- and nanostripes of self-assembled Au nanocrystal superlattices by direct micromolding. *Nano Research* 3: 537.

36. Shibu, E.S., Radha, B., Verma, P.K., Bhyrappa, P., Kulkarni, G.U., Pal, S.K., and Pradeep, T. 2009. Functionalized Au_{22} clusters: Synthesis, characterization and patterning. *ACS Applied Materials & Interfaces* 1: 2199.

37. Naik, R.R., Stringer, S.J., Agarwal, G., Jones, S.E., and Stone, M.O. 2002. Biomimetic synthesis and patterning of silver nanoparticles. *Nature Materials* 1: 169.

38. Lee, J.-H., Kim, H.-E., Im, J.H., Bae, Y.M., Choi, J.S., Huh, K.M., and Lee, C.-S. 2008. Preparation of orthogonally functionalized surface using micromolding in capillaries technique for the control of cellular adhesion. *Colloids Surfaces B* 64: 126.

39. Göbel, O.F., Branfield, T.E., Stawski, T.M., Veldhuis, S.A., Blank, D.H.A., and ten Elshof, J.E. 2010. Stacks of functional oxide thin films patterned by micromolding. *ACS Applied Materials & Interfaces* 2: 2992.

40. Chiu, D.T., Jeon, N.L., Huang, S., Kane, R.S., Wargo, C.J., Choi, I.S., Ingber, D.E., and Whitesides, G.M. 2000. Patterned deposition of cells and proteins onto surfaces by using three-dimensional microfluidic systems. *Proceedings of the National Academy of Sciences of the United States of America* 97: 2408.

41. He, Q., Sudibya, H.G., Yin, Z., Wu, S., Li, H., Boey, F., Huang, W., Chen, P., and Zhang, H. 2010. Centimeter-long and large-scale micropatterns of reduced graphene oxide films: Fabrication and sensing applications. *ACS Nano* 4: 3201.

42. Park, M.J., Choi, W.M., and Park, O.O. 2006. Patterning polymer light-emitting diodes by micromolding in capillary. *Current Applied Physics* 6: 627.

43. Tsay, C., Zha, Y., and Arnold, C.B. 2010. Solution-processed chalcogenide glass for integrated single-mode mid-infrared waveguides. *Optics Express* 18: 26744.

44. Blümel, A., Klug, A., Eder, S., Scherf, U., Moderegger, E., and List, E.J.W. 2007. Micromolding in capillaries and microtransfer printing of silver nanoparticles as soft-lithographic approach for the fabrication of source/drain electrodes in organic field-effect transistors. *Organic Electronics* 8: 389.

45. Rogers, J.A., Bao, Z., and Raju, V.R. 1998. Nonphotolithographic fabrication of organic transistors with micron feature sizes. *Applied Physics Letters* 72: 2716.

46. Deng, T., Goetting, L.B., Hu, J., and Whitesides, G.M. 1999. Microfabrication of half-wave rectifier circuits using soft lithography. *Sensors and Actuators A* 75: 60.

47. Duan, X., Zhao, Y., Berenschot, E., Tas, N.R., Reinhoudt, D.N., and Huskens, J. 2010. Large-area nanoscale patterning of functional materials by nanomolding in capillaries. *Advanced Functional Materials* 20: 2519.

48. Radha, B. and Kulkarni, G.U. 2009. Dewetting assisted patterning of polystyrene by soft lithography to create nanotrenches for nanomaterial deposition. *ACS Applied Materials & Interfaces* 1: 257.

49. King, E., Xia, Y., Zhao, X.M., and Whitesides, G.M. 1997. Solvent-assisted microcontact molding: A convenient method for fabricating three-dimensional structures on surfaces of polymers. *Advanced Materials* 9: 651.

50. Lee, S.-H., Kim, H.-N., Kwak, R.-K., and Suh, K.Y. 2009. Effects of mold rising angle and polymer concentration in solvent-assisted molding. *Langmuir* 25: 12024.

51. Yu, X., Wang, Z., Xing, R., Luan, S., and Han, Y. 2005. Solvent assisted capillary force lithography. *Polymer* 46: 11099.

52. Rogers, J.A. and Nuzzo, R.G. 2005. Recent progress in soft lithography. *Materials Today* 8: 50.

53. Kumar, A. and Whitesides, G.M. 1993. Features of gold having micrometer to centimeter dimensions can be formed through a combination of stamping with an elastomeric stamp and an alkanethiol "ink" followed by chemical etching. *Applied Physics Letters* 63: 2002.

54. Schmid, H. and Michel, B. 2000. Siloxane polymers for high-resolution, high-accuracy soft lithography. *Macromolecules* 33: 3042.

55. Jacobs, H.O. and Whitesides, G.M. 2001. Submicrometer patterning of charge in thin-film electrets. *Science* 291: 1763.

56. Gates, B.D., Xu, Q., Stewart, M., Ryan, D., Willson, C.G., and Whitesides, G.M. 2005. New approaches to nanofabrication: Molding, printing, and other techniques. *Chemical Reviews* 105: 1171.
57. Kim. T.I., Kim, J.H., Son, S.J., and Seo, S.M. 2008. Gold nanocones fabricated by nanotransfer printing and their application for field emission. *Nanotechnology* 19: 295302.
58. Felmet, K., Loo, Y.-L., and Sun, Y. 2004. Patterning conductive copper by nanotransfer printing. *Applied Physics Letters* 85: 3316.
59. Hur, S.H., Khang, D.Y., Kocabas, C., and Rogers, J.A. 2004. Nanotransfer printing by use of noncovalent surface forces: Applications to thin-film transistors that use single-walled carbon nanotube networks and semiconducting polymers. *Applied Physics Letters* 85: 5730.
60. Meitl, M.A., Zhu, Z.T., Kumar, V., Lee, K.J., Feng, X., Huang, Y.Y., Adesida, I.,Nuzzo, R.G., and Rogers, J.A. 2006. Transfer printing by kinetic control of adhesion to an elastomeric stamp. *Nature Materials* 5: 33.
61. Chou, S.Y., Krauss, P.R., and Renstrom, P.J. 1996. Imprint lithography with 25-nanometer resolution. *Science* 272: 85.
62. Resnick, D.J., Sreenivasan, S.V., and Willson, C.G. 2005. Step and flash imprint lithography. *Materials Today* 8: 34.

17

Fabrication of Microelectronic Devices

Monica Katiyar, Deepak, and Vikram Verma
Indian Institute of Technology Kanpur

CONTENTS

17.1 Introduction

The second part of the past century is known as the *silicon age*, and the backbone of this era is the precise manufacturing of circuit elements on a single-crystal Si wafer. This particular technology of semiconductor processing is known as *planar technology*, and it makes the integration of a large number of electronic components on a single wafer possible [1]. The miniaturization of individual devices to make large-scale integrated circuits would not have been possible without the advances in processes, techniques, and equipment. The availability of cheap electronics today is a testimony to the advances that have taken place over the past 50 years. Advances in microelectronic device fabrication have led to new fields such as microelectromechanical systems (MEMS) [2]. In microelectronic circuits, there is no physical displacement between components, while MEMS processing is done to allow physical displacement of components. To a very large extent, today's nanofabrication technology also has its lineage in microelectronics fabrication as many tools developed to work at the microrange have been improved and enhanced to provide nanolevel observation and manipulation.

To build a very complicated part, the first step is to break it down into various monolithic components, cast or machine them, and finally join them using a mechanical or metallurgical joining technique. Microelectronics fabrication is done in such a way that the joining between components is done by directly building one component on the other. In other words, joining is at the atomic level (nanoscale) between the interfaces of two components. The interface itself is a component, for example, in a p–n junction; we see later that it is a requirement that we have ultraclean joints between two materials. To attain this precision, a clean room environment is required for microelectronic fabrication.

To illustrate the working principle of microelectronic processing, let us consider the part in Figure 17.1 that we wish to fabricate—a simple p–n diode in silicon. We make ohmic contacts (this means the resistance of the contact is zero) to this junction, using metal junctions on both sides. In this chapter, all examples are taken from Si technology, because it has the most advanced processing technology and it also has a large share of the integrated circuit market. The processing steps to make this diode are shown in Figure 17.2.

We start with a single-crystal silicon wafer of the required doping, which is most likely made by pulling a single crystal from a melt and then having it cut and polished to the right size [3]. On this Si wafer, approximately 500 mm thick and between 50 and 300 mm in diameter, we first deposit a layer of silicon oxide (SiO_2); this is generally done by thermal oxidation or chemical vapor deposition (CVD). The thickness of this layer is decided by the requirement of the subsequent processes. A window is opened in the oxide. This process is called *photolithography*, and it is subdivided into many subprocesses as shown in Figure 17.2b. In photolithography, we spin-coat a photosensitive polymer (photoresist (PR)) on the substrate. Typically, we bake the wafer before and after spin coating (prebake and postbake). This is followed by exposure of the coated wafer to UV light. However, the PR experiences exposure to UV light only in the desired areas because it is protected by a patterned mask (e.g., a glass plate on which chromium is coated in a pattern, preventing the UV light from passing through). Following the development of the PR, the surface that needs to be removed is exposed and the one that needs to be retained is protected by the PR. We use an etching agent that will remove the undesired part of the SiO_2. Finally, the PR is also etched away from the top of the layer and we get the net shape desired on the SiO_2. In the next step, a p-type dopant is introduced into the silicon through the window using the diffusion or ion implantation process. The thickness of the masking oxide should be enough to ensure that dopants are not introduced in the covered portion of the silicon wafer. The p–n junction is created by this process. Once again, we deposit a SiO_2 layer; next, we open two windows in this layer as shown in (vi) of Figure 17.2a, using the previously

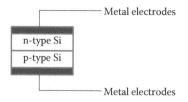

FIGURE 17.1
A p–n diode to be fabricated.

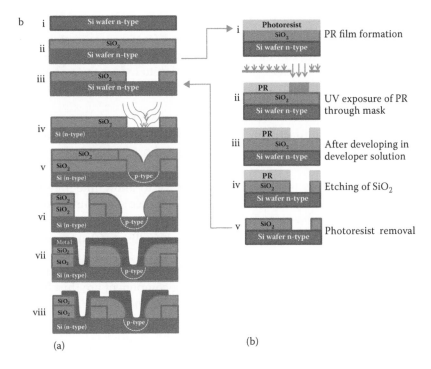

FIGURE 17.2
(a) Processing steps in the fabrication of a p–n junction diode: (i) a single crystal silicon wafer of the required doping; (ii) deposition of a layer of silicon oxide (SiO_2); (iii) photolithography to open a window; (iv) introduction of a p-type dopant in silicon through the window; (v) once again, deposition of a SiO_2 layer; (vi) photolithography to open another window; (vii) metal deposition; (viii) photolithography to define contacts to the p and n regions. (b) Subprocesses involved in the photolithography step: (i) spin coating of a photosensitive polymer (photoresist) on the substrate; (ii) exposure of the coated wafer by UV light through a patterned mask; (iii) development of the photoresist; (iv) etching to remove the undesired part of the SiO_2 layer; (v) etching of the remaining photoresist, resulting in the desired window in the SiO_2 layer.

explained photolithography process. A thin film of metal electrode is deposited (sometimes, additional processing may be needed to ensure ohmic contact) to make contact with the n and p portions of the Si wafer. The metal film will again need to be patterned using photolithography. Now, the device is ready. However, in practice, another passivating layer is deposited and patterned to protect the devices. The final device in Figure 17.2 may not look the same as that in Figure 17.1 from the geometry point of view, but it has exactly the same electrical characteristics. For microelectronics, this is more important than the physical appearance. Another interesting aspect of this transformation of the component shape is that we can take both contacts from the top of the surface; this makes integration very easy. In integrated circuits, millions of such devices need to be fabricated and connected on a single wafer. All fabrication steps happen in a single plane—hence it is called *planar technology*. After fabrication, the chip is tested, wire bonded, and packaged; these are also very important aspects of microelectronics fabrication, which are not discussed in this chapter.

17.1.1 Power of Large-Scale Integration in Microelectronics

We have briefly seen the processes involved in making a discrete device on silicon, but the power of planar technology lies in its capability to integrate a large number of discrete devices (transistors, capacitors, and resistors) to make circuits on a silicon wafer. Further, with the starting size of the silicon wafer increasing from approximately 2 to 12 in., a large number of such circuits can be fabricated on it simultaneously. The advantage of micromanufacturing is that the yield is sufficiently high to reduce the cost of a single chip. There are many factors that contribute to the success of semiconductor manufacturing technology:

1. Cleanliness during manufacturing is critical to maintain low contamination of the surface. Therefore, clean rooms are a must for good process control and high yield in manufacturing.

2. Development of process technology is necessary to allow integration of larger number of devices every year. Moore (Figure 17.3) predicted this advancement in 1970, and his prophecy has been fulfilled.

3. Use of automation in microelectronics fabrication facilities (fabs) improves yield.

In summary, we have created a component by adding the material and patterning it. The two processes that add a new material to an existing component are thin film deposition and diffusion/ion implantation (this modifies the existing material). Processes that pattern the components are photolithography and etching. Each of them is discussed separately in subsequent sections. Since this information is meant for engineers/researchers not working in the field, the level of details has been kept to a minimum for the sake of clarity.

17.2 Thin Film Deposition/Growth

Advances in microelectronics and optoelectronics have driven the development of thin film science and technology, and vice versa [4]. To add a new material, thin film deposition

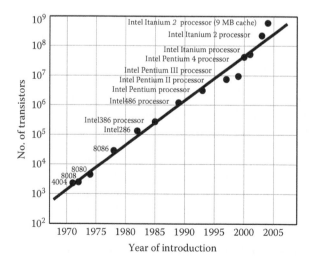

FIGURE 17.3
Moore's law for microprocessors—showing an exponential increase in the number of devices for an application.

is used where the atomic flux, in vapor phase, of the material is brought to the surface and deposited. (There are also liquid-phase-based techniques, but these are used in special cases.) There are two main categories here, physical vapor deposition (PVD) and CVD. We can deposit almost any material, metals, insulators, and semiconductors, as needed in the device structure.

During the thin film deposition process, vapor-phase precursors (growth species) are created, and deposition techniques generally are named after the method used for creating the growth species, for example, thermal evaporation, e-beam deposition, and CVD. Growth species could be atoms, ions, molecules, or clusters; they are characterized by their composition, flux, and energy. These are then transported to the substrate where they are deposited by means of processes taking place at the surface—accommodation, adsorption followed by surface diffusion and incorporation in the film, or reevaporation. Here, it is important to mention that the use of vacuum is required during deposition to maintain the cleanliness of the surface on which the film is deposited and to minimize impurity incorporation in the film. Processes occurring at the surface are important in deciding the properties of the film, hence, substrate cleanliness and temperature are important. The thickness of the film is monitored in real time using a deposition rate monitor (DTM). It is based on the piezoelectric properties of a quartz crystal. While the film is deposited on the substrate, it is also deposited on the crystal. By measuring the resonance frequency of the quartz crystal, the amount of material deposited on it can be inferred. After calibration, this information is used to monitor the film thickness on the substrate.

A typical system, shown in Figure 17.4, consists of a vacuum chamber (bell jar, vacuum pumps, and pressure gauges), a substrate holder and a heater that holds the substrate (it can also rotate the substrate), a thermocouple to monitor the temperature of the substrate, the source of growth species, and a DTM. Vacuum is helpful in all the three steps of film deposition—generation of the vapor phase (e.g., evaporation rate depends on the level of vacuum in the chamber), mass transport from the source to the substrate (line-of-sight from the source to the substrate is possible in the absence of scattering by ambient gas molecules), and low contamination of the substrate surface.

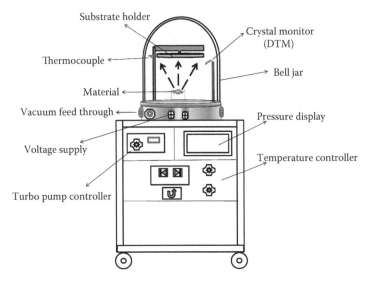

FIGURE 17.4
A typical thin film deposition system.

17.2.1 Importance of Vacuum

Using the kinetic theory of gases, we get the expression, Equation 17.1, for the rate of arrival of atoms or molecules (F) on the substrate surface from the gaseous environment:

$$F = \frac{P}{\sqrt{2\pi M k_B T}}$$

(17.1)

where P is the partial pressure of the gas species having atomic or molecular weight M and temperature T, and k_B is the Boltzmann constant. Assuming that all incident atoms stick to the surface, the time to deposit one monolayer of atoms on the surface can be calculated. At normal pressure, it will take less than a microsecond for the surface to get contaminated. To avoid this contamination, film deposition is done in vacuum. Also, from the point of view of joining the two materials at the atomic level, a small amount of contamination at the interface will lead to poor adhesion between the layers. If the partial pressure of the growth flux is much higher than the base pressure, there will be less contamination of the film during growth. Therefore, the desired purity of the thin film determines the required vacuum level in the chamber. For example, ultrahigh vacuum levels ($< 10^{-8}$ bar) are required for depositing silicon, whereas a medium level ($\sim 10^{-6}$ bar) is sufficient for a metal layer.

17.2.2 Surface Processes

Once the growth flux arrives on the surface, surface processes decide the film formation (Figure 17.5). To get deposited on the surface, the incident atom must transfer all its kinetic energy to the substrate. The lateral or horizontal component of the velocity will keep the atom mobile on the surface. If the incident atom is not accommodated, it is reflected or bounced back. This is determined by the incident energy of the growth species, thin film, and substrate materials. This phenomenon, which takes place in a very short time, is called *thermal accommodation*. Once accommodated, the atom can move around on the surface owing to surface diffusion. Eventually, the atom gets adsorbed on the surface through either physical or chemical adsorption. In chemisorption, electron transfer takes place between a film and substrate atoms, whereas in physisorption, there is no electron transfer and a van der Waal type of bond exists. An adsorbed atom does not necessarily become part of the film; it can thermally desorb after some time. Relative rates of these surface processes are important in deciding the film structure.

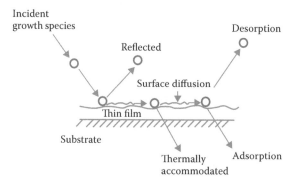

FIGURE 17.5
Schematic of surface processes on the substrate leading to film formation.

17.2.3 Thermal Evaporation

The two common PVD processes used in the semiconductor industry are thermal evaporation and sputtering. The difference between these processes is in the method of producing growth flux from the source material. In the case of thermal evaporation, the source material is heated to a temperature at which there is enough partial pressure generated to get the desired deposition rate (Equation 17.1). The source material can be heated by resistance, arc, induction, or e-beam. In resistance heating, the source is heated by a resistive filament; sometimes, the material to be evaporated is kept directly on the filament or in a crucible heated by the filament. In arc heating, a high current discharge is used for heating. In e-beam deposition, electron beams from a thermionic gun or a plasma gun are used to heat the material.

There are several issues with thermal evaporation—temperature control, uniformity and surface coverage that vary with distance and angle from the source, and contamination from the crucible or other components in the system. Temperature control can be attained by using a thermocouple and more accurate feedback loops to the filament. Uniformity can be improved by rotating the wafers using a large evaporation source and appropriate design of the deposition chamber.

17.2.4 Sputtering

In the PVD process, the source material is physically sputtered from the target by energetic ions. The bombarding ions are created by glow discharge (diode, triode, and magnetron) or an ion beam. Plasma or glow discharge is a mixture of ionized gas, electrons, ions, and neutral atoms, created by applying voltage to the working gas present between the two electrodes. Electrons are the dominant charge carriers. Figure 17.6 shows a schematic of this process. Plasma is quasineutral as perturbation from neutrality occurs at the edge. Accelerating the electrons between the electrodes will generate energetic and metastable species from the gas, for example, excited states of atoms or ions, active chemical species from molecules, and photons. Positive ions are accelerated toward the cathode, and they sputter the target material. Sputtering yield is defined as the number of atoms ejected from the target surface per incident ion. From kinematics, the yield is related to the momentum transfer from the energetic particles to the target surface atom. The threshold energy for sputtering is approximately equal to the heat of sublimation of the target material.

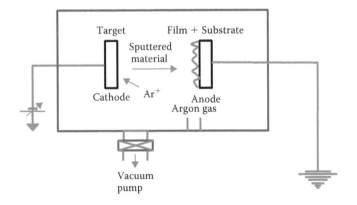

FIGURE 17.6
Schematic of a sputtering system for film deposition.

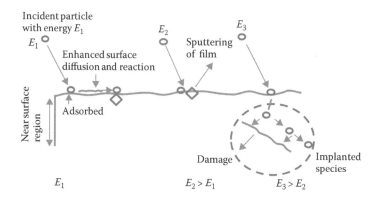

FIGURE 17.7
Schematic of additional surface processes due to the presence of high energy growth flux in sputtering.

The extra energy provided by the accelerating ions to the sputtered atoms results in a growth flux with higher energy. The additional surface processes that occur during the sputter deposition of a film owing to the presence of energetic flux are shown in Figure 17.7. As we increase the energy of the incident flux, a higher fraction of atoms is deposited as a film. If this energy becomes too high, energetic growth flux starts sputtering the substrate instead of causing deposition (This etching capability is used in dry etching as discussed in Section 17.5). Implantation of the incident atom below the surface is also possible, and this leads to altered subsurface layers and enhanced diffusion. For the correct range of energetic bombardment, it is observed that the film density is improved in sputtering as opposed to that of films deposited by thermal evaporation.

It would seem from Figure 17.6 that only conducting materials can be used as target, but that is not true. Insulating targets with radiofrequency (rf) power supply can be used for creating the plasma. Another variation of sputtering is magnetron sputtering, where the plasma can be sustained at lower pressures and electric field with the help of a magnetic field.

17.2.5 Chemical Vapor Deposition

In CVD, the source material is in the gaseous form and is brought in contact with the surface, where it adsorbs and reacts to form the film. The by-products of this reaction are desorbed and exhausted from the chamber. The deposition system is simple compared to the PVD techniques, as shown in Figure 17.8a. The deposition rate depends on the

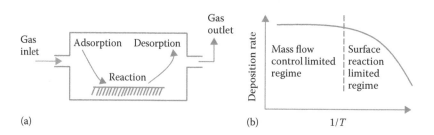

FIGURE 17.8
Schematic of (a) a CVD system and (b) different regimes for CVD.

availability of the precursor gas at the reactant surface and on the reaction rate. At high temperatures, the reaction rate is fast and film deposition is in a flow-controlled regime. In this region, temperature dependence is small. At low temperatures, deposition is controlled by the reaction rate and exhibits a strong dependence on temperature as shown in Figure 17.8b.

There are many variations of this process; for example, depending on the pressure at which CVD is carried out, we have atmospheric pressure chemical vapor deposition (APCVD) and low pressure chemical vapor deposition (LPCVD). LPCVD provides better uniformity in a batch type reactor where many wafers are stacked and processed together. The low pressure allows uniform distribution of the reactants throughout the chamber. We also have plasma-enhanced chemical vapor deposition (PECVD), which is often used to increase the reaction rate. This allows film deposition at lower substrate temperatures. Metalorganic chemical vapor deposition (MOCVD) uses metalorganic precursors during deposition that dissociate at lower temperatures to form the film. This allows film deposition to take place at a lower temperature.

17.2.6 Thermal Oxidation

Instead of depositing the material, we can also grow a film by oxidizing the substrate; for example, SiO_2 films are grown on Si by thermal oxidation. These films are free of contamination and have an excellent Si/SiO_2 interface because the surface is never exposed to the atmosphere. Furthermore, thermal oxidation only requires oxygen gas (dry oxidation) or water vapor (wet oxidation) and a high-temperature furnace. This process is preferred over electrochemically grown or CVD-deposited oxides to deposit SiO_2 as it exhibits superior dielectric property. Therefore, for the gate insulator in a metal oxide semiconductor transistor, we use thermal oxidation.

Figure 17.9 shows the processes involved in the oxidation of silicon. First, the oxidizing species adsorb at the surface, followed by diffusion through the oxide layer and reaction at the Si/SiO_2 interface, contributing to growth of oxide thickness. In the initial period of oxidation, when oxide thickness is negligible, the oxide growth is linear with time and controlled by the surface reaction rate. As the oxide thickness increases, diffusion of the oxidizing species through the oxide becomes the rate-controlling step and the growth rate is parabolic with time. The oxidation rate is dependent on the oxidizing species, orientation of the silicon, doping of silicon, flow rate of oxygen, and temperature.

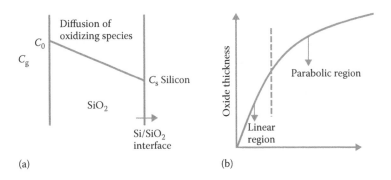

(a) (b)

FIGURE 17.9
(a) Schematic of the SiO_2 growth mechanism on silicon and (b) the thickness versus time curve.

17.2.7 Thin Film Growth Models

Depending on the technique in use and the processing parameters, it has been observed that thin films grow in three modes. These are referred to as *island-like growth*, *layer-by-layer growth*, and *mixed growth* models, as shown in Figure 17.10. Island growth takes place when the interaction between film atoms is greater than that between the film and substrate atoms. Layer-by-layer growth takes place when the interaction between the substrate and film atoms is stronger. A mixed model is one in which the first few monolayers are grown layer by layer, after which the influence of the substrate wanes, leading to island type growth.

17.2.8 Film Structure

Finally, irrespective of the technique used, processing parameters are optimized to get the desired film property and microstructure. One of the important properties is the film macrostructure; we need films that have smooth morphology and are free from voids and pinholes. They should adhere well and provide good step coverage; stress in the film should be controlled to maintain adhesion. The various structural defects that can affect thin films are shown schematically in Figure 17.11. Uniformity of film thickness is another important parameter. For better uniformity, growth flux must impinge on the substrate from many directions. A large area source and a high surface mobility promote this.

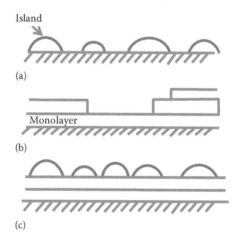

FIGURE 17.10
Schematic of film growth models.

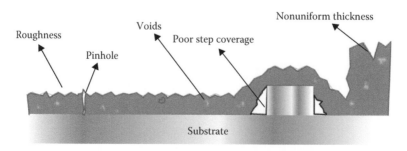

FIGURE 17.11
Schematic of various structural defects in thin films.

17.3 Dopant Incorporation

Silicon as a semiconductor by itself has limited utility. Its power in devices is harnessed when a p–n junction is created by locally modifying the conductivity using n- and p-type dopants. Typical n-type dopants are P and As, whereas p-type doping can be achieved with B. To appreciate the quantities in which the dopant atoms are present in Si, first consider that the number of Si atoms in a cubic centimeter is approximately 10^{22}. Typical doping levels, amounts by which the Si atoms in the lattice are replaced by the dopant atoms, are 10^{16}–10^{20}/cm^3; the amount at the lower end of the range is for creating junctions and that at the higher end is for realizing ohmic contacts. This doping in microelectronic fabrication is carried out mostly by *diffusion* or *ion implantation*, both of which are now described.

17.3.1 Diffusion

The diffusion method of doping combines several steps. The basic approach involves (1) exposing the Si surface to a dopant source, (2) carrying out a low-temperature and short-duration diffusion of dopant, incorporating it only in a thin surface layer of Si, followed by (3) removing the dopant source introduced in the first step, and finally, (4) driving-in or redistributing the dopant that was incorporated in the thin surface layer during the second step deeper into the Si by the diffusion process under appropriate temperature and time combinations [5].

The dopants are introduced at the surface using solid sources or gas sources. A common and practical solid source is a spin-on-glass, which, as the name suggests, is coated on the Si wafer by spin coating. The dopant is mixed in an organosilane, which is a liquid of viscosity suitable for spin coating. After coating a thin layer, heat treatment converts the organosilane to a SiO$_2$ matrix, within which lie the dopant atoms. For example, to dope with B, carborane may be used, which by heat treatment converts to B$_2$O$_3$ in a SiO$_2$ matrix. With this type of source, boron concentration at the Si surface is established by the reaction of this oxide at the Si surface according to the equation 2B$_2$O$_3$ + 3Si = 4B + 3SiO$_2$.

Alternative solid, liquid, or gas sources are possible. In most cases, though, the dopant is finally transported to the wafer in gaseous form. If the source were solid, it is heated to provide enough vapor pressure. Then the vapors of the dopant are transported with a carrier gas to the Si wafer. Similarly, for a liquid source, for example, trimethylborate for B doping, a carrier gas is bubbled through the liquid. This carrier gas saturated with the dopant then forms the source of the dopant. If a gas source of the dopant is available, it obviously can be transported in a similar manner. The problem with this approach in contrast to spin-on glass is that it is difficult to maintain uniformity in a long diffusion furnace in which hundreds of wafers may be stacked from one end to the other. Thus, a spin-on source may be more attractive.

The dopant brought in contact with the Si surface is then allowed to diffuse into the Si wafer. The governing diffusion equation for establishing the concentration of the dopant at any location and time is

$$D\frac{\partial^2 C}{\partial x^2} = \frac{\partial C}{\partial t}$$

(17.2)

In this, we have assumed that the diffusion coefficient of the dopant, D, is a constant and that its concentration, $C(x,t)$, varies with the position x, measured from the surface, into the depth of the silicon wafer, and time, t; that is, Si lies in $x > 0$. Step 2, the predeposition step, represents a case of diffusion in which the surface concentration of the dopant is fixed. In practice, the diffusion source is so high in dopant concentration that the solubility of the dopant species in the silicon, C_s, determines the surface concentration. Hence, for the boundary condition, $C(0,t) = C_s$ and $C(\infty,t) = C_0$ and initial condition $C(x > 0,0) = C_0$, the solution to the diffusion equation is

$$\frac{C(x,t) - C_0}{C_s - C_0} = erfc\,\frac{x}{2\sqrt{Dt}}$$

(17.3)

This C_0 is the background concentration of dopant already present in the wafer, if any; otherwise, it will be zero. Figure 17.12 shows the concentration profile for increasing timescales.

When the timescale is kept short, as in the predeposition step, for all practical purposes, the dopant is confined to only a few monolayers at the surface. And in this surface layer, the total quantity of the dopant incorporated is

$$Q_0 = \int_0^\infty c(x,t)\,\mathrm{d}x = 2C_s\left(\frac{Dt}{\pi}\right)^{1/2}$$

(17.4)

When the dopant source is removed in step 3, for example, by etching off the spin-on-glass, the amount of dopant, Q_0, remains within a small depth at the surface. In step 4, this retained dopant is redistributed by another drive-in diffusion. In this case, the concentration of the dopant in the Si changes spatially by diffusion, but in such a manner that the total amount of the dopant remains Q_0; hence, the terminology used is *redistribution*.

The solution of the diffusion equation for the boundary condition changed to $\frac{\partial c(0,t)}{\partial x} = 0$ (impermeable barrier at $x = 0$) is Gaussian, as given below:

$$C(x,t') = \frac{Q_0}{\sqrt{\pi D't'}}\,e^{\frac{-x^2}{4D't'}}$$

(17.5)

in which D' is the diffusivity at the temperature of redistribution for time t'.

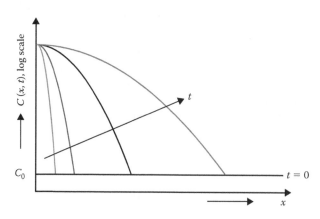

FIGURE 17.12
Concentration profile after the dopant incorporation step.

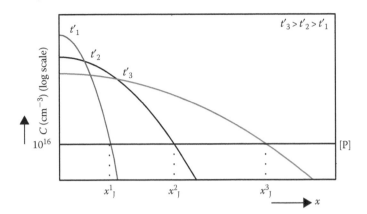

FIGURE 17.13
Concentration profile after the drive-in step.

If we started with an n-type Si wafer uniformly containing $10^{16}/cm^3$ of phosphorus ([P]) dopant, a p–n junction can be obtained by diffusing boron ([B]). The junction depth, x_j, is located where the [B] also becomes $10^{16}/cm^3$. In Figure 17.13, the Gaussian profile of the diffused dopant B is shown at various times. For all timescales, the total amount of dopant remains Q_0, but the longer the diffusion time, greater is the junction depth.

This method of incorporating dopants in a semiconductor is a powerful one. Essentially, a batch type process that lasts for up to 10 h can handle a few hundred wafers at a time. This process was the workhorse for doping until the 1980s and continues to be used in the fabrication of power devices. However, in modern integrated circuits, the junction depths and distances separating two p–n junctions may be as low as 10–100 nm. Such small dimensions are difficult to achieve by the diffusion process. Therefore, an alternative in the form of doping by ion implantation is practiced in the fabrication of highly complex integrated circuits.

17.3.2 Ion Implantation

In ion implantation, the dopant is introduced by exposing a wafer to a flux of ions. The technology for creating the ion beam is borrowed from accelerator physics, which involves using an appropriate gas source containing the dopant atoms, such as boron fluoride for boron. This gas source is first ionized by, for example, energetic electrons coming off a hot filament. These ions are then accelerated to high energies, ten to a few hundred kiloelectron volts, mostly in the form of kinetic energy. These accelerated ions are a mixture of many types. Thus, they are subjected to an analyzing magnet that selects only a beam of specific ions, such as B^+ for p-type doping or As^+ for n-type doping. We describe the ion implantation process in semiconductors assuming that such an ion beam of flux F_i (ions per unit area per unit time), or the corresponding ion beam current I_b, of known energy between ten and a few hundred kiloelectron volts is available.

As shown in Figure 17.14, an ion beam is rastered on the surface of the wafer to cover the entire area. Essentially, it means that only one wafer can be processed at a time. However, in a modern implanter, with automatic loading and unloading, the wafer throughput can be speeded up. In any case, demand for high process capability in integrated circuits that can only be met by ion implantation at an acceptable production rate has made ion implantation the workhorse of doping technology in integrated circuit fabrication.

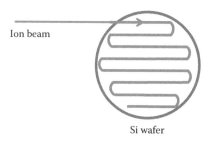

Ion beam

Si wafer

FIGURE 17.14
Rastering of the wafer by an ion beam to achieve uniform doping across the wafer.

The energetic ions that impinge on the wafer surface come to rest after approximately 10–100 nm depth. This penetration depth is referred to as the *range* (R) of the ions. The range of a given ion in a specific semiconductor primarily depends on the energy of the ions. Thus, by changing the energy, it is possible to implant the dopants at a prescribed depth, unlike in the diffusion method, which always forces the highest concentration of dopant at the surface. This is the first advantage of the ion implantation process over the diffusion process.

Further, ion flux can be controlled independently so that the amount of dopant introduced is adjustable. If the implant time for ions of charge state Z is t, the dose of the dopant (number of dopant atoms per unit area of the wafer) introduced is $Q_0 = {}^{I_b \times t}/_{Zq \times A}$, where q is the electronic charge and A is the implanted area. This dose, available in the range 10^{10}–10^{17} ions/cm^2, is distributed in the wafer such that the peak concentration of the dopant in the wafer can be in the range 10^{14}–10^{20}/cm^3. It may also be noted that the control in achieving this concentration is extremely good, within a few percentage.

The second major advantage over the diffusion method of doping is that these two independent "handles" control the depth at which the dopant is implanted and its amount extremely well.

The schematic in Figure 17.15 is useful to understand this further. In a vacuum chamber, the Si wafer is placed on a holder, which itself is externally grounded. The ion beam introduced into this vacuum chamber is made to fall on the wafer. Since the wafer is grounded, the compensating charge (electrons) is provided by the ground; that is, for example, for every ion of B$^+$ implanted, the ground supplies one electron. Thus, the current between the wafer and the ground is monitored, rather than the ion beam current, which tells us the dose of ions the wafer has received. If $I(t)$ is the instantaneous current ($Z = 1$ for B$^+$ or As$^+$),

$$\text{Number of dopant atoms implanted} = \frac{\int_0^t I(t)dt}{q}$$

(17.6)

and the dose in a wafer of area A is

$$Q_0 = \frac{\int_0^t I(t)dt}{qA}$$

(17.7)

Since monitoring the ion flux would be difficult, this current integrator provides a convenient method for real-time control of the ion dose, or equivalently, the number of dopant atoms implanted per unit area of the wafer.

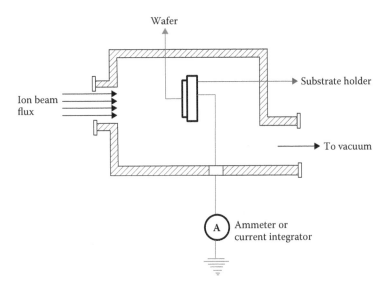

FIGURE 17.15
Schematic of an ion implanter.

The ions that impinge on the wafer are scattered by the lattice. The primary mechanisms are nuclear stopping and electronic stopping. The former is due to the collision of ions with the target (wafer) atoms and the latter is due to the transfer of energy to the electrons by coulombic interaction. In the process, excitation or even electron ejection (ionization) is possible. By either mechanism, as the energetic ions entering the substrate come to rest, the distance an ion has traveled is called the *range*. Being a probabilistic event, some ions undergo more collisions than others and also recoil differently in directions perpendicular to the direction of the beam. Hence, there is a distribution in positions where these ions stop. For example, Figure 17.16 shows several ions that have come to rest at different locations in the wafer. Rather than the range, we are more interested in the projection of the ions in the $-x$ direction, the depth into the wafer. Corresponding to the way these ions stop, the concentration of dopant (cm^{-3}) in the $-x$ direction is also shown. The peak of dopant atoms lies at the projected range, R_P, and the standard deviation in the dopant concentration is ΔR_p. Unfortunately, the same randomness in process leads to a lateral distribution of the dopant perpendicular to the implant direction, that is, there is also a lateral straggle of the dopant. That distribution has a standard deviation of ΔR_d.

Fundamentally, the depth to which the ions are implanted, R_P, is dependent on the energy of the ions; it is greater for larger implant energies. Thus, in Figure 17.17, the dopant concentration profile marked "A" can be shifted deeper into the wafer to the position marked "B" by increasing the ion energy. The shape, especially in the tail region, may also change. On the other hand, by changing the dose, by either changing the ion beam flux or the time for implantation, the concentration profile "A" can be shifted up to higher concentrations (marked "C"). In all cases,

$$\int_0^\infty C(x)\,\mathrm{d}x = Q_0$$

$$(17.8)$$

Ion implantation has additional advantages over the diffusion method of doping. Because diffusion is mostly an isotropic process, there is as much lateral diffusion as there

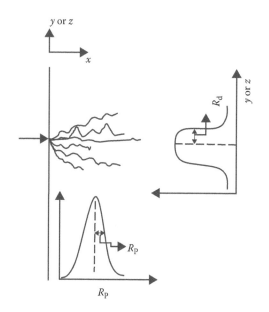

FIGURE 17.16
Projected range and lateral straggle of ions during implantation.

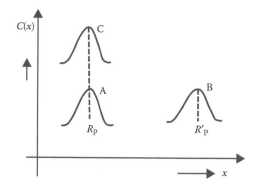

FIGURE 17.17
Effect of ion beam energy and dose on the implantation profile.

is diffusion in the depth. In ion implantation, the lateral spread of the dopant is much less. Apart from that, diffusion is also a high-temperature process, whereas ion implantation is performed at room temperature. An important consequence of high temperatures in the diffusion method is that the mask that protects the regions on the wafer where dopants are not desired will have to be protected by a hard layer such as that of silicon nitride or oxide. However, because of low temperatures in ion implantation, a 2- to 5-μm thick PR is enough to stop the ions.

However, ion implantation does pose several problems. The matter of throughput has already been discussed. However, most important is the safety issue. Radiation hazard, high voltages, and toxic gases require elaborate safety arrangements. Also, the equipment takes up a large area and is expensive.

Apart from equipment-related issues, there is an important process- and materials-related issue. The implantation process leads to damage in the target crystal, the Si substrate.

This damage could be the simple creation of point defects or dislocation loops and could even cause amorphization. Thus, it is necessary to anneal the material to repair the damage, which brings back the high-temperature process. However, since the objective is to only anneal the damage, it is possible to carry this out within a short duration. This annealing of Si at elevated temperatures may not cause much of a problem, but in materials such as GaAs, where preferential loss of As may occur, additional provisions to cap the wafer have to be made. Further, the dopants that have been introduced tend to diffuse at annealing temperatures, and this diffusion is often accelerated by the presence of excess defects, leading to what is known as *transient-enhanced diffusion.*

17.4 Photolithography/Patterning

A semiconductor device on a Si wafer is fabricated by adding many layers in specific patterns [6]. These layers may be insulators, such as SiO_2 or Si_3N_4, or metals of many kinds. Unfortunately, although we desire a patterned addition of these layers, most processes are designed to add layers on the entire area of the wafer. Thus, after adding a layer on a wafer, portions of these deposited layers must be removed to achieve the desired pattern.

Two strategies are possible. The first and the more common one involves addition of a layer on the whole wafer followed by the development of a protective pattern of PR on the wafer and then the etching off of the underlying layer by a chemical that attacks it where the protective PR is not present. Ultimately, the PR is also removed, with the net result that a patterned layer has been added on to the wafer. Since the role of the PR is to protect the underlying layer from etching, it should be less susceptible to chemical attack than the layer that is being patterned. Often, this requirement is met by the PR. But, when the etchant is likely to attack the PR also, then alternatives are possible; those specific cases are not discussed here, but in general, this requires depositing an additional hard layer that is first patterned by the PR, which then plays the same role as the PR, when the underlying original layer is being patterned. In both these cases, however, first a layer is deposited and then patterned, in contrast to the second strategy described below.

In some cases, a process called *lift-off* is also practiced in which the order of layer deposition and PR patterning is reversed; that is, we develop instead a pattern of PR on the wafer first and then deposit the layer on the entire surface of the wafer. The strategy is to ensure discontinuities in the deposited layer at the edges of the preexisting PR pattern. The idea is that when the PR is dissolved in an appropriate chemical, the discontinuities allow the chemical to access the PR; this chemical attacks the PR until all of it is removed. What is then left on the wafer is a patterned layer of the material deposited.

In both these strategies, photolithography plays an important role. The term *photo* implies that a light source is used in the process. But, when the pattern dimension becomes smaller, below even the wavelength of UV light, then the light source is replaced by an e-beam or x-rays. In those cases, the terminology becomes *e-beam lithography* or *x-ray lithography*. Both these methods have their own fabrication peculiarities, but fundamentally they are similar to photolithography; therefore, we discuss only photolithography here.

At the heart of photolithography is the PR material, which is sensitive to light. A negative PR hardens and a positive PR softens when exposed to light of specific wavelengths. In both tones, the PR essentially contains three components, a matrix material as a binder, a photoactive compound for photosensitivity, and a solvent for good film-forming properties.

The main ingredient is, of course, the photoactive compound, but the matrix acts as a binder that governs the mechanical properties of the film and is responsible for resistance to etching by chemicals employed in patterning the underlying layers. In contrast, the solvent in the resist does not remain in the PR film, rather it maintains the PR in liquid state during its shelf life and makes it amenable to spin coating on the wafer; it is only when the solvent is driven off after spin coating that the PR takes the form of a mechanically stable film on the wafer.

The main photoactive material is termed *positive* or *negative*. The reason for this terminology will become apparent once their application is described. At this stage, let us understand that once a film of positive PR is formed on a wafer and the wafer is subjected to a chemical called *developer*, the PR will not dissolve. But, the photoactive compound in a positive PR is such that when it is exposed to UV light, the PR turns soluble in the same developer. An example of such a substance is diazonaphthaquinone, although for most commercially available PRs the chemicals used are not disclosed. The matrix material by itself may have been soluble in the developer, but addition of the photoactive substance makes it insoluble until it is exposed to UV light. In contrast, the photoactive agent and the matrix material in a negative PR are normally soluble in a developer. But, when it is exposed to UV light, the cross-linking reaction in the photoactive substance, such as bis-arylazide, leads to a network that makes the PR film insoluble in a developer.

The use of the photolithography process is described with the example of a desired pattern of SiO_2 on a Si wafer, as in Figure 17.18h. The schematic in Figure 17.18 illustrates the process step by step. (a) A layer of SiO_2 is deposited on Si wafer by one of the several available methods, thermal oxidation, sputtering, or CVD. This is the layer to be patterned. (b) A PR film, negative or positive, approximately 2–3 μm thick, is spin-coated on the wafer, that is, a measured quantity of PR is dispensed on the wafer and the wafer is spun fast so that the PR spreads out before the solvent dries off. In order to achieve good adhesion to the wafer, it may be necessary to first bake the wafer to remove the adsorbed moisture and/or to give a surface treatment. Negative PRs usually provide good adhesion, but as feature sizes have become smaller, positive PRs have become more important where much more care is necessary for good adhesion. (c) During spin coating, most of the solvent is already removed. But, in a soft bake process, by heating at a low temperature around 100°C, all of the solvent is driven off. (d) The wafer is then exposed to UV light through a mask, parts of which are meant to block the UV light. Note the opaque and transparent regions of the mask, chosen according to the resist used. The mask could be a glass plate on which the opaque region, for example, could be due to Cr coating. The negative resist, which is normally soluble in a developer, when exposed to UV light has turned insoluble, whereas the positive resist, normally insoluble, has turned soluble wherever it is exposed to UV light. (e) The PR is subjected to a developer that removes its soluble portion. Before this step, sometimes another soft bake may be required. The solvents used in the developer are important. Environmentally harmful chemicals such as toluene and xylene were used at one time for negative PRs. There has been a gradual shift toward positive PRs that can be developed by mild inorganic aqueous solutions. However, nowadays, PRs are available in both tones, which can be developed by the same solvent. (f) Hard baking of the PR is carried out to provide rigidity before etching. (g) SiO_2 is etched by a dilute HF aqueous solution, which is chemically insensitive to the underlying Si. The etchant attacks both SiO_2 and the PR. The question is of selectivity. When the etch rate of the PR is much less than that of SiO_2, the etchant is highly selective and the process will be successful. (h) The PR is stripped by using chemicals known as *strippers* or by plasma ashing. The result is that both with positive and negative PRs, the same pattern is achieved. The positive PR

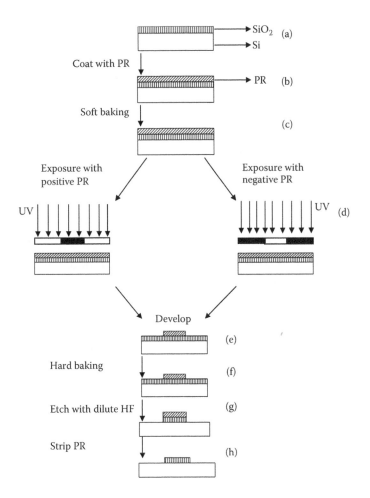

FIGURE 17.18
Schematic of photolithography steps.

is so named because it generates a positive image of the mask. The same logic explains the terminology for negative PR.

17.5 Etching

Etching is the process by which a part is machined. This is required to remove various layers from selected areas that are exposed using photolithography as described in the previous section. Sometimes, there is also a need to remove a layer partially or completely for better interfaces. We can use either wet etching or dry etching. In wet etching, the etchant used is one in which the material to be etched can be dissolved, but the masking layer (PR/oxide layer) is not dissolved. If the part to be etched is amorphous or polycrystalline, then wet etching is isotropic. If it is single crystalline, the differential etch rates of different planes can lead to profiles that depend on the orientation of the silicon wafer.

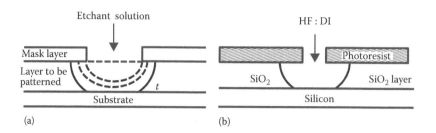

FIGURE 17.19

(a) Isotropic wet-etching profile and (b) selective etching of SiO_2 using HF-based etchant solution.

For more details on this part, which is more relevant to MEMS, the reader is referred to Reference 2. Figure 17.19 shows the etch profile development with time in an amorphous layer during wet etching. Owing to isotropic etching, there is a possibility of the formation of a ledge, which can be troublesome for later processing. Also, the width of the etched trench changes with depth (Figure 17.19b), which may not be desirable.

Process parameters that can change the kinetics of dissolution are the etchant solution and its concentration; temperature, stirring, or aeration of the bath; and etching time. These are to be carefully monitored in order to get the desired profile in the fabricated part. Etching also has an effect on the roughness of the surface, which needs to be minimized; otherwise, it can deteriorate the adhesion with the next layer that is deposited. High etch rates lead to rough surfaces. Therefore, to improve surface finish, lower concentrations of the etchant and lower temperatures are used. If any material needs to be used in the microelectronics industry, it is required to have a well-controlled etching process. For a very long time, this was a problem with using copper metal in the microelectronics industry.

In silicon technology, selectivity of HF : DI solution to etch SiO_2 is used to open windows in the SiO_2 mask. SiO_2 has a very high etch rate in HF : DI solution compared to silicon and this becomes a self-terminating machining process (Figure 17.19). In the semiconductor industry, we can wet etch Si_3N_4, Si, and metals. More details about various etching chemistries can be found in Reference 4.

The problem of dimensional accuracy exists in wet etching. The dimension of the opening at the silicon surface is not the same as the opening of the windows. The extent of underetching and overetching can also decide the final opening on the silicon surface. Therefore, design rules are developed to correct for these parameters when designing the photolithography mask.

In addition, because of the isotropic nature of wet etching, the machined walls are not vertical. Dry etching eliminates this problem since it has a highly anisotropic etch rate, as shown in Figure 17.20. Similar to mechanical drilling, this process also uses energetic directional ions, which remove the material to create a hole. This process involves the physical removal of material. These ions can be from the ion beam of an inert gas such as argon. The process is also known as *argon ion milling*. Sometimes, the plasma of a reactant gas is used, which additionally removes the material by reacting with it; the by-products are subsequently removed. This process is also called *reactive ion etching (RIE)*. Generally, fluoro- and chlorocarbon gases are used in dry etching; however, there are other options available, depending on the layers to be etched.

Dry-etching processes involving physical sputtering are not highly selective of different layers; hence, end detection is necessary to decide when to stop etching. This is often done by monitoring the reflectance of the surface or the chemical composition of the gases in the

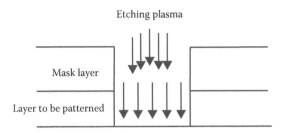

FIGURE 17.20
Vertical etch profile in dry etching.

chambers, which changes as the process proceeds from one layer to the next. In addition, the mask layer needs to have enough thickness to survive the full duration of the etching process.

The advantages of dry etching are in providing vertical walls in the etched holes and better dimensional accuracy. Moreover, as the dimensions of the holes are reduced, the wet etchant may not reach the bottom of the hole, but gas atoms, ions, or molecules having dimensions on the order of atomic scale are likely to reach there. Finally, dry-etching processes reduce the wet chemical waste from the microelectronics industries.

The disadvantage of dry etching pertains to the damage caused by energetic ions to the layer, which, at times, can change the device characteristics by introducing defects in the material.

17.6 Cleaning Process

This is a process necessary between the processes discussed in the previous sections. Although it does not change the shape or properties of different parts, it is very important to ensure low contamination levels of the various interfaces. Contamination can come from the process itself (e.g., purity of different etchants used during etching) and from the storage and handling equipment. Typically, the cleaning process is done using wet chemicals. Depending on the surface to be cleaned, the chemistry of the cleaning solution is decided. In addition to chemical contamination, particle contamination is also very important. Since the part to be fabricated is in the microscale range, particles of similar size will lead to the production of a defective device and low yield of the production line (Figure 17.21). Therefore, cleaning is very important to the overall process yield.

The cleaning process to be used depends on the materials on the surface. For example, we can use a HF-containing process for cleaning a silicon surface, but not a silicon oxide surface as HF etches the oxide surface. A typical cleaning process that is used in silicon processing is RCA cleaning (named after Radio Corporation of America). The following steps are used in the cleaning process; the concentration and temperature are varied by the user, depending on the requirement.

- **Piranha etch: H_2SO_4, H_2O_2, and DI water**

  In this step, the organic contamination on the surface is oxidized.

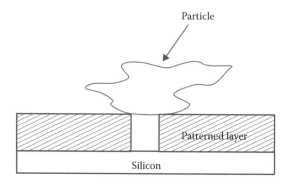

FIGURE 17.21
Blocking of the patterned window by a particle of similar dimension.

- **RCA-1: NH$_4$OH, H$_2$O$_2$, and DI water**

  In this step, the oxidized layer is removed and the particles are also removed from the surface.

- **RCA-2: HCL, H$_2$O$_2$, and DI water**

  In this step, metallic impurities are removed from the surface.

- **Dilute HF etch**

  Often, a thin silicon oxide layer is deposited on the surface of the wafer, which can be removed by this process.

All cleaning processes are terminated with a final DI water rinse followed by N$_2$ gas drying. This is a seemingly simple process, but it needs to be controlled as water marks can lead to a defective device.

In order to reduce the chemical and DI water consumption, industries are also developing dry-cleaning processes where the vapors of the cleaning agents are brought in contact with the surface.

17.7 Future of Semiconductor Technology

The development in this industry is well organized. The semiconductor industry associations of five different countries—the United States, Taiwan, Japan, Korea, Europe—sponsor an organization called *International Technology Road Map for Semiconductors* (*ITRS*). The chip makers, equipment makers, consortia, and universities of these countries are the members of this organization. It gives the future requirement of the semiconductor industries for next 15 years, and these tables are updated every year. The road map is divided among different technology groups. This exercise has been instrumental in realizing the predictions of Moore's law and has provided directions for research and development worldwide.

References

1. Grove, A.S. 1967. *Physics and Technology of Semiconductor Devices*. Singapore: John Wiley and Sons.
2. Sze, S.M. 1994. *Semiconductor Sensors*. New York: John Wiley & Sons, Inc.
3. S. Wolf and Tauber, R.N. 1986. *Silicon Processing for the VLSI Era*. Vol. 1, *Process Technology*. California, USA: Lattice Press.
4. Gandhi, S.K. 1994. *VLSI Fabrication Principles*. New York: John Wiley and Sons, Inc.
5. Mayer, J.W. and Lau, S.S. 1990. *Electronic Materials Science: For Integrated Circuits in Si and GaAs*. New York: Macmillan publishing company.
6. Wolf, S. 1990. *Silicon Processing for the VLSI Era*. Vol. 2, *Process Integration*. California: Lattice Press.

18

An Integrated Wafer Surface Evolution Model for Chemical Mechanical Planarization (CMP)

Abhijit Chandra, Ashraf F. Bastawros, Xiaoping Wang, Pavan Karra, and Micayla Haugen

Iowa State University

CONTENTS

18.1 Introduction

Chemical mechanical planarization (CMP) is recognized as a promising planarization method in integrated circuit (IC) manufacturing. A detailed description of the CMP process can be found in several references (e.g., Steigerwald et al., 1997; Luo and Dornfeld, 2004; Oliver, 2004). CMP is a necessary process step for <250-nm line width oxide layers and is a critical operation in the Cu dual damascene (CuDD) technology because the yields are extremely sensitive to the CMP performance. By removing all previous metal-layer deposits, CMP represents a kind of reverse processing. The product quality resulting from a Cu-CMP process depends largely on the spatial (local and global) uniformity of the material removal rate (MRR) and the associated defect density, while the process efficiency is largely dependent on the magnitude of the mean MRR.

CMP is performed by sliding a wafer surface on a relatively soft polymeric porous pad that is flooded with a chemically active slurry containing abrasive particles of submicron diameter. The chemical properties of the slurry interact with the mechanical properties of the abrasive particles, and the polishing pad and wafer at the nano- and macroscales respectively, and with the surface morphologies at the microscale. As a result, the wafer surface, pad surface, and slurry characteristics *coevolve*. This evolution controls the quality and effectiveness of the CMP process. The finished wafer surface, however, is prone to CMP-process-induced scratches. The existence of a single scratch whose depth is greater than a critical threshold can render the chip unusable, and this threshold continues to decrease with successive generations of IC manufacturing.

A wide range of studies on the CMP process have been reported. For example, previous work investigated MRR (Komanduri, 1996; Evans et al., 2003) and the effects of the pad and slurry properties (Bastawros et al., 2002) on the process. Wang et al. (2005a,b) introduced the effects of pad wear and its evolution in an effort to extend the pad response model developed by Bastawros et al. (2002) and Luo and Dornfeld (2004) to assess the propensity to scratching. It is also well known that the slurry gradually evolves with time, with and even without continued processing, and that there is a strong correlation between slurry evolution and the generation of scratches on the finished wafers. In the fields of physics and colloidal chemistry, a variety of modeling efforts as well as experimental investigations (Lin et al., 1989, 1990) of slurry agglomeration have been reported. There also exists a wide body of literature in which the interactions between mechanical and chemical evolution of slurry properties have been investigated (Komulski, 2001; Che et al., 2005).

In traditional CMP, a wafer is held upside down by a rotating wafer carrier and is brought into contact with a polishing pad mounted on a rotating platen. During the CMP process, a slurry flows into the interface between the wafer and the pad. A combination of the mechanical down force applied by the wafer carrier as well as the chemical action of the slurry is used to polish the wafer surface. The workpiece material can be a metal or a dielectric. CMP is very effective at reducing local step height, which is the difference in thickness of the current layer across feature scale length.

Between 1999 and 2008, there has been a tremendous increase in the volume of usage of CMP (percentage of wafers requiring CMP increased from <15% to >70%) and the number of CMP steps per wafer (increasing from <10 to >23). During the same period, however, process complexity in terms of the number of CMP step applications and the uniqueness of each application also increased by orders of magnitude (Rhoades and Murphy, 2006; ITRS, 2007, 2009).

However, *CMP is expensive both in terms of capital cost* (a typical CMP machine costs several million dollars) and *cost of operation (CoO) or cost of consumables such as pad and slurry.* Recently, *with shrinking feature size (currently <15 nm), CMP also has been mired by defectivity (e.g., scratch and film delamination) concerns* (e.g., ITRS, 2007, 2009) *at a multiplicity of length scales* ranging from within wafer nonuniformity (WIWNU) at wafer scale and within die nonuniformity (WIDNU) at die scale to defectivity concerns at feature or nanoscale. According to this roadmap, defectivity tolerance levels are expected to drop further because of the continued reduction in feature sizes. Reliability of the CMP process remains a primary goal, and currently, the CMP industry is attempting to reduce defectivity levels from all possible angles. As recent news articles in *Semiconductor International* point out, efforts are underway to modify all consumables, such as pad and slurry, and the *possibilities to modify the CMP platform or machine* are also being investigated.

Under the current state of the art, however, a global (die-scale) nonplanarity may be generated because of the differences in feature scale parameters (e.g., pattern density, line width) at different locations. The global thickness variation has a significant impact on subsequent process steps such as lithography and etching. Global thickness variation also impacts circuit performance: long-range clock wires passing through regions of different thicknesses result in different capacitances and may result in clock skew (Stine et al., 1997). Thus, the development of an understanding of the effects of feature-scale parameters on global planarization rates plays an important role in the optimization of the CMP process and layout density design.

Cale et al. (2002) developed a model for CMP. In this model, they considered the contact model of Greenwood and Williamson to develop an expression for the contact pressure between the wafer and the pad that occurs at microscale since the pad asperities are on the order of microns. Using this expression, an equation for the hydrodynamic layer is derived, which is the same as the distance d used in this chapter. Feature-scale and wafer-scale models are also discussed in this work. Another hydrodynamic model for polymer asperities has been proposed by Kim et al. (2003).

Seok et al. (2003) developed a model that involves wafer-scale effects such as wafer flexibility and bulk pad deformation, along with particle-scale effects such as material removal by particles. Slurry layer thickness is also calculated using the contact stresses and hydrodynamic fluid pressures. Material removal is developed as a function of radius across the wafer.

Warnock (1991) proposed a phenomenological model that allows the quantitative prediction of wafer topography. In this model, the effects of the surrounding features on the removal rate at a given point were considered. Higher surrounding features decrease the removal rate, while lower surrounding features enhance it. Three adjustable parameters were used to describe the pad deformation, the pad roughness, and the relative velocity. The model predicted the experimental data very well. However, these parameters have no direct physical meanings.

A contact wear model by Chekina et al. (1998) was used to predict wafer topography evolution at feature scale. The pressure distribution along the wafer surface was studied using contact mechanics, taking into account the elastic deformation of the polishing pad. Vlassak (2004) developed another contact-mechanics-based model, which takes into account both the roughness and elastic deformation of the polishing pad in computing the pressure distribution. Roughness of the pad was considered by the contact model of Greenwood and Williamson (1966). This model is able to evaluate the evolution of the wafer profile during the CMP process. It depends on an iterative solution procedure, in which two equations describing the pressure distribution between the polishing pad and wafer and the pad deformation need to be iteratively solved. This procedure requires a significant amount of computation to reach convergence.

Ouma et al. (2002) developed a model to predict the wafer surface profile across a die. The concept of "effective pattern density (EPD)" was introduced in this model. The removal rate was assumed to depend on the EPD and the blanket wafer polish rate (BWPR). The BWPR was determined experimentally. The calculation of EPD was based on pad deformation characteristics. In this model, it was assumed that the down area was not polished until the up area was worn out completely.

Fu and Chandra (2003) presented an analytical dishing and step height reduction model for CMP. They assumed that (1) the pad contacts with the wafer at any point of the interface, (2) the higher area releases the pressure on the adjacent lower area, and (3) the force redistribution due to pad bending is proportional to the dishing/step height. This model gives an analytical prediction for step height reduction and dishing. The bending factor α that the authors introduced in their model to describe the force redistribution due to pad bending is briefly described and used subsequently in this chapter with some modifications.

Requirements of such nonuniformity avoidance entail lower defectivity and improved predictability across the board for CMP applications. Thus, *it has become critical to focus on improving the reliability of the CMP* process. This is compounded by the ITRS (2007, 2009) *mandate of lowering cost of ownership (CoO)*. To overcome this hurdle, a simulation software (www.cmpsim.com) for rapid process planning activities, *CMPSim*, is necessary for adhering to the ITRS (2007, 2009) road map.

CMP plays a central role in micromanufacturing. The development of appropriate micromanufacturing and nanomanufacturing processes is critical to realizing ICs with multilevel metallization designs. For commercial success, however, this development needs to occur in a timely manner. Faster development translates to faster realization and faster time to revenue. Over the past 35 years, through significant improvements in design and manufacturing, the IC industry has been able to adhere to Moore's law of doubling transistor density every 18 months. This law has held through significant variations in business cycles as well as market and economic conditions.

Three manufacturing processes of key importance to IC fabrication are (1) deposition (e.g., physical vapor deposition (PVD), chemical vapor deposition (CVD)), (2) planarization (CMP), and (3) patterning via etch (e.g., photolithography with deep ultraviolet steppers). Photolithography enables shrinks in feature sizes (horizontal integration), while deposition and CMP enable more complex stacks (vertical integration (ITRS, 2007)). The inability of photolithography equipment to etch (print) on nonplanar surfaces necessitates a planarization process—a need that is filled by CMP because of its ability to provide surfaces with both local and global planarity over an entire stepper field.

However, adhering to Moore's law has become increasingly more difficult in recent years as gate widths have shrunk below 100 nm and are projected to reach 35 nm. Owing to higher performance thresholds (e.g., for defect propensity), process development cycles have stretched. This stretching of process development cycles is significant in planarization technologies such as CMP. During the period 1999–2004, the volume of usage (percentage of wafers requiring CMP increased from <15% to >50%) and the number of CMP polishes per wafer (increasing from <10 to >23) have increased. During the same period, however, process complexity in terms of the number of CMP polish applications and uniqueness of each application has also increased by orders of magnitude (Rhoades and Murphy, 2006; ITRS, 2007). In particular, these requirements entail lower defectivity and improved predictability across the board for various CMP applications. Thus, it has become critical to focus on *a transformative* rapid CMP process development scheme and to attempt to shrink the cycle time as the IC industry moves forward. A typical CMP process development sequence is shown in Figure 18.1. *Steps highlighted in dark gray can be replaced by a predictive modeling capability.*

In this chapter, some salient features of the integrated model embodied in the *CMPSim* software are presented to describe the evolution of the wafer surface during a CMP process. The model assumes direct contact between a patterned wafer and a rough polishing pad. It ignores the effects of abrasive particles at the feature scale. However, these effects may be introduced at the particle (nano) scale (Wang et al., 2005a,b). The pad asperities are assumed to have random height distribution, spherical tips, and periodical spacing across the pad surface. A brief description of pad topography characterization is provided in Section 18.2.1. In Section 18.2.2, a solid–solid contact model based on the work of Greenwood and Williamson (1966) is utilized to determine the mean pressure at any

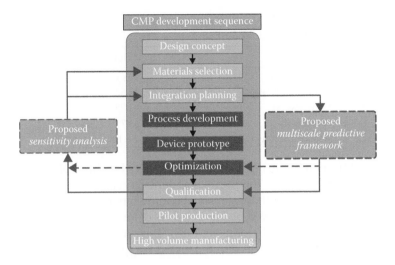

FIGURE 18.1
CMP process development sequence, showing the impact of the proposed predictive framework in shortening the development cycle.

point on the patterned wafer surface during a time step. A methodology is proposed to redistribute the pressure due to the effect of the surrounding topography at a given location. The modified pressure is then used to determine the local MRR using Archard's wear law (1953). The model is validated in Section 18.3 against the experimental observations of Ouma et al. (2002). In Section 18.4, a parametric study is conducted to investigate the effects of the polishing pad properties on the post-CMP wafer topography. Insights gathered from this investigation are discussed in Section 18.5. A comprehensive table of variables is provided at the end of the chapter under the section Nomenclature.

18.2 Model Development

18.2.1 Pad Characterization

The topography of polishing pads plays a significant role in CMP processes. Some characterization of the polishing pad is given on the basis of a representative IC 1000 pad (Bastawros et al., 2002; Borucki, 2002; Guo et al., 2004; Wang et al., 2005a,b). The pad surface has numerous pores with an average pore diameter of 30–50 μm. These pores typically occupy about 35% of the volume of the material and are separated by polyurethane membranes, which form asperities/roughness on the polishing pad. During the polishing process, it is these asperities that actually contact with the wafer surface. The real contact area is usually very small and is about 1% or less of the nominal contact area between the wafer and the pad (Shan, 2000). The real contact pressure is then much higher than the nominal pressure and directly affects the MRR in CMP. The pad topography can be characterized by the probability density function (PDF) of pad asperity heights. Yu et al. (1993) suggested a Gaussian distribution to describe the height distribution of pad asperities. Borucki (2002) and Wang et al. (2005a,b) used a Pearson IV PDF to model the rough surface of the pad. Vlassak (2004) applied an exponential distribution to reduce

TABLE 18.1

Parameter Values for Pad Asperity Heights

Parameters	Values
Asperity radius (R_p)	50 μm
Standard deviation of asperity height (σ)	15.625 μm
Skewness (γ)	−1.25
Kurtosis (β)	6.875
Asperity density (η_s)	$2 \times 10^8/m^2$

the computational burden in the model. The pad asperity heights PDF can be determined by statistical parameters such as the mean, standard deviation, skewness, and kurtosis. Table 18.1 presents the values of these statistical parameters that will be used in the simulation (Stein, 1996; Wang et al., 2005a,b). These values determine if the pad asperity height is distributed as a Pearson IV PDF (Johnson, 1985). The asperity tip radius, assumed to be constant, and the asperity density used in the model are given in Table 18.1.

18.2.2 Contact Model Description

Greenwood and Williamson (1966) developed a model (Figure 18.2) for contact between a smooth flat surface and a rough surface using Hertzian contact theory (Johnson, 1985). Adapting this model to the CMP process simulation, the wafer surface is assumed to be smooth and flat on the global scale, while the pad surface is assumed to be rough and contains asperities with a given height PDF $\phi(Z_p)$ relative to the mean plane of surface. The asperities are assumed to be uniformly spaced across the pad surface with an area density η_s. All asperities are assumed to have spherical tips, each with an identical radius R_p.

The separation distance d in Figure 18.2 is defined as the distance between the mean plane of the pad and the smooth wafer surface. An asperity with height Z_p larger than Zd will carry a load L (according to the Hertzian contact theory)

$$L(d) = \frac{4}{3} E^* R^{*\frac{1}{2}} (Z_p - d)^{\frac{3}{2}}$$

(18.1)

where E^* is the equivalent Young's modulus and R^* is the equivalent radius (Equations 18.5 and 18.6). The contact area A between the asperity and the wafer is

$$A(d) = \pi R^* (Z_p - d)$$

(18.2)

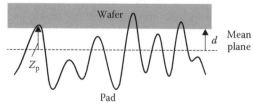

FIGURE 18.2
Illustration of the Greenwood and Williamson model (1966).

The total load L_{total} and the real contact area A_{re} of all pad asperities are given as the mean load/mean contact area of one single asperity multiplied by the total number of the pad asperities with η_s as the asperity area density:

$$L_{\text{total}}(d) = A_o \eta_s \int\limits_{Z_p - d > 0} \frac{4}{3} E^* R^{*\frac{1}{2}} \left(Z_p - d \right)^{\frac{3}{2}} \phi\left(Z_p \right) dZ_p$$

(18.3)

$$A_{\text{re}}(d) = A_o \eta_s \int\limits_{Z_p > d} \pi R^* \left(Z_p - d \right) dZ_p$$

(18.4)

where A_o is the nominal area and R^* is the equivalent radius given by

$$\frac{1}{R^*} = \frac{1}{R_p} + \frac{1}{R_w}$$

(18.5)

R_p and R_w are the asperity radius of the pad and the wafer, respectively. The wafer surface is assumed to have a large radius compared to the pad asperity. Thus, we may assume $R^* \approx R_p$.
E^* is the equivalent modulus, given by

$$\frac{1}{E^*} = \frac{1 - v_w^2}{E_w} + \frac{1 - v_p^2}{E_p}$$

(18.6)

E_w and E_p are the Young's moduli of the wafer and the pad respectively; v_w and v_p are Poisson's ratios of the wafer and the pad, respectively. Usually, $E^* \approx E_p/(1 - v_p^2)$ since E_w is much larger than E_p.

According to the Greenwood and Williamson (1966) model, the load-balancing equation between the nominal pressure $\bar{P}$ applied on the wafer surface and the separation distance d is given by

$$\bar{P} = \frac{L_{\text{total}}(d)}{A_o} = \frac{4\eta_s E^* R^{*\frac{1}{2}}}{3} \int\limits_{Z_p - d > 0} \left(Z_p - d \right)^{\frac{1}{2}} \phi\left(Z_p, t \right) dZ_p$$

(18.7)

The ratio A_f of the real contact area A_{re} to the nominal area A_o is

$$A_f = \frac{A_{\text{re}}(d)}{A_o} = \pi \eta_s R^* \int\limits_d^{\infty} \left(Z_p - d \right) \phi\left(Z_p \right) dZ_p$$

(18.8)

Following Archard's law (1953), it was assumed that the MRR is proportional to the average real contact pressure P_{re}, that is,

$$P_{\text{re}} = \frac{L_{\text{total}}(d)}{A_{\text{re}}(d)} = \frac{\bar{P} A_o}{A_{\text{re}}(d)} = \frac{\bar{P}}{A_f}$$

(18.9a)

$$\overline{MRR} = C_w P_{\text{re}}$$

(18.9b)

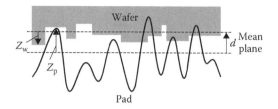

FIGURE 18.3
Schematic of the contact model between a rough pad and patterned wafer.

where C_w is a coefficient of proportionality. It is proportional to the relative velocity between the wafer and the pad and related to their material properties. Equation 18.9a,b is a form of Archard's law (1953). It is similar to the Preston equation (Preston, 1927).

The model described earlier assumed that the wafer surface is smooth. However, in most situations, the wafer has a patterned surface instead of a smooth surface. Figure 18.3 shows the schematic representation of the CMP model described in this chapter. The pad is full of asperities and has the properties described in Section 18.2.1. The wafer surface has different heights Z_w, referred to as its *mean plane*. It then follows that the load-balancing Equation 18.7 can be modified by considering an equivalent rough pad surface and a smooth wafer surface (Greenwood and Tripp, 1967) as

$$\bar{P} = \frac{4\eta_s E^* R^{*\frac{1}{2}}}{3} \int_{Z_p - d > 0} (Z - d)^{\frac{3}{2}} \varphi(Z) dZ$$

(18.10)

where $Z = Z_p + Z_w$ with a PDF $\varphi(Z)$. If Z_w is assumed to have a distribution $\Psi(Z_w)$, $\varphi(Z)$ can be estimated (Freund, 1992). From Equation 18.10, the load-balancing separation d can be solved, provided $\bar{P}$, $\varphi(Z)$, and η_s, E^*, R^* are given and estimated. In the actual CMP process, the wafer surface height variation is much smaller than the height of the pad asperity. Thus the effect of wafer surface height on the separation distance d can be neglected. Then, Equation 18.10 can be simplified as Equation 18.7.

After knowing the load-balancing separation distance d, the mean pressure $\tilde{P}(x)$ at any point on the wafer during the time step Δt can be calculated.

During the time step Δt, many pad asperities are in contact with the wafer at the point x_i one after another, assuming that Δt is sufficiently large for these contacting pad asperities to follow the height distribution $\phi(Z_p)$. d is assumed to be constant during Δt. If the effect of the surrounding wafer topography is not considered, the mean contact pressure $\tilde{P}$ at point x_i during Δt is equal to the real contact pressure between a rough pad and a local smooth wafer with the separation distance $\tilde{d} = d - Z_w(x_i)$:

$$\tilde{P}(x_i) = \frac{L_{\text{total}}(\tilde{d})}{A_{\text{re}}(\tilde{d})}$$

(18.11)

18.2.3 Effects of the Surrounding Topography

18.2.3.1 Pressure Redistribution

Warnock (1991) points out that depending on the pad flexibility, the polishing rate at any given point is affected by its surrounding topography. Fu and Chandra (2003) proposed

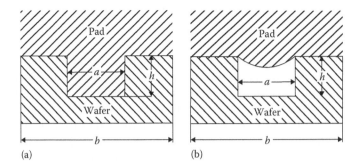

FIGURE 18.4
Schematics of the pattern wafer/pad contact interface: (a) pad in contact with the wafer completely and (b) pad not in contact with the down area of the wafer.

an approach for pressure redistribution to consider the surrounding topography effect. A brief description of their approach is given in Figure 18.4a. h is the step height between the up area and the down area. They assumed that the pad contacts with the wafer completely and deforms in a manner similar to a set of linear elastic springs. The pressures for the down area and the up area were $\tilde{P}_{down}$ and $\tilde{P}_{up}$ respectively, without considering the effects of the surrounding features. As a result of pad bending, there is force redistribution between the up area and the down area. The force change ΔF was assumed to be proportional to the height difference

$$\Delta F = \alpha \cdot h \tag{18.12}$$

where α is defined as the bending factor in units of newtons per square meter. In Fu and Chandra (2003), ΔF was a line load (assuming the uniform state along the direction orthogonal to the page). Then the pressures $\tilde{P}_{down}$ and $\tilde{P}_{up}$ were changed to the equivalent pressures $\tilde{P}^*_{down}$ and $\tilde{P}^*_{up}$ by

$$P^*_{up} = \tilde{P}_{up} + \frac{\Delta F}{(b-a)} \tag{18.13}$$

$$P^*_{down} = \tilde{P}_{down} - \frac{\Delta F}{a} \tag{18.14}$$

Here, b is the pitch, and the ratio a/b is the pattern density. In Fu and Chandra (2003), the pressure on the up area and the down area was changed evenly neglecting the fact that a point on the up area nearer to the down area should carry more load and a point on the down area nearer to the up area should be released of more load (Choi et al., 2006). In other words, the effect of the surrounding topography on a given point was the same if the surrounding points had the same height difference with the given point, regardless of the effect of the distance between the given point and the surrounding points. In fact, for the surrounding points with the same height difference, the point at a smaller distance from the given point will have a larger influence. An appropriate weight function may be introduced to address this influence of proximity.

The model of Fu and Chandra (2003) assumes that the pad contacts with the wafer surface completely. This is true when the step height between the up area and the down area is small enough. However, at the beginning of the CMP process, the pad may not contact completely with the wafer. Accordingly, two approaches are used in the present work to redistribute the pressure due to the pad bending.

The first approach is used when the step height h is larger than a critical contact height (H) and the pad does not contact with the down area of the wafer. This contact height (H) depends on the width of the down area (Smith et al., 1999) for a specified pressure. When the down area is wider, the critical contact height (H) is larger.

The second approach is a modified version of Fu and Chandra (2003). It is used when the pad contacts with the wafer completely, that is, the step height h is lower than H. The case in which the pad contacts with the down area partially is not considered here for simplicity.

A combination of these two approaches is used in the current work to consider the influence of the surrounding topography.

1. The pad does not contact with the down area of the wafer because of the surrounding up area as in Figure 18.4b. This is the case if the step height h is large and the space width a is small such that the up area relieves the load on the down area completely. Pressure redistribution is done in the following way: the load carried by the down area is assumed to be supported evenly by the up area. After calculating the average pressure $\tilde{P}_{\text{down}}$ without considering the effect of the surrounding topography, we simply release the load carried by the down area a and completely transfer it to the surrounding up area ($b - a$) by

$$\Delta P = \frac{\tilde{P}_{\text{down}} \cdot a}{(b - a)}$$

(18.15)

Then

$$P^*_{\text{down}} = 0$$

(18.16a)

$$P^*_{\text{up}} = P_{\text{up}} + \Delta P$$

(18.16b)

where $\tilde{P}_{\text{down}}$ and $\tilde{P}_{\text{up}}$ are the contact pressures calculated by Equation 18.11 without considering the effect of the surrounding topography.

2. As the step height h decreases, the pad contacts with the down area of the wafer completely, as in Figure 18.4a. Then the pressure redistribution method of Fu and Chandra (2003) is modified to obtain the pressure redistribution. The pressure change at a given point due to a neighboring point with the height difference h is given by

$$\Delta P = \hat{\alpha} \cdot h$$

(18.17)

The definition of $\hat{\alpha}$ here is the pressure change at a given point due to a neighboring point with a unit height difference. The unit of measurement of $\hat{\alpha}$ is newtons per cubic meter. It is different from that defined by Fu and Chandra (2003) for the plane strain condition but represents a similar idea. So we still call it the *bending factor* and denote it by α.

In this model, it is assumed that the pressure at any location on the wafer is affected by its surrounding topography within an influence length l over which the pad bends. So, the pressure could be changed by

$$\Delta P(x_i) = \sum_{x_i-l/2}^{x_i+l/2} w(x_i - x) \cdot \alpha \cdot \left[Z_w(x_i) - Z_w(x) \right]$$

(18.18)

$$P^*(x_i) = \tilde{P}(x_i) + \Delta P(x_i)$$

(18.19)

where $Z_w(x_i) - Z_w(x)$ is the height difference and $x_i - x$ is the distance between two points (within the influence zone) on the wafer. $\tilde{P}(x_i)$ is the pressure calculated using Equation 18.11. $w(x_i - x)$ is a weight function that addresses the effect of the distance between the two points. The weight function should decrease monotonically with increase in the distance, that is, when a point is farther away from the point x_i, its influence on point x_i should become less significant. The weight function assumed here is a standardized Gaussian function as shown in Equation 18.20, where σ_1 is the standard deviation, the value of which is decided by the influence length l. From the property of the Gaussian function, when $x_i - x$ is beyond $(-3\sigma_1, 3\sigma_1)$, $w(x_i - x)$ is almost zero as shown in Figure 18.5. We set $\sigma_1 = l/6$ in the simulation.

$$w(x_i - x) = \exp\left(\frac{-(x_i - x)^2}{2\sigma_1^2} \right)$$

(18.20)

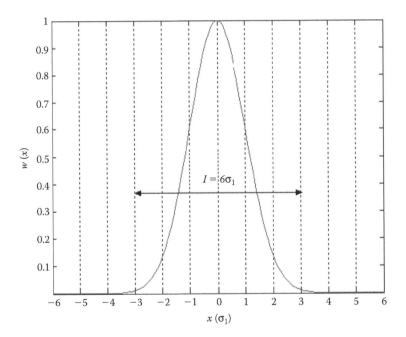

FIGURE 18.5
Profile of the standardized Gaussian function.

18.2.3.2 Interscale Consistency

From Equations 18.15 through 18.18, the pressure decrease (increase) at a given point x_i due to a higher (lower) neighboring point will be compensated by the pressure increase (decrease) at this neighboring point due to the point x_i. Thus, the requirement of global force equilibrium at the die scale is satisfied.

The moment balance also needs to be satisfied. After calculating the pressure change $\Delta P(x_i)$ using Equations 18.15 and 18.18, the moment balance is checked over the die scale. If the moment change ΔM due to the pressure change $\Delta P(x_i)$ is less than a specified small tolerance value (e.g., $10^{-3}P_oA_o L$, where A_o and L are the die area and length), it can be assumed that the moment balancing requirement is met. If the extra moment is larger than the specified tolerance, the pressure needs to be redistributed. In doing this, we add a pressure correction $\Delta P'(x_i)$ such that it generates a negative moment $-\Delta M$ and $S\Delta P'(x_i) = 0$. For simplicity, the pressure correction $\Delta P'(x_i)$ is linearly distributed along the die surface and is calculated as follows:

$$\sum_{0}^{L}\Delta P'(x_i)\cdot x_i = -\Delta M$$

(18.21)

$$\Delta P'(x_i) = \frac{x_i}{\frac{L}{2}}\Delta P'\left(\frac{L}{2}\right) \qquad 0 \le x_i \le \frac{L}{2}$$

(18.22a)

$$\Delta P'(x_i) = -\frac{L - x_i}{\frac{L}{2}}\Delta P'\left(\frac{L}{2}\right) \qquad \frac{L}{2} \le x_i \le L$$

(18.22b)

where $\Delta P'(\frac{L}{2})$ is the pressure correction at the center point. By substituting Equation 18.22 into Equation 18.21, $\Delta P'(\frac{L}{2})$ is given as

$$\Delta P'\left(\frac{L}{2}\right) = \frac{-\Delta M}{\left[\sum_{0}^{L/2}\frac{x_i}{\frac{L}{2}}x_i - \sum_{L/2}^{L}\frac{L - x_i}{\frac{L}{2}}x_i\right]}$$

(18.23)

The pressure distribution $\Delta P'(x_i)$ is then obtained. The effective local contact pressure is corrected as

$$P^*(x_i) = \tilde{P}(x_i) + \Delta P(x_i) + \Delta P'(x_i)$$

(18.24)

The resulting net force and moments are checked against the tolerance value.

If we look at a zone with the influence length l, the center point will affect its surrounding points within this zone. A point at the zone edge has a distance of $l/2$ from the center point. So, the pressure change at an edge point due to the center point has a ratio

$$w\left(\frac{l}{2}\right) = \exp\left(\frac{-(\frac{1}{2})^2}{2(\frac{1}{6})^2}\right) \approx 0.01$$

(18.25)

This indicates that, in the simulation, the pressure effect on the edge point due to the center point has an approximate error of 1%. If needed, a different distribution function may be chosen to reduce this error. This selection of the distribution function is equivalent to the selection of shape functions in a finite element analysis scheme.

18.2.4 Local Material Removal Rate (MRR)

With modified pressure P^* at a point x_i on the wafer surface, Archard's law (1953) can be used to get the mean MRR of the wafer at the point x_i during the time step Δt.

$$MRR(x_i) = C_w \cdot P^*(x_i) \tag{18.26}$$

The total thickness ΔZ_w removed from the wafer surface at the point x_i during the time step Δt is computed by

$$\Delta Z_w(x_i) = MRR(x_i) \cdot \Delta t \tag{18.27}$$

By updating $Z_w(x_i, t_0 + \Delta t) = Z_w(x_i, t_0) - \Delta Z_w(x_i)$ for every time step, a time history of the wafer profile can be obtained.

18.2.5 Simulation Process at the Die Scale

In order to obtain the wafer profile evolution at the die scale, a two-scale modeling approach combining the effects due to the die-scale profile and those due to the feature-scale pattern structures are used. In Part A, the die-scale (global) profile is extracted to calculate the pressure distribution along the wafer while neglecting the detail pattern structure at the feature scale. In Part B, further pressure redistribution is incorporated because of the finer feature-scale structure. The simulation process is shown schematically in Figure 18.6 and may be described as follows:

18.2.5.1 Part A—Die Scale

1. The global profile at the die scale is obtained by extracting the wafer thickness on the up area and neglecting the pattern structure at the feature scale (as shown in Figure 18.7).
2. The local contact pressure distribution is evaluated on the basis of the global profile using Equation 18.11.
3. Owing to the surrounding topography effect, the local contact pressure from step (2) is redistributed using Equations 18.18 and 18.19 by assuming that the pad contacts with the wafer up area everywhere during the polishing process.

18.2.5.2 Part B—Feature Scale

1. The final effective local contact pressure is obtained by redistributing the local contact pressure in step (3) because of the feature-scale structure effect. Equations 18.15 and 18.16 are used when the pad does not contact with the down area at the feature scale. Equations 18.18 and 18.19 are used when the pad contacts with the down area.

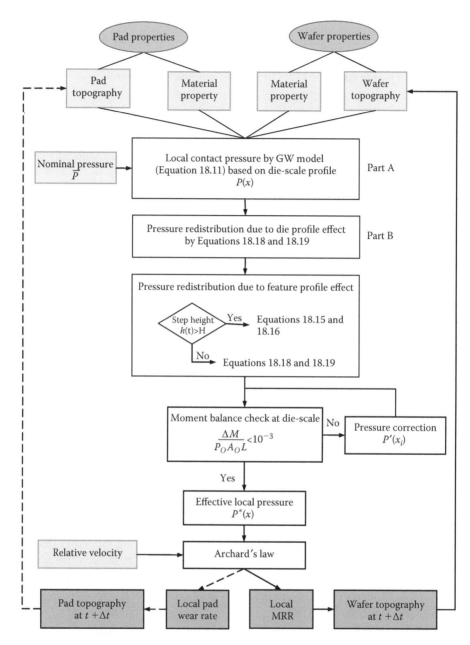

FIGURE 18.6
Simulation process flow.

2. The pressure change due to the surrounding topography is corrected to satisfy the moment balance by using Equations 18.21 through 18.23.

3. The local MRR is calculated from the effective local contact pressure using Equations 18.26 and 18.27. The wafer topography is updated on the basis of the local MRR.

4. The updated wafer profile is applied and the procedure is repeated for the next time step (as shown in Figure 18.6).

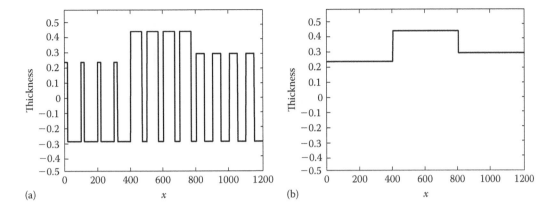

FIGURE 18.7
(a) Wafer profile and (b) extracted wafer up area profile at die scale.

The model is developed to evaluate the wafer profile evolution by integrating the pattern density and the die-scale profile. It can be extended to investigate the step height reduction and dishing at the feature scale by considering the global profile used in Part A as a flat surface.

Pad topography plays an important role in MRR. It was found that MRR drops rapidly if the pad is not conditioned (Stein et al., 1996). Pad wear effect on the wafer surface evolution is not considered in this chapter.

18.3 Model Verification

In this stage, the model predictions of the die-scale global step formation due to variations in feature-scale parameters (e.g., pattern density, line width, pitch) are validated against experimental observations in Ouma et al. (2002). The experiment is done by an Ebara polisher with IC1000 pads used over a subpad of fixed thickness. The mask layout of one die used in the experiment is shown in Figure 18.8. It is 20 mm × 20 mm and discretized into many blocks. Each block is 4 mm × 4 mm with different pattern structures. The prefix D denotes pattern density. Each density structure consists of vertical lines and spaces of 100-μm pitch (the length of the line width and the space width). For example, D10 means that the line width is 10 μm and the space width is 90 μm. The prefix P denotes pitch. The density for each pitch structure is fixed at 50% (equal line width and space width). For example, P20 means that both the line width and the space width are 10 μm each. Along the line L3 and L4, the pattern density gradually increases from block to block. The step pattern density is defined along the line L5. L6 has a constant pattern density, but the pitch length for each block is different. The pre-CMP wafer has a patterned surface with the initial step height (the height difference between the up area and the down area) of 0.6 μm. In their experiment, the thickness was measured on the up area (line) and the down area (space) at several locations.

18.3.1 Parameter Estimation

In the simulation, the nominal pressure is set as 34 kPa under the assumption that the nominal pressure applied to each die across the wafer is constant. The properties of pad

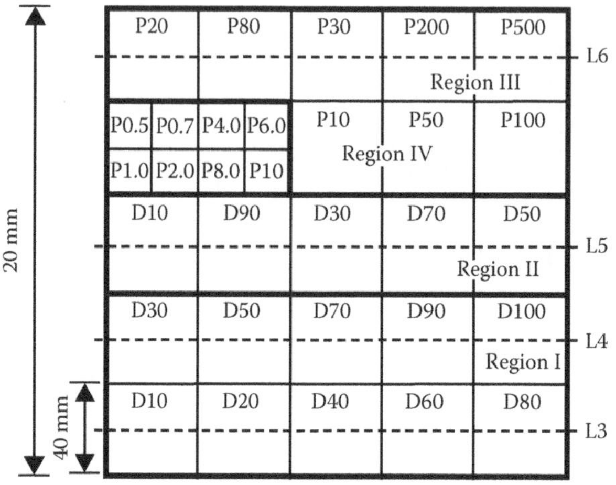

FIGURE 18.8
Layout of the oxide CMP characterization mask. (From Ouma, O., et al., Characterization and modeling of oxide chemical–mechanical polishing using planarization length and pattern density concepts, *IEEE Transactions on Semiconductor Manufacturing* © 2002 IEEE.)

asperities given in Stein (1996) are used in the simulation. The parameter values for pad asperity properties are presented in Table 18.1. The distribution of pad asperity heights is characterized by a Pearson IV distribution. The procedure used to estimate the pad asperity heights PDF can be found in Wang et al. (2005a,b) and Johnson (1985). The Young's modulus E_p for the IC1000 pad is assumed to be 29 MPa (Ouma et al., 2002).

Other parameters including l (the influence length), C_w (the coefficient of proportionality), α (the pad bending factor), and critical contact height (H) needed in the simulation are estimated as follows:

1. The coefficient C_w

    The coefficient C_w in Equations 18.9 and 18.26 is related to the relative velocity between the wafer and pad as well as to the pad and wafer material properties. In the model, the MRR is proportional to the local contact pressure, with the coefficient C_w being the proportionality constant. Thus, if the local MRR and local contact pressure are known, C_w can be estimated.

    For a blanket wafer CMP, the contact pressure can be easily calculated by the model of Greenwood and Williamson by assuming that the wafer surface is smooth. For a pad with the statistical parameters given in Table 18.1, the ratio of the real contact area to the nominal contact area calculated using Equation 18.8 is about 0.8%. Thus, the real contact pressure calculated using Equation 18.9a is about 4.2 MPa when the nominal pressure is 34 kPa.

    The MRR estimation for a blanket wafer surface can be found in Ouma (1998) and Xie et al. (2006). Here, the MRR is simply estimated as the average wafer thickness reduced during unit time in the experiment. From the experimental data for L5 at 29- and 89-s polishing, the MRR is estimated as the mean thickness reduced from the wafer profile at 29 s to the wafer profile at 89 s for the corresponding locations. The mean MRR turns out to be 318 nm/min approximately. So, the coefficient C_w is estimated as 1.25×10^{-9} μm/s/Pa.

2. Influence length at feature scale (l_f) and at die scale (l_d)

The influence length is defined as a distance over which the pad bends. At the feature scale, pad bending is influenced by the local pattern structure. On the mask layout shown in Figure 18.8, the pattern structure is periodical in each block. Therefore, the feature-scale influence length l_f may be assumed to be the pitch length. In Part B of the simulation where the feature scale is considered, the influence length l_f is set to be the pitch length.

As the polishing time increases, a variation of the die-scale (global) profile will be generated since the MRR is different in different regions of the wafer owing to variations in pattern structures. Then the pad also bends to conform to the die-scale (global) profile. The influence length at the die scale can be characterized by the deformation length within which the pad surface will deform to follow the deformation of a loaded point. In other words, the MRR at a given point is influenced by its surrounding topography within this deformation length. The deformation length is defined as the planarization length in Ouma (1998). The approach to obtain this length is also investigated there. Therefore, the pad influence length (l_d) in Part A of the simulation is approximated as 3 mm, and the average planarization length calculated from Table 18.1 in Ouma et al. (2002).

3. The pad bending factor α

When the pad contacts the patterned wafer, the contact pressure at a given point depends not only on the pad deformation at that point but also on the pad deformation at the neighboring points. The zone of influence over which the neighboring points affect a given point, has already been considered by the influence length. But the extent or magnitude of the influence of the neighboring points for a given location is governed by the pad bending factor. The estimation of the pad bending factor can be done as follows:

For the wafer profile evolution (actually, the step height reduction) simulation at feature scale, the influence length can be set as the pitch length, and the coefficient C_w can be estimated by the blanket wafer polishing. Only one parameter remains to be estimated, that is, the pad bending factor. The feature-scale model prediction (step height reduction) is fitted to a single experimental data point to estimate the value of the pad bending factor.

In the simulation for the experimental data in Ouma et al. (2002), the step height reduction versus time for the center point in each block on L5 may be used to estimate the pad bending factor since global pad bending has the smallest effect there. From the experimental data for L5, we choose the data point at the location of 10 mm. The step height at 29 s for this location is about 0.33 μm, and it reduces to about 0.021 μm. The pattern density is 0.3 and the pitch length is 100 μm at this location. Using the feature-scale simulation to evaluate the step height reduction at the selected point, the pad bending factor turns out to be 0.55×10^{12} N/m³ if C_w is set to be 1.25×10^{-9} μm/s/Pa.

4. The critical contact height H

The critical contact height H determines the start of polishing for down area on the wafer. It is different for different pattern structures and different applied pressures for a specified pad. In Cu CMP, dishing occurs because of the local pad bending. The saturation dishing depth (the asymptotic value of dishing over a long duration) for a nonselective slurry and the elastic pad property are assumed to reflect the maximum pad bending depth. So it may be assumed to be the critical contact height *H*. The values for each pattern structure are obtained from Laursen and Grief (2002).

18.3.2 Comparison of Simulation Results to Experimental Data

The values for parameters used in the simulation are summarized in Table 18.2. Figures 18.9 and 18.10 shows the model predictions (denoted by lines) compared to the experimental data (denoted by points). The solid line and the solid points are the data for 29 s of polishing. The dashed line and the open points are for 88 s of polishing. The root mean square (RMS) of the prediction error is given in Table 18.3.

From Figure 18.9 and the prediction error shown in Table 18.3, it is observed that the simulation reliably captures the effects of the pattern density on the post-CMP wafer surface. At the region with lower pattern density, the up area thickness reduces faster, whereas at the region with higher pattern density, the up area thickness reduces slower. It is interesting to note that the predicted MRR results are a little faster than the experimental measurement at 88 s. Pad wear is a likely cause for this discrepancy. No pad wear is considered in our die-scale model. It is also noted that the down area surface for 29 s from simulation slightly underestimates the experimental data. This may be due to the initial roughness of the pad. Before the step height reaches the contact height *H*, at which the pad contacts with the wafer surface completely, some asperities with small radius and greater height may have already contacted the down area and polished it in the experiment. In our model, this situation is simplified.

18.4 Parametric Study

A model integrating the effects of both die-scale and feature-scale profiles is developed and is verified against the experimental data in Section 18.3.2. On the basis of this model,

TABLE 18.2

Parameter Values for the Simulation

Parameters	Values
Young's modulus E_p	29 MPa
Coefficient C_w	1.25×10^{-9} μm/Pa/s
Die-scale influence length l_d	3 mm
Feature-scale influence length l_f	Pitch length
Pad bending factor α	0.55×10^{12} N/m³
Critical contact height *H* (for L5)	0.11, 0.068, 0.1, 0.082, 0.09 μm

TABLE 18.3

Prediction Error

Line No. on Wafer	RMS Prediction Error (Å)
L3	396
L4	418
L5	434
L6	416

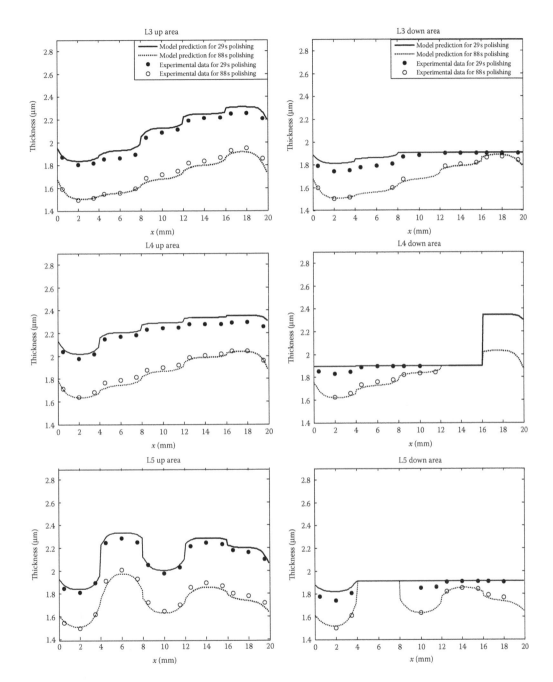

FIGURE 18.9
Predictions for up area and down area across line L3, L4, L5, and L6, respectively. (From Ouma, O., et al., Characterization and modeling of oxide chemical–mechanical polishing using planarization length and pattern density concepts, *IEEE Transactions on Semiconductor Manufacturing* © 2002 IEEE.)

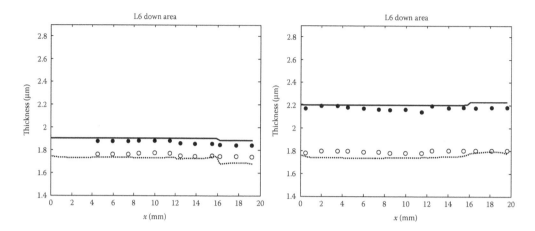

FIGURE 18.9 (continued)

impacts of several parameters, such as pad roughness, the pad global influence length l_d, and pad bending factor α, are schematically studied using the wafer profile shown in Figure 18.11. There are two different pattern structures on the wafer. The pattern density in the center zone (pitch length is 100 μm) and the edge zone are 0.3 and 1.0, respectively.

1. Effect of pad roughness on mean MRR

   In this study, the parameters in Table 18.1 are used and the standard deviation for pad surface properties is varied. The nominal pressure $\bar{P}$, coefficient C_w, the influence length l_d, and the pad bending factor are fixed during the simulation. It is assumed that the pad contacts with the down area of the wafer surface. Figure 18.12 shows the mean MRR during 1 min of polishing for pads with different standard deviation values. The simulation results show that a pad with a higher standard deviation provides a larger MRR. For a pad with a standard deviation of 25 μm, the MRR is about two times that obtained from the pad with a standard deviation of 5 μm. The standard deviation of pad asperity height usually indicates the pad surface roughness. A pad with a higher standard deviation indicates a rougher pad surface. A rougher pad has a smaller real contact area with the wafer. This results in higher real contact pressure. Thus, a higher MRR is obtained by such a pad. During the CMP process, a freshly conditioned pad usually produces a high MRR. However, it drops rapidly with time if the pad is not conditioned. The mean MRR decay has been observed by experimental investigations in Stein et al. (1996). This phenomenon can be explained by this model. The initially higher pad asperities in contact with the wafer are worn down with continued polishing. The pad surface becomes smoother and the standard deviation of pad asperity heights will get smaller with the polishing process if the pad is not conditioned. With the evaluation of pad surface evolution, the MRR decay can be simulated by the model.

2. Effect of the die-scale influence length l_d on global planarity of post-CMP wafer surface

   The influence length at feature scale l_f only affects the local planarity. The effect of the influence length on the global planarity is studied by changing the die-scale influence length l_d in the simulation. The pad bending factor is fixed as 0.3×10^{12} N/m^3.

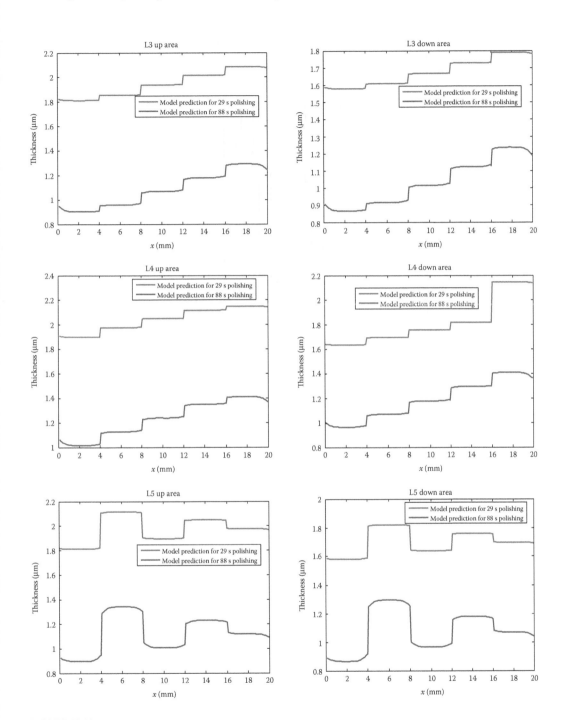

FIGURE 18.10
Wafer surface evolution with a dry (200 MPa) pad.

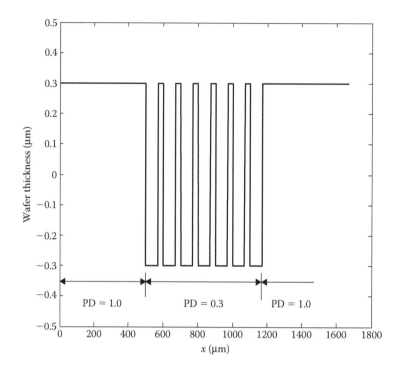

FIGURE 18.11
The wafer profile before CMP.

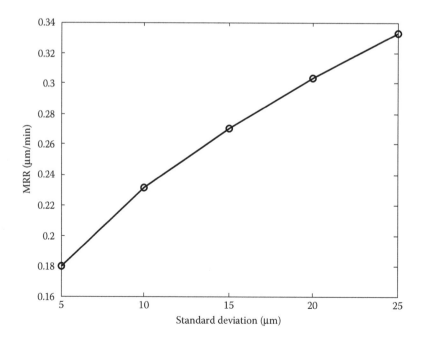

FIGURE 18.12
MRR with the pad roughness.

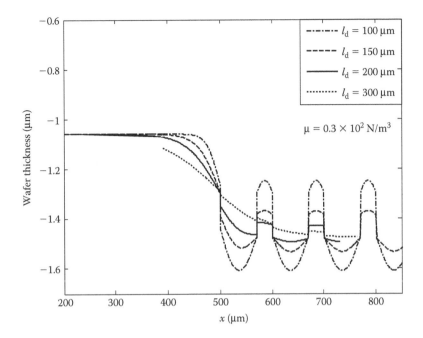

FIGURE 18.13
Wafer profile after 10 minutes of polishing with different influence lengths.

Figure 18.13 presents the wafer profile after 10 min of polishing using different influence lengths l_d. It is found that a longer influence length l_d results in a more uniform post-CMP wafer profile. The standard deviation of the wafer thickness is used to describe the planarity of the wafer surface. The post-CMP profile with an influence length of 300 μm has a standard deviation 20% less than with the influence length of 100 μm. If we assume the pad to be an elastic (Winkler) foundation, the influence length indicates that a given spring is affected by the neighboring springs within this distance. A pad with a larger influence length would mean that more springs affect each other. The MRR at a given point will be affected by the surrounding topography over a longer distance. The pad bending effect will increase the MRR at a high point and reduce the MRR at a low point. If the pad-bending factor is spatially uniform, the total pad bending effect will be higher for a pad with a longer influence length. It can be conjectured that the polished wafer surface will be more uniform with a longer influence length. An influence length of 0 indicates that those springs become uncoupled. When they are pressed onto the patterned wafer surface, the force applied at a given point only depends on the deformation of the corresponding spring and is not affected by its neighbors. Therefore, the MRR at a given point is not affected by its surrounding topography. The MRR at a high point will not be enhanced by its surrounding down area, while the MRR at low point will not be reduced by its surrounding up area. The post-CMP wafer profile in such a case will have lower planarity.

3. Effect of the bending factor α on global planarity of post-CMP wafer surface.

The effect of the bending factor on the global planarity of post-CMP wafer surface is investigated by keeping the influence length l_d constant at 300 μm. Figure 18.14

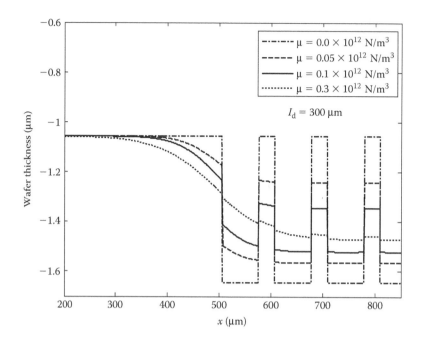

FIGURE 18.14
Wafer profile after 10-minute polishing with different bending factors.

shows the wafer profile after 10 min of polishing with different bending factors. It may be observed that the larger bending factor gives better global planarity if l_d is constant. The post-CMP profile with the bending factor of 0.3×10^{12} N/m^3 has a standard deviation 34% less than with the bending factor of 0. The bending factor describes the pad bending ability. The greater bending ability associated with higher bending allows the higher area on the wafer to effectively shade its neighboring down areas and releases the load from the down area. Thus, the MRR is enhanced at a up area and reduced at a down area. The wafer surface will be polished more uniformly. In order to get a better planar surface, a pad with a higher bending factor should be used. The pad bending factor is significantly influenced by the pad surface microstructure. An approach to manufacturing such a pad can be found in Fu and Chandra (2003).

Both the bending factor and the influence length affect the post-CMP wafer surface. However, they contribute in different ways. The bending factor considers the height difference effect, while the influence length considers the zone of influence of a given point on its surrounding topography.

18.5 Conclusion

An integrated model is proposed to evaluate the wafer surface evolution during a CMP process. The interface pressure between a patterned wafer and a rough pad is evaluated on the basis of the work of Greenwood and Williamson (1966). A methodology is proposed to take into account the effects of the surrounding topography due to pad bending. The model predictions are verified against the experimental data of Ouma et al. (2002). The estimation procedure for the parameters used in the model is discussed. The effects of pad roughness, die-scale influence length, and the bending factor on the wafer surface evolution are studied. It is observed that a rougher pad gives a higher MRR. Both the influence length l_d and the pad-bending factor α contribute to the planarity of the wafer surface. The higher values of the influence length and the bending factor give a better global planarity of the post-CMP wafer surface.

It is important to recognize that a simple "hand-shake" arrangement or superposition of results from two disparate scales (e.g., die scale and feature scale) may produce erroneous results (Saari, 2009). Accordingly, a multiplicative decomposition (Chandra et al., 2008, 2010) is assumed here.

Nomenclature

L	Load
E^*	Equivalent Young's modulus
R^*	Equivalent radius
Z_p	Asperity height
D	Separation distance
A	Contact area
R^*	Equivalent radius
L_{total}	Total load
A_0	Nominal area
η_s	Asperity area density
A_{re}	Real contact area
A_f	Ratio of real contact area to nominal area
R_p	Pad radius
R_w	Wafer radius
v_w	Poisson's ratio of wafer
v_p	Poisson's ratio of pad
E_w	Young's modulus of wafer
E_p	Young's modulus of pad
P_{re}	Average real contact pressure
T	Time
$\bar{P}$	Nominal pressure
$\tilde{P}_{up}$	Mean pressure of up area
$\tilde{P}$	Mean pressure
$\tilde{P}_{down}$	Mean pressure of down area
F	Force
α	Bending factor
H	Step height
P^*_{up}	Equivalent pressure of up area
P^*_{down}	Equivalent pressure of down area
(a/b)	Pattern density, where b is the pitch
P	Pressure

H	Critical contact height
$\hat{\alpha}$	Pressure change at a given point due to unit height difference
X	A given point
Z_w	Height on wafer surface
Z	Total height
P^*	Equivalent pressure
L	Influence length
σ_1	Standard deviation
P'	Corrected pressure
M	Moment
MRR	Material removal rate
$\overline{MRR}$	Average material removal rate
C_w	Coefficient of proportionality for relative velocity between wafer and pad
$\tilde{d}$	Corrected separation distance

Acknowledgments

This material is based on work supported by the U.S. National Science Foundation Grant Nos. CMMI-0323069 and CMMI-0900093. The authors gratefully acknowledge this support. Any opinions, findings, and conclusions or recommendations expressed in this material are those of the authors and do not necessarily reflect the views of the sponsoring agencies.

References

Archard, J.F. 1953. Contact and rubbing of flat surfaces. *Journal of Applied Physics* 24: 981–988.

Bastawros, A.-F., Chandra, A., Guo, Y., and Yan, B. 2002. Pad effects on material-removal rate in chemical–mechanical planarization. *Journal of Electronic Materials* 31(10): 1022–1031.

Borucki, L. 2002. Mathematical modeling of polish-rate decay in chemical-mechanical polishing. *Journal of Engineering Mathematics* 43: 105–114.

Cale, T.S., Bloomfield, M.O., Richard, D.F., Soukan, S., Janseny, K.E., Tichy, J.A., and Gobbert, M.K. 2002. Integrated multiscale process simulation in microelectronics. *Computational Material Science* 23(1): 3–14.

Chandra, A., Karra, P., Bastawros, A.F., Biswas, R., Sherman, P.J., Armini, S., and Lucca, D.A. 2008. Prediction of scratch generation in chemical mechanical planarization. *Annals of CIRP* 57(1): 559–562.

Chandra, A., Karra, P., and Bastawros, A.F. 2010. Defectivity avoidance in chemical mechanical planarization: Role of multi-scale and multi-physics interactions. *ECS Conference*. Las Vegas, NV, Paper: E3-1457.

Che, W., Bastawros, A.-F., and Chandra, A. 2005. Synergy between chemical dissolution and mechanical abrasion during chemical mechanical polishing of copper. *Materials Research Society Symposium Proceedings* 867, 275–280.

Chekina, O.G., Meer, L.M., and Liang, H. 1998. Wear-contact problems and modeling of chemical mechanical polishing. *Journal of the Electrochemical Society* 145(6): 2100–2106.

Choi, J., Tripathi, S., Hansen, D., and Dornfeld, D.A. 2006. Chip scale prediction of nitride erosion in high selective STI CMP. *2006 CMP-MIC Conference*. Santa Clara, CA, USA, pp. 160–167.

Evans, C.J., Paul, E., Dornfeld, D., Lucca, D.A., Byrne, G., Tricard, M., Klocke, F., Dambon, O., and Mullany, B.A. 2003. Material removal mechanisms in lapping and polishing. *Annals of CIRP* 52(2): 611–634.

Fu, G. and Chandra, A. 2003. An analytical dishing and step height reduction model for chemical mechanical planarization. *IEEE Transactions on Semiconductor Manufacturing* 16(3): 477–485.

Freund, J.E. 1992. *Mathematical Statistics*. New Jersey: Prentice Hall.

Greenwood, J.A. and Williamson, J.B.P. 1966. Contact of nominally flat surfaces. *Proceedings of the Royal Society of London* A295: 300–319.

Greenwood, J.A. and Tripp, J.H. 1967. Elastic contact of rough spheres. *Journal of Applied Mechanics* 34(1): 153–159.

Guo, Y., Chandra, A., and Bastawros, A.-F. 2004. Analytical dishing and step height reduction model for chemical mechanical planarization (CMP) with a viscoelastic pad. *Journal of the Electrochemical Society* 151(9), G583–G589.

ITRS, 2007. International technology roadmap for semiconductors. http://www.itrs.net/ (accessed 2011).

ITRS, 2009. International technology roadmap for semiconductors. http://www.itrs.net/ (accessed 2011).

Johnson, K.L. 1985. *Contact Mechanics*. Cambridge, New York: Cambridge University Press.

Kim, A.T., Seok, J., Tichy, J.A., and Cale, T.S. 2003. A multiscale elastohydrodynamic contact model for CMP. *Journal of the Electrochemical Society* 150(9): G570–G576.

Komanduri, R. 1996. On material removal mechanisms in finishing of advanced ceramics and glasses. *Annals of CIRP* 45: 509–514.

Komulski, M. 2001. *Chemical Properties of Material Surfaces*. New York: Marcel Dekker.

Laursen, T. and Grief, M. 2002. Characterization and optimization of copper chemical mechanical planarization. *Journal Electronic Material* 31(10): 1059–1065.

Lin, M.Y., Lindsay, H.M., Weitz, D.A., Klein, R., Ball, R.C., and Meakin, P. 1989. Universality of fractal aggregates as probed by light scattering. *Proceedings of the Royal Society of London A* 423/1864: 71–87.

Lin, M.Y., Lindsay, H.M., Weitz, D.A., Klein, R., Ball, R.C., and Meakin, P. 1990. Universal diffusion-limited colloid aggregation. *Journal of Physics: Condensed Matter* 2: 3093–3113.

Luo, J. and Dornfeld, D.A. 2004. *Integrated Modeling of Chemical Mechanical Planarization for Sub-Micron IC Fabrication*. Berlin Heidelberg, Germany: Springer.

Oliver, M.R., ed. 2004. *Chemical Mechanical Planarization of Semiconductor Materials*. Berlin, New York: Springer.

Ouma, O. 1998. Modeling of chemical mechanical polishing for dielectric planarization, PhD Thesis, MIT.

Ouma, O., Boning, D.S., Easter, W.G., and Saxena, V. 2002. Characterization and modeling of oxide chemical—Mechanical polishing using planarization length and pattern density concepts. *IEEE Transactions on Semiconductor Manufacturing* 15(2): 232–244.

Preston, F.W. 1927. The theory and design of plate glass polishing machines. *Journal of Society Glass Technology* 11: 214–256.

Rhoades, R.L. and Murphy, J. 2006. Adopting CMP processes to achieve new materials integration. *MICRO* 24(6): 51–54.

Saari, D.G. 2010. Mathematical snapshots motivated by "dark matter", *American Inst. Physics Conf. Proc.* 1283, 75, DOI: 10.1063/1.3506083.

Seok, J., Sukam, C.P., Kim, A.T., Tichy, J.A., and Cale, T.S. 2003. Multiscale material removal modeling of chemical mechanical polishing. *Wear* 254: 307–320.

Shan, L. 2000. Mechanical interactions at the interface of chemical mechanical polishing, PhD Thesis, GIT.

Smith, T.H., Fang, S.J., Boing, D.S., Shinn, G.B., and Stefani, J.A. 1999. A CMP model combing density and time dependencies. *1999 CMP-MIC*. pp. 97–104.

Steigerwald, J.M., Murarka, S.P., and Gutmann, R.J. 1997. *Chemical Mechanical Planarization of Microelectronic Materials*. New York: John Wiley & Sons.

Stein, D., Hetherington, D., Dugger, M., and Stout, T. 1996. Optical interferometry for surface measurement of CMP pads. *Journal of Electronic Materials* 25, 1623–1627.

Stine, B., Mehrotra, V., Boning, D., Chung, J., and Ciplickas, D. 1997. A simulation methodology for assessing the impact of spatial/pattern dependent interconnect parameter variation on circuit performance. *IEDM Technical Digest*. Santa Clara, CA, USA, pp. 133–136.

Vlassak, J.J. 2004. A model for chemical—Mechanical polishing of a material surface based on contact mechanics. *Journal of the Mechanics and Physics of Solids* 52: 847–873.

Wang, C., Sherman, P.J., and Chandra, A. 2005a. A stochastic model for the effects of pad surface topography evolution on material removal rate decay in chemical-mechanical planarization. *IEEE Transactions on Semiconductor Manufacturing* 18(4): 695–708.

Wang, C., Sherman, P.J., Chandra, A., and Dornfeld, D. 2005b. Pad surface roughness and slurry particle size distribution effects on material removal rate in chemical mechanical planarization. *CIRP Annals—Manufacturing Technology* 54(1): 309–312.

Warnock, J. 1991. A two-dimensional process model for chemical mechanical polish planarization. *Journal of the Electrochemical Society* 138(8): 2398–2402.

Xie, X., Boning, D., Meyer, F., Rzehak, R., and Wagner, P. 2006. Analysis and modeling of nanotopography impact in blanket and patterned silicon wafer polishing. *2006 CMP-MIC Conference*. Santa Clara, CA, USA, pp. 243–253.

Yu, T.K., Yu, C.C., and Orlowski, M. 1993. A statistical polishing pad model for chemical–mechanical polishing. *Proceedings of the IEEE International Electron Devices Meeting*. Santa Clara, CA, USA, pp. 35.4.1–35.4.4.

Index